Proceedings of the 1st International Conference TENDEV 2023

Jean-Vasile Andrei / Nicoleta Mateoc-Sîrb /
Andrea Feher (eds.)

Proceedings of the 1st International Conference TENDEV 2023

Challenges and Strategies for Sustainable Development facing the Climate Change

Berlin - Bruxelles - Chennai - Lausanne - New York - Oxford

Library of Congress Cataloging-in-Publication Data
A CIP catalog record for this book has been applied for at the
Library of Congress.

Bibliographic information published by the Deutsche Nationalbibliothek.
The German National Library lists this publication in the German National
Bibliography; detailed bibliographic data is available on the Internet at
http://dnb.d-nb.de.

ISBN 978-3-631-91332-1 (Print)
E-ISBN 978-3-631-92298-9 (E-PDF)
E-ISBN 978-3-631-92299-6 (EPUB)
DOI 10.3726/ b22066

© 2024 Peter Lang Group AG, Lausanne
Published by Peter Lang GmbH, Berlin, Germany

info@peterlang.com - www.peterlang.com

All rights reserved.

All parts of this publication are protected by copyright.
Any utilization outside the strict limits of the copyright law, without the
permission of the publisher, is forbidden and liable to prosecution.
This applies in particular to reproductions, translations, microfilming, and
storage and processing in electronic retrieval systems.

Foreword

The main object of activity of the Research Center for Sustainable Rural Development of Romania, the initiator and organizer of the International Conference "*Multidimensional Tendencies in Sustainable Development*" is the research of sustainable rural development, a complex problem that requires multidisciplinary studies, given the achievement of a balance between the need to preserve rural space and the tendency, lately, of modernization of agricultural, non-agricultural, cultural activities and rural life as a whole.

The Research Center for Sustainable Rural Development of Romania carries out scientific research topics that meet the needs of knowledge, research, improvement and development of rural space, according to the current requirements of rural life evolution. The dimension of rural space and the problem of sustainable rural development in Romania generate complex themes, research and actions due to the need to maintain a balance between the requirement of preserving traditional rural space and the trend of urban expansion, both economically and ecologically, socially and culturally.

The research fields of the Center are as follows: local and regional rural development in Romania and in the European Union; rural economy – fundamental component of Romania's rural development; assessing the impact of applying the Common Agricultural Policy on the rural economy in Romania; financing systems of agriculture and rural development, the researchers of this center carry out research and studies of the evolution of rural space, under the conditions of integration into the European Union, compatibility of Romanian rural policies with European ones.

The Research Center for Sustainable Rural Development of Romania of the Romanian Academy – Timisoara Branch initiated the organization of a multiannual international conference with the generic name *"Multidimensional Tendencies in Sustainable Development"*, whose content in the first edition organized in 2023 was *"Challenges and Strategies for Sustainable Development facing the Climate Change"*, with the main themes: sustainable rural development; circular economy; environmental economics and policies; food safety and security; EU Common Agricultural Policy; smart agriculture; tourism, etc.

During the conference, attended by researchers in the field of sustainable development from Romania, Republic of Moldova, Hungary, Republic of Serbia, Ukraine, Bulgaria, Poland, Slovakia, Nigeria, Azerbaijan and India, 67 scientific papers were presented, of which 21 are included in this volume *"Proceedings of the 1st International Conference TENDEV 2023"*.

The conference had a multidisciplinary character, addressing fundamental issues in different fields and perspectives related to sustainable development, the scientific works of sustainable development assuming simultaneous research themes of economic development, environmental protection and social welfare.

Climate change, being one of the greatest challenges of mankind globally, represents a major problem of sustainable development. In order to address the key issues of the strategies that are needed to be implemented, the papers of this volume deal with opportunities for adaptation to climate change, with the aim of opening new pathways of compatibility of agriculture and the environment.

Academician Păun Ion Otiman

Preface

The selected papers for the current volume cover a wide area of sustainable rural development with insights for multiple fields of expertise. One set of deteailed research results deal with bioeconomy either as a more systemic approachs of the agro-based sectors or the biocapacity and ecological foot print, either more punctual at the elevel of biomass production. Risk management attitudes in small scale operations or analysis of structural changes reflected by the international agri-food trade add to the side of economic assessments. Financial analyses at institutional and system levels introduce another set of research findings displaying results from Ukraine, Republic of Moldavia or adding the longterm evolution of land market in Hungary. Food safety and security is explored from a dual perspective, as a public good with social impact and as background stability factor in time of crisis. Hospitality industry benefits from the results regarding the sustainable travel investigated in a multinational approach for customers using contemporary instruments of reservation.

The most consistent part of research is represented by the sector in charge of producing the bioresources, the agricultural production, introducing findings and results from applied new technologies such as in the fields of plums, vegetables, maize, soyabean, wheat or millet. Scientific investigations cover a wide range from the new technology application to the exploration models and techniques. Management of natural resources, such as the soil, equal part in ecosystem services as support and central resource in agricultural production, are introduced. Findings regarding education and training for the agricultural production sector has a sustainability check and performance assessment modelled for the rural environment. Social and cultural aspects of rural development are screened for a contemporary issue linked to labour availability and behaviour, gender equity and equality or transition of cultural major events towards the rural area.

The entire set of papers bringing up research results and findings answering to several of the Sustainable Development Goals has a sense of complementarity integrating elements that could lead to synergies when coupled and approached jointly. The multidisciplinary and pluri-disciplinary character of the scientific event created the background premises for the expression of this rich and divers set of research novelties. The sustainable development millennium goals benefit from contributions originating on three different continents covering most fields of rural development, incorporating circular economy and promoting sustainability as a must particularly in trans-sectoral approaches.

Dr. Cosmin Sălășan

Acknowledgement

The *Proceedings of the 1st International Conference TENDEV 2023* includes the scientific papers presented at the international conference "**Multidimensional Tendencies, in Sustainable Development**", the first edition "**Challenges and Strategies for Sustainable Development facing the Climate Change**". The conference enjoyed a remarkable national and international presence, being an activity included in the **cultural project "Cultural Connection through Conference and Artistic Acts" – CULTARTCONEX,** beneficiary Romanian Academy – Timişoara Branch, Research Center for Sustainable Rural Development of Romania, funded by the Grow Timişoara 2023 program (Timişoara – European Capital of Culture in 2023) from the state budget of the Ministry of Culture and developed through the Project Center of Timisoara.

This proceeding is carried out with the financial support of the following partners, to whom we would like to thank: Association for the Support of Rural Space Research and Development (ASCDSR); SC Agroindustriala PETIM SRL, Peciu Nou; SC JOLTA RADU CONSULTING SRL, Marghita; SC GAPA SRL, Timisoara; "Pogorârea Sfântului Duh" Parish, Lugoj.

Contents

List of contributors 15

Radu-Lucian Blaga, Eugenia Țigan, Simona Gavrilaș
Economic and environmental approaches for agro-based sectors
sustainable development 23

Karoly Bodnar, Zoltan Istvan Privoczki, Ioan Brad
Development of the price of agricultural land in the South-Eastern
region of Hungary 41

Sorinel Ionel Bucur
The sustainability of rural area – An approach from the perspective
of biocapacity and ecological footprint 59

Lavinia Denisia Cuc, Dana Rad, Silviu Gabriel Szentesi, Gabriel Croitoru, Gavril Rad
Evaluation of safety and hygiene measures in Romanian hospitality
industry in the context of the COVID-19 pandemic and customer profile 71

Nadiia Davydenko, Oleksandra Smirnova, Zoia Titenko, Alina Buriak
The state of formation of the state financial control system in the field
of land relations in Ukraine 91

Camelia Anișoara Gavrilescu
Romania's international agri-food trade – Why in a permanent deficit?
A post-accession analysis 103

Marko Jeločnik, Lana Nastić, Boris Kuzman
Improving the vegetable growing by the use of new technologies 123

Andreea Lidia Jurjescu, Alina Șimon, Florin Sala
Modeling and multicriteria analysis of soybean production
variation – Case study in Romania 145

Ildikó Kolozsvári, Árpád Székely, Noémi Valkovszki, Ágnes Kun, Mihály Jancsó, György Dajcs
Efficiency of slurry application in winter wheat with special reference to earthworm population trends 161

Bashiru Dahiru Magaji, Yusuf Usman Oladimeji, Ado Yakubu, Henry Egwuma, Benjamin Ahmed, Abubakar Abdullahi Hassan, Miroslav Raicov
Risk attitudes of micro, small and medium agribusiness enterprises in Nigeria 177

Nicoleta Mateoc-Sîrb, Teodora Mateoc-Sîrb, Ariana Velciov, Cosmin Lădariu, Zeno Gârban
Food safety and security from public desideratum to agrobiological and social aspects 197

Natalia Mocanu, Vasile Secrieru
The role of financial audit in public institutions in the Republic of Moldova 211

Marieta Nesheva, Leyda Todorova
Assessment of plum hybrids based on their viral symptoms in field conditions 225

Bianca-Florentina Nistoroiu, Ragif Huseynov, Ştefan Laurenţiu Prahoveanu
Sustainable development through gender equality 235

Andreea Adriana Petcov, Manuela Dora Orboi, Ana Mariana Dincu, Andreia Sasu, Raul Pascalau
Study on the ammonium impact in the deep waters of some villages in Caras Severin county, Romania 247

Veronica Prisacaru, Alina Caradja
Modeling the relationship between the performance of agricultural vocational education and the sustainable development of the rural environment 257

Valentina Ofelia Robescu, Valentina Nicoleta Florea, Gabriel Croitoru, Vasile Cumpănaşu
Theoretical aspects regarding the use of biomass for a sustainable economy 275

Srdjan Šljukić, Milovan Mitrović
Sociological aspects of sustainable development: Cultural capital in rural settlements in Serbia 289

Cosmina-Simona Toader, Ciprian Ioan Rujescu, Andrea Ana Feher, Małgorzata Zajdel, Małgorzata Michalcewicz-Kaniowska, Iveta Ubrežiová, Levente Komarek
Investigating tourists' interest in sustainable travel when using online booking accommodation – A multinational approach 305

Sanyam Varma, Manish Sen, Anush Jain, Laura Iosefina Smuleac, Sorin Mihai Stanciu
Cross cultural management: Challenges and strategies for managing a global workforce 323

Vishal Singh Varma, Reshu Gupta Singh, Ravi Kumar Goyal, Kritika Tekwani, Ramnika Kaur, Liana Mihaela Fericean
The economic potential of millet farming in Rajasthan (India): Opportunities and challenges 337

List of contributors

Benjamin Ahmed
Department of Agricultural Economics, Faculty of Agriculture/Institute for Agricultural Research, Ahmadu Bello University, Zaria, Nigeria, benujah@yahoo.com

Radu-Lucian Blaga
"Aurel Vlaicu" University of Arad, Faculty of Economics, Arad, Romania, radu.blaga@uav.ro

Karoly Bodnar
Hungarian University of Agriculture and Life Sciences, Hungary, bodnar.karoly.lajos@uni-mate.hu

Ioan Brad
University of Life Sciences "King Mihai I" from Timisoara, Romania, ioanbrad@usvt.ro

Sorinel Ionel Bucur
Institute of Agricultural Economics, Bucharest, Romania, bucursorinelionel@yahoo.com

Alina Buriak
State Tax University, Irpin, Ukraine, alina.v.bu@onlbne.ua

Alina Caradja
Technical University of Moldova, Chisinau, Republic of Moldova, alinacaradja@gmail.com

Gabriel Croitoru
"Valahia" University of Târgoviște, Faculty of Economics, Târgoviște, Romania croitoru.gabriel2005@yahoo.com

Lavinia Denisia Cuc
"Aurel Vlaicu" University of Arad, Faculty of Economics, Arad, Romania laviniacuc@yahoo.com

Vasile Cumpănașu
"Constantin Brâncuși" University of Târgu Jiu, Romania, vasilecumpanasu01@gmail.com

György Dajcs
Kardoskúti Agrár PLC, Kardoskút, Hungary

Nadiia Davydenko
State Tax University, Irpin, Ukraine, davidenk@ukr.net

Ana Mariana Dincu
University of Life Sciences "King Mihai I" from Timișoara, Faculty of Management and Rural Tourism, Romania, anamariadincu@usvt.ro

Henry Egwuma
Department of Agricultural Economics, Faculty of Agriculture/Institute for Agricultural Research, Ahmadu Bello University, Zaria, Nigeria, henry4him@gmail.com

Andrea Ana Feher
University of Life Sciences "King Mihai I" from Timisoara, Faculty of Management and Rural Tourism, Romania, andreafeher@usvt.ro; Romanian Academy –Timisoara Branch, Research Center for Sustainable Rural Development, Romania.

Liana Mihaela Fericean
University of Life Sciences, "King Mihai I" from Timisoara, Romania, liana.fericean@gmail.com

Valentina Nicoleta Florea
"Valahia" University of Targoviste, Romania, floreanicol@yahoo.com

Simona Gavrilaș
"Aurel Vlaicu" University of Arad, Faculty of Food Engineering, Tourism and Environmental Protection, Arad, Romania, simona.gavrilas@uav.ro

Camelia Anișoara Gavrilescu
Institute of Agricultural Economics, Romanian Academy, Bucharest, Romania, cami_gavrilescu@yahoo.com

Zeno Gârban
University of Life Sciences "King Mihai I" from Timisoara, Faculty of Food Engineering Romania, zeno.garban@yahoo.com; Romanian Academy-Timişoara Branch, Working Group for Xenobiochemistry, Romania.

Ravi Kumar Goyal
CCS NIAM, Jaipur/Rajasthan, India, coo.rkvy@ccsniam.ac.in

Abubakar Abdullahi Hassan
Department of Agricultural Economics, Faculty of Agriculture/Institute for Agricultural Research, Ahmadu Bello University, Zaria, Nigeria, abuifadhilah@gmail.com

Ragif Huseynov
Azerbaijan Technological University, Department of Economics, Azerbaijan, ragif1984@gmail.com

Anush Jain
NIMA Institute of Management, Pune, Maharashtra, India, anushjain1917@gmail.com

Mihály Jancsó
Research Center for Irrigation and Water Management, Institute of Environmental Sciences, Hungarian University of Agriculture and Life Sciences, Szarvas, Hungary.

Marko Jeločnik
Institute of Agricultural Economics, Belgrade, Serbia, marko_j@iep.bg.ac.rs

Andreea Lidia Jurjescu
Agricultural Research and Development Station, Lovrin, Romania, andreea-lidia19@gmail.com

Ramnika Kaur
Rajasthan Grameen Aajeevika Vikas Parishad, Jaipur/Rajasthan, India, ramnika-kaur2010@gmail.com

Ildikó Kolozsvári
Research Center for Irrigation and Water Management, Institute of Environmental Sciences, Hungarian University of Agriculture and Life Sciences, Szarvas, Hungary, kolozsvari.ildiko@uni-mate.hu

Levente Komarek
University of Szeged, Faculty of Agriculture, Institute of Economics and Rural Development, Hungary, komarek.levente@szte.hu

Ágnes Kun
Research Center for Irrigation and Water Management, Institute of Environmental Sciences, Hungarian University of Agriculture and Life Sciences, Szarvas, Hungary.

Boris Kuzman
Institute of Agricultural Economics, Belgrade, Serbia, kuzmanboris@yahoo.com

Cosmin Lădariu
Timiş County Health Directorate – Ambulance Service, Ministry of Health, Department for Emergency Situations, Romania, cosmin_ladariu@yahoo.com

Bashiru Dahiru Magaji
Department of Agricultural Economics, Faculty of Agriculture/Institute for Agricultural Research, Ahmadu Bello University, Zaria, Nigeria, bashirudhr@yahoo.com

Nicoleta Mateoc-Sîrb
Romanian Academy –Timisoara Branch, Research Center for Sustainable Rural Development, Romania; University of Life Sciences "King Mihai I" from Timisoara, Faculty of Management and Rural Tourism, Romania, nicoletamateocsirb@usvt.ro

Teodora Mateoc-Sîrb
University of Medicine and Pharmacy, Faculty of Medicine, Timisoara, Romania, teodora.mateoc@gmail.com

Małgorzata Michalcewicz-Kaniowska
Bydgoszcz University of Science and Technology, Faculty of Management, Poland, malgorzata.michalcewicz-kaniowska@pbs.edu.pl

Milovan Mitrović
University of Belgrade, Faculty of Law, Republic of Serbia, milovanm@ius.bg.ac.rs

Natalia Mocanu
State University of Moldova, Chisinau, Republic of Moldova, mocanunatalia@gmail.com

Lana Nastić
Institute of Agricultural Economics, Belgrade, Serbia, lana_n@iep.bg.ac.rs

Marieta Nesheva
Agricultural Academy, Fruit Growing Institute, Plovdiv, Bulgaria; marieta.nesheva@abv.bg

Bianca-Florentina Nistoroiu
Bucharest University of Economic Studies, Doctoral School Economics II, Bucharest, Romania, nistoroiubianca@yahoo.com

Yusuf Usman Oladimeji
Department of Agricultural Economics, Faculty of Agriculture/Institute for Agricultural Research, Ahmadu Bello University, Zaria, Nigeria, dr.oladimejiyusuf@gmail.com

Manuela Dora Orboi
University of Life Sciences "King Mihai I" from Timișoara, Faculty of Management and Rural Tourism, Romania, orboi@usvt.ro

Raul Pascalau
University of Life Sciences "King Mihai I" from Timișoara, Faculty of Agriculture, Romania, raul.pascalau@usvt.ro

Andreea Adriana Petcov
University of Life Sciences "King Mihai I" from Timișoara, Faculty of Engineering and Applied Technologies, Romania, andreeapetcov@usvt.ro

Ștefan Laurențiu Prahoveanu
School of Advanced Studies of the Romanian Academy, Bucharest, Romania, stefanprahoveanu@gmail.com

Veronica Prisacaru
Moldova State University, Chisinau, Republic of Moldova, veronica.prisacaru@usm.md

Zoltan Istvan Privoczki
Agro-Assistance Kft., Csongrad, Hungary, agrarpalyazat@gmail.com

Dana Rad
"Aurel Vlaicu" University of Arad, Faculty of Educational Sciences Psychology and Social Work, Arad, Romania, dana@xhouse.ro

Gavril Rad
"Aurel Vlaicu" University of Arad, Faculty of Educational Sciences Psychology and Social Work, Arad, Romania, radgavrilarad@gmail.com

Miroslav Raicov
Romanian Academy-Timisoara Branch, Research Center for Sustainable Rural Development, Romania, mikiraicov@gmail.com

Valentina Ofelia Robescu
"Valahia" University of Targoviste, Romania, robescu_ofelia@yahoo.com

Ciprian Ioan Rujescu
University of Life Sciences "King Mihai I" from Timisoara, Faculty of Management and Rural Tourism, Romania, rujescu@usvt.ro

Florin Sala
Agricultural Research and Development Station, Lovrin, Romania; University of Life Sciences "King Mihai I" from Timisoara, Romania, florin_sala@usvt.ro

Andreia Sasu
University of Life Sciences "King Mihai I" from Timişoara, Faculty of Engineering and Applied Technologies, Romania, andreia_botos@yahoo.com

Vasile Secrieru
Academy of Economic Studies of Moldova, Chisinau, Republic of Moldova, secrieruvasile@mail.ru

Manish Sen
AISECT, Jaipur, Rajasthan, India, manish.sen@aisect.org

Alina Şimon
Agricultural Research Development Station, Turda, Romania, alina.simon@scdaturda.ro

Reshu Gupta Singh
Poddar International College, Jaipur/Rajasthan, India, reshugupta111@gmail.com

Srdjan Šljukić
University of Novi Sad, Faculty of Philosophy, Republic of Serbia, srdjan.sljukic@ff.uns.ac.rs

Oleksandra Smirnova
State Tax University, Irpin, Ukraine, o.m.smirnova@dpu.edu.ua

Laura Iosefina Smuleac
University of Life Sciences, "King Mihai I of Romania" from Timisoara, Faculty of Agriculture, Romania, laurasmuleac@usvt.ro

Sorin Mihai Stanciu
University of Life Sciences, "King Mihai I of Romania" from Timisoara, Faculty of Management and Rural Tourism, Romania, sorinstanciu@usvt.ro

Árpád Székely
Research Center for Irrigation and Water Management, Institute of Environmental Sciences, Hungarian University of Agriculture and Life Sciences, Szarvas, Hungary.

Silviu Gabriel Szentesi
"Aurel Vlaicu" University of Arad, Faculty of Economics, Arad, Romania, silviuszentesi@yahoo.com

Kritika Tekwani
Indian Institute of Management, Ahmedabad/Gujrat, India, kritikat@iima.ac.in

Eugenia Țigan
"Aurel Vlaicu" University of Arad, Faculty of Food Engineering, Tourism and Environmental Protection, Arad, Romania, eugenia.tigan@uav.ro

Zoia Titenko
National University of Life and Environmental Sciences of Ukraine, Kyiv, Ukraine, zoyateslenko@ukr.net

Cosmina-Simona Toader
University of Life Sciences "King Mihai I" from Timisoara, Faculty of Management and Rural Tourism, Romania, cosminatoader@usvt.ro

Leyda Todorova
Agricultural Academy, Fruit Growing Institute, Plovdiv, Bulgaria; l.todorowa@abv.bg

Iveta Ubrežiová
Catholic University in Ružomberok, Faculty of Education, Slovakia, iveta.ubreziova@ku.sk

Noémi Valkovszki
Research Center for Irrigation and Water Management, Institute of Environmental Sciences, Hungarian University of Agriculture and Life Sciences, Szarvas, Hungary.

Vishal Singh Varma
Rajasthan Grameen Aajeevika Vikas Parishad, Jaipur/Rajasthan, India, vishal.singh.varma17@gmail.com

Sanyam Varma
Bio-Technologist, Jaipur, Rajasthan, India, verma.sanyam@gmail.com

Ariana Velciov
University of Life Sciences "King Mihai I" from Timisoara, Faculty of Food Engineering, Romania, arianavelciov@usvt.ro

Ado Yakubu
Department of Agricultural Economics, Faculty of Agriculture/Institute for Agricultural Research, Ahmadu Bello University, Zaria, Nigeria, adoydogua@gmail.com

Małgorzata Zajdel
Bydgoszcz University of Science and Technology, Faculty of Management, Poland, malgorzata.zajdel@pbs.edu.pl

Radu-Lucian Blaga, Eugenia Țigan, Simona Gavrilaș

Economic and environmental approaches for agro-based sectors sustainable development

Abstract: *Since the farming areas and the natural resources are limited in space and time, novel solutions must be implemented. The present article targets current views on integrating the relatively new concept of high-value natural farming areas into future agriculture. The extensive analysis of the current state of the art highlights the advantages of implementing the mentioned concept for these zones. We used data from the most comprehensive databases to conduct this systematic literature review, focusing primarily on the economic and environmental aspects. The preferred reporting items for methodic evaluation and meta-analyzes guided the data selected for this paper. The views discussed, and the tools analyzed through the article are aligned with the current trends of having a minimal environmental impact and obtaining healthier products. The concrete points addressed respond to the contemporary specific market demands. The analysis results show that the high-value natural farming areas offer several economic and ecological advantages and disadvantages that must be considered when making management and conservation decisions and assessing their sustainability. Moreover, a sustained investment approach is essential for the world's agrifood systems to become more resilient, efficient, sustainable, and inclusive. Social capital in local communities supports the symbiosis of agriculture and tourism. However, optimal regional linkages are hampered by the quantitative and qualitative mismatch between locally supplied products and the tourism sector requirements. Based on these premises, we substantiated our investigation, which aimed to determine the degree of interest of the residents of the Lipovi Hills area for ecological label guesthouses. The research results highlight that most respondents need to be more familiar with what the European ecological label or ecological product means. These elements lead us in the direction of recommending continuous training and the idea of developing new information projects and awareness in the field.*

Keywords: high nature value farming area, agrifood systems, ecotourism, sustainable land use, agriculture funding, environmental impact, local communities.

Introduction

Building a sustainable society can only be done by implementing durable principles in all sectors. Agriculture is no exception. It has to be viewed as a joint able to sustain the area from the financial, ecosystemic, public, and nutritional points of view (Soussana, 2014). The vision of rural development and its alignment with local planning can create meaningful change with sufficient funding and support (Park and Lee, 2019).

The article comprises two main parts. One *presents the current state of the art regarding the possibilities of designing and developing eco-friendly economic activities*. The focus is on the rural plain and hilly areas. The second is a *survey implemented in a specific part of West Romania, targeting the ecolabel guest houses.*

The review part of the article aims to resume, identify, and understand new insights into agri-environmental systems that promote sustainable and efficient eco-friendly farming practices that ensure long-term food security. It was also important to highlight how conservation agriculture can help maintain soil fertility, reduce erosion, and conserve water resources in the *high nature value farming areas* (HNVF) systems. Such a concept is closely related to the routes for involving local communities and small farmers in sustainable agricultural practices and how awareness of the importance of agri-environment vision can be promoted among the farmers. These were the fundaments considered initially for this study.

The second aspect of interest was evaluated through a survey. Its purpose was to quantify the population situated in the Lipova Hill area's interest in ecologically labelled products, targeting mainly the guest houses. Such facilities can be developed in locations relatively close to where the analysis was carried out. Two hypotheses were formulated. The first (**H1**) from which the study started considered *that the inhabitants of the area prefer to stay in hotels during a stay to the detriment of boarding houses*. The second (**H2**) assumes *that education, in this sense, is necessary for requirements for guesthouses with the European ecological label or the development of ecotourism.*

Literature review

Agro systems with high natural value

Adopting the proper direction to ensure greeneries agrochemical growth conditions and minimizing the ecological impact will contribute to sustainable agricultural development (Sheoran et al., 2022).

The *high nature value farming area* (HNVF) concept was considered an integrative vision. Low-intensity agricultural landscapes (HNVF) are characterized, according to Lomba et al. and Mäkeläinen et al., by the prevalence of high levels of natural and semi-natural habitats (Lomba al., 2014; Mäkeläinen et al., 2019). The local communities can access funding grants that sustain their daily life wellbeing. The ones that targeted the HNVF showed some possible improvements in this sector (Šatalová et al., 2021). The production activities tend to have a minimum environmental impact due to applying bio-economy principles, the target being zero waste (Torres-Miralles et al., 2022).

Financing systems for high nature values farming areas

HNVF areas have unique ecological, landscape, and cultural features that depend on traditional farming practices. Funding schemes for them are essential to support sustainable agriculture and biodiversity conservation. The financial support offered to these farmers and communities helps ensure their survival while promoting sustainable land use, biodiversity, and cultural heritage conservation. There are various funding options for HNVF.

Financial institutions may offer unique loan products with lower interest rates and more extended repayment periods for farmers in HNVF areas. Credit plays a vital role in purchasing inputs among farmers to improve agricultural productivity (Dawuni, Mabe and Tahidu, 2021). Farmers must understand better lending principles and credit risk assessment to use their inherent business strengths in competing for loans with conventional farms (Jones, Escalante and Rusiana, 2015).

Banks rely on qualitative factors and excessive collateral to overcome information asymmetry when making lending decisions (Sandhu, 2021). Adequate training and skills development is needed for all, especially young people and entrepreneurs involved in agricultural management and marketing, good farming, and organic practices (Bhinekawati, 2016; Jones, Escalante and Rusiana, 2015).

Another global phenomenon noted at the beginning of the 21st century is "agricultural finance", the involvement of finance companies in purchasing and leasing agricultural land. While it is financially beneficial for companies to invest in agriculture, they do so in locations where people want to live in a healthy environment without intensive farming (Sippel, Lawrence and Burch, 2017).

Economic and ecological advantages/disadvantages for high nature value farming areas – Economic benefits: Tourism

Tourists can be attracted to the HNVF area to engage in various recreational activities: hiking, flora and wildlife viewing, and cultural tours, contributing to the local economy by spending on accommodation, meals, entertainment, and local product purchases. Table 1 summarizes some main advantages the population may consider, whether countryside or town residents.

Table 1. The high-natural-value farmland areas drafted primary interests

Main interest	Reference
Revalorize the traditions that were lost due to society upgrading data	(Lee and Kim, 2010)
Remediation impact on farming regions that were subject to many transformations in the past years	(Fleischer and Tchetchik, 2005; Schermer et al., 2016)
Adding value by sensitizing the population concerning their significance	(Tew and Barbieri, 2012)
Farm's social resilience, enabling farmers to use local food supply chains	(Stotten, 2021)

Reconversion

This idea is another economic advantage offered by HNVF areas that can reduce the risk of farm business breakdown due to crop failure and market fluctuations, thus providing farms with more stable incomes. These are among the diversification efforts initiated to increase HNVF resilience. The principles of such an approach have been less considered till now (Hill, 2017).

Economic disadvantages: Low productivity

Farming practices in HNVF areas can generate lower productivity than intensive. Musafili et al. discussed a holistic perspective in their study. The approach targeted cultural tourism and local craft valorization (Musafili et al., 2019).

Enhanced rates

Strict Sanitary and Phyto-Sanitary (SPS) actions and Maximum Residue Levels (MRL) related to such agricultural practices affect consumer behavior and market prices (Crivelli and Gröschl, 2016; Drogué and DeMaria, 2012; Fontagné et al., 2015; Li and Beghin, 2014; Xiong and Beghin, 2014). These Non-Tariff Measures (NTMs) regulate international trade in agrifood commodities, sometimes even preventing it, by imposing trade protectionism.

Over time, it became a competitive advantage for farmers committed to adopting organic production and its quality certification, allowing their integration into value chains and increasing the viability and resilience of farms in these areas (Bolwig, Gibbon and Jones, 2009; Chiputwa, Spielman and Qaim, 2015).

Minimal commercial entree

The impact differs as markets and standards become more stringent, posing a barrier to integrating low-income countries into global agrarian export markets (Ehrich and Mangelsdorf, 2018; Herzfeld, Drescher and Grebitus, 2011; Melo et al., 2014). A remarked tendency, especially for small producers, is to implement regulations to access international markets individually (Olper, Curzi et and Pacca, 2014). Such an approach might have a positive influence on export trends.

Constrained supplies

HNVF areas may need more resources, such as the technical, marketing, and management knowledge of farmers, water, and arable land, which can reduce agriculture's productivity and economic viability. Individual farms often need more skills and resources to market the business effectively (Embacher, 1994).

The perceived yield of green manufacturing and the negative environmental effect of conventional agriculture on the soil positively influenced the intensity of adopting these measures (Dapaah Opoku, Bannor and Oppong-Kyeremeh, 2020).

Environmental benefits: Ecosystems preservations

It is considered that the HNVF areas cover more than a quarter of the European Union's (E.U.) agricultural land. Their main attributes are biodiversity saving, countryside view, regional cohesiveness, characteristic products, and/or local recruitment (Lomba et al., 2020).

The importance of the landscape-territory vision as an entry point for modelling HNVF systems toward socially desirable scenarios is noted (Bernard et al., 2023). The changes are part of the trend highlighted by Lomba et al. of moving HNVF towards social-ecological sustainability, such as improving rural services, encouraging technological innovation, and rewarding the provision of ecosystem services (Lomba et al., 2020).

Soil, water, and landscapes are biodiversity hotspots and habitats for rare and protected species (Moreno et al., 2016; Pereira et al., 2019). These elements provide multiple ecosystem services by participating in water and nutrient cycling, contributing to its overall regulation (Guerra, Pinto-Correia and Metzger, 2014; Moreno et al., 2018).

To better manage agricultural landscapes, it is essential to understand the relationship between contrasting farmland practices, resulting landscape patterns, biodiversity, and the delivery of ecosystem services (E.S.) (Buchadas et al., 2022).

Carbon segregation

Several conventional agricultural practices damage the environment and are a significant source of anthropogenic greenhouse gases (GHGs) such as CO_2 and NO_2. Waheed et al., modelling data between 1990 and 2014, found that carbon dioxide emissions can be reduced by increasing the use of renewable energy and forest land while decreasing the use of this land in agriculture (Waheed et al., 2018).

The SMART-type farm preserves the environment by using clean energy, emitting less carbon, and reducing water waste. The results are powered by solar energy, and recycling water containing waste feeds the hydroponics. In conclusion, smart farming is possible, feasible, and sustainable (Musa and Basir, 2021).

Environmental disadvantages: Ground relinquishment

HNVF areas can become unsustainable over time, leading to land abandonment and the loss of ecological and cultural value attributed to agricultural landscapes (MacDonald et al., 2000). Many HNVFs are widely used in natural and semi-natural grasslands in mountainous areas, often managed as shared property (Keenleyside et al., 2014; O'rourke, Charbonneau and Poinsot, 2016). Due to their relatively low agricultural value and current farming technologies, these marginal and step-grazed land types are among the agrarian landscapes most prone to abandonment (Plieninger and Bieling, 2013; Verburg et al., 2010).

Land erosion and parcelling

Conventional management includes grazing, mowing, burning, and removing trees and shrubs but excludes plowing or substantial fertilization (Norderhaug and Johansen, 2011). Intensively managed farming systems provide mainly provisioning services (e.g., food and fibre), while low-intensity farming systems can support a broader range of ecosystem services (E.S.) and higher levels of biodiversity (Foley et al., 2007; Power, 2010; Rockström et al., 2017; Swinton et al., 2007). To better manage High Nature Value Farmland (HNVF), it is essential to understand the relationship between contrasting farmland practices, resulting landscape patterns biodiversity, and the delivery of these E.S. (Buchadas et al., 2022).

To sufficiently address the threats facing natural and semi-natural habitats, policy and research frameworks in the European Union should pay more attention to the socioecological complexity inherent in their management and support the involvement of different actors in participatory governance processes. It is consistent with a whole-farm approach implicit in high-nature-value farming systems (Herzon et al., 2021).

Methodology

In agriculture, environment, business, and management, knowledge is being generated at an enormous pace while staying fragmented and interdisciplinary simultaneously. The main objective of the systematic literature review is to observe the trends of integrating the relatively new concept of high-value natural farming areas into future agriculture, having a minimal environmental impact and obtaining healthier products. As a result, we conducted a systematic literature review of previous studies on economic and ecological approaches for the sustainable development of the agro-based sector.

The research area considered is geographically, socio-economically, and historically delimited of the Lipova Hills. One of the reasons for choosing it is its location near the Natura 2000 Natural East Zarandul Site. There are optimal conditions for ecotourism and the possibility of creating guesthouses with the European ecological label.

The quantitative research method was applied to empirically validate the proposed hypotheses, with the questionnaire as the research tool. The research consisted of the field application of questionnaires with semi-open, scaled, and factual questions to 180 respondents between October and November 2022. Later, the SPSS program was used for data analysis and interpretation. The sampling procedure used was non-random (non-probability) with independent quotas. The stratification factors chosen were the environment of origin of the respondents (the urban environment of Lipova and Recaș, respectively, the rural environment communes adjacent to the two cities) and their gender (male, female), the limitation imposed by the established variables facilitating the effectiveness of surveying in the direction of identifying people which correspond to the indicated quotas, at the same time increasing the subjectivity of the analysis.

After collecting the information, databases were created, and subsequent data analysis was performed in the SPSS program-IBM Statistics-Version 23, provided by IBM Corp. The expected error was ±5 %, with a probability threshold of 95 %. The respondent's demographical aspects are presented in Table 2.

Table 2. Respondents' socio-demographic aspects

Feature	Characteristics	Share [%]
Age	20–30	6.7
	31–40	66.7
	41–50	16.6
	51–60	10.0
Gender	Male	53.3
	Female	46.7
Training Level	vocational school	13.3
	high school	43.3
	post-high school	6.7
	higher education	36.7
Social status	employee	63.3
	student	16.7
	retired	6.7
	unemployed/	3.3
	long-term leave employer	10.0
Environment	rural	53.3
	urban	46.7

Source: Authors' own research.

Results and discussions

The studied references have highlighted that implementing a *sustained investment approach* and *preserving high nature-value areas* (maintaining biodiversity, conserving soil, contributing to carbon sequestration) will contribute to more sustainable and climate-resilient agriculture. From economic and environmental perspectives, the importance of exploring more sustainable alternatives in agricultural practices was underlined. Ecotourism is dominant in many areas of less developed countries. It allows tourists to interact with rural, agrarian communities. Maintenance and preservation of distinct landscape features can be assessed by tracking the diversity of peculiar elements and their state of conservation (Pinto-Correia et al., 2022).

For which products/services did you first hear the term eco-labelling? The question was asked to familiarize the respondents with the specifics of the topic chosen for the research. To this question, an equal proportion of 40 % of the respondents claimed household electrical or food products. Only 6.7 % consider this European ecological label to be applied to tourist services, Figure 1. This first answer highlights two aspects: *most respondents need to differentiate between products with European ecological labels*, the label that does not apply to food products, and *biological products*, which come from biological agriculture. This question already *confirms the second hypothesis of the current study.*

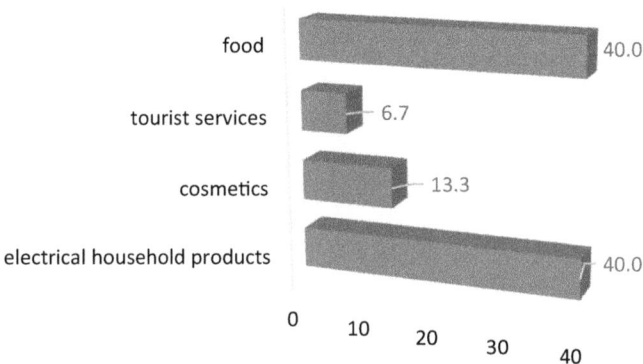

Figure 1. Respondent's answers regarding the first type of eco-labelled product they heard of
Source: Data from field quizzes 2022.

The field research being carried out in the area with small towns, such as the cities of Lipoca and Recas, highlights that the respondents and, respectively, the residents of this area, and probably not only them, are eager for vacations spent in hotels, Figure 2. Thus, 66.7 % of the respondents chose this option when asked *Where they usually stay in Romania when they travel for relaxation outside the locality.* Only 10 % prefer guesthouses, and 23.3 % prefer villas.

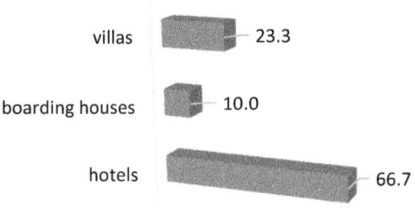

Figure 2. Respondents preferences for the accommodation units when they travel outside the home town for relaxation purposes
Source: Data from field quizzes 2022.

A direct correlation can also be observed between the choice of a hotel as a vacation location and the respondent's residence (Table 1), with 53.3 % of them coming from rural medium and 46.7 % from urban. The fact that most of the respondents prefer to stay in hotels confirms the first hypothesis of this research, which *assumes that the residents of the researched area choose to stay in hotels to the detriment of boarding houses.*

Although they prefer hotels, they want to fit into the allocated budget, 33.3 %, but also to be different vacation options compared to previous destinations, 43.3 %, and 13.3 % prefer to be a destination recommended by acquaintances, Figure 3.

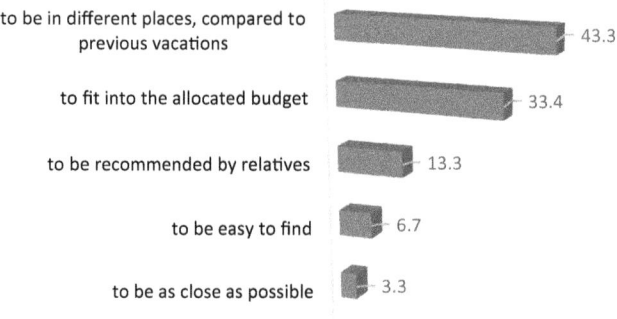

Figure 3. Respondent's criteria when choosing an ecotourism destination for a holiday
Source: Data from field quizzes 2022.

The vast majority of respondents claim that they get information about a vacation destination on the Internet, 46.7 %, or on the recommendation of friends, 30 %, while 20 % turn to agencies in this regard, Figure 4.

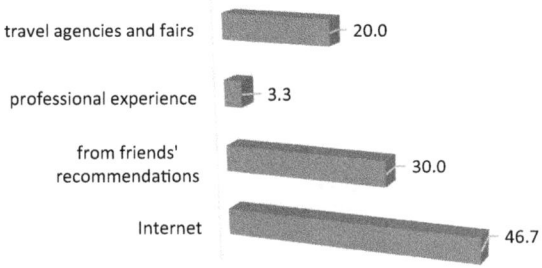

Figure 4. Respondent's sources of information about vacation destination and the services needed to spend it
Source: Data from field quizzes 2022.

Although the respondents' age pattern has a reasonably balanced breakdown, those aged between 31 and 40 are predominant at 66.7 %. This aspect is correlated with 50.6 % of them saying they want an exotic destination when choosing a vacation destination; local traditions attract 20 %, Figure 5. These elections also highlight that the second hypothesis is confirmed.

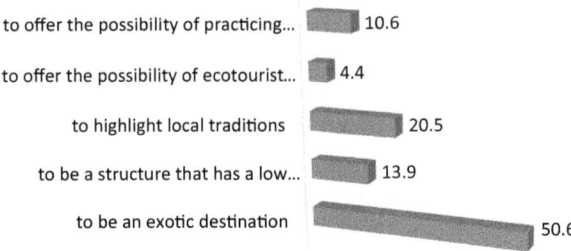

Figure 5. What respondents look when they choose a specific vacation destination
Source: Data from field quizzes 2022.

Another aspect that was asked of the respondents was to express their opinion regarding what they consider to be lacking in the development of ecotourism in Romania. In this regard, most respondents think there is a lack of specialized staff, 50 %, followed by those who consider that there is a lack of interest in promoting this type of tourism, 30 %, Figure 6.

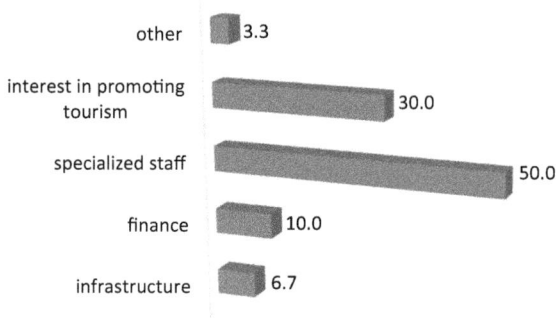

Figure 6. What respondents think is missing in the development of Romanian ecotourism
Source: Data from field quizzes 2022.

Conclusions

The failure of many HNVF initiatives to address them creates a need to empower local communities to take advantage of travel opportunities. One solution is to link the two domains as a promising way to make sightseeing more economically inclusive by using the value chain approach in this linkage. In such a way, the local farmers are incorporated into tourism food supply chains ethically and beneficially.

The low productivity of agricultural practices in HNVF areas generates delinquent behaviors among farmers. On the one hand, the acceptance of low farm yields complemented by diversification of non-agricultural activities valorizes the natural and cultural heritage of the area. On the other hand, the intensification of agricultural activities results in a gradual loss of the traditional meaning of agriculture and orientation towards mass food production.

In addition to confirming the assumptions from which it started, the research also brings us a plus, highlighting that ecotourism development in Romania needs more qualified personnel. These answers can be considered an alarm signal for ecotourism development in Romania.

One of the limitations of the current research is that the area under study is geographically extended and could not be included in its entirety, proposing that the cities be more eloquent. Another limitation was that many respondents were sceptical about answering these questionnaires even though they were anonymous.

References

Bernard, C., Poux, X., Herzon, I., Moran, J., Pinto-Correira, T., Dumitras, D.E.,Ferraz-de-Oliveira ... Vlahos, G. 2023. Innovation brokers in high nature value farming areas: A strategic approach to engage effective socioeconomic and agroecological dynamics. *Ecology and Society*, 28(1), p. 20. https://doi.org/10.5751/ES-13522-280120.

Bhinekawati, R. 2016. *Corporate social responsibility and sustainable development: Social capital and corporate development in developing economies.* 1st ed. London: Gower. https://doi.org/10.4324/9781315395463.

Bolwig, S., Gibbon, P. and Jones, S. 2009. The economics of smallholder organic contract farming in tropical Africa. *World Development*, 37(6), pp. 1094–1104. https://doi.org/10.1016/j.worlddev.2008.09.012.

Buchadas, A., Moreira, F., McCracken, D., Santos J. L. and Lomba, A. 2022. *Ecology Society*, 27(1), p. 5. https://doi.org/10.5751/ES-12947-270105.

Chiputwa, B., Spielman, D.J. and Qaim, M. 2015. Food standards, certification, and poverty among coffee farmers in Uganda. *World Development*, 66, pp. 400–412. http://dx.doi.org/10.1016/j.worlddev.2014.09.006.

Crivelli, P. and Gröschl, J. 2016. The impact of sanitary and phytosanitary measures on market entry and trade flows. *The World Economy*, 39(3), pp. 444–473. http://hdl.handle.net/10419/73846.

Dapaah Opoku, P., Bannor, R.K. and Oppong-Kyeremeh, H. 2020. Examining the willingness to produce organic vegetables in the Bono and Ahafo regions of Ghana. *International Journal of Social Economics*, 47(5), pp. 619–641. https://doi.org/10.1108/IJSE-12-2019-0723.

Dawuni, P., Mabe, F.N. and Tahidu, O.D. 2021. Effects of village savings and loan association on agricultural value productivity in Northern Region of Ghana. *Agricultural Finance Review*, 81(5), pp. 657–674. https://doi.org/10.1108/AFR-02-2020-0024.

Drogué, S. and DeMaria, F. 2012. Pesticide residues and trade, the apple of discord? *Food Policy*, 37(6), pp. 641–649. https://doi.org/10.1016/j.foodpol.2012.06.007.

Ehrich, M. and Mangelsdorf, A. 2018. The role of private standards for manufactured food exports from developing countries. *World Development*, 101 pp.16-27. https://doi.org/10.1016/j.worlddev.2017.08.004.

Embacher, H. 1994. Marketing for Agri-tourism in Austria: Strategy and realisation in a highly developed tourist destination. *Journal of Sustainable Tourism*, 2(1–2), pp. 61–76. https://doi.org/10.1080/09669589409510684.

Fleischer, A. and Tchetchik, A. 2005. Does rural tourism benefit from agriculture?, *Tourism Management*, 26(4), pp. 493–501. https://doi.org/10.1016/j.tourman.2003.10.003.

Foley, J.A., Asner, G.P., Costa, M.H., Coe, M.T., DeFries, R., Gibbs, H.K., Howard, E.A., Olson, S., Patz, J., Ramankutty, N. and Snyder. P. 2007. Amazonia revealed: forest degradation and loss of ecosystem goods and services in the Amazon Basin. *Frontiers in Ecology the Environment*, 5(1), pp. 25–32.

Fontagné, L., Orefice, G., Piermartini, R. and Rocha, N. 2015. Product standards and margins of trade: Firm-level evidence. *Journal of International Economics*, 97(1), pp. 29–44. https://doi.org/10.1016/j.jinteco.2015.04.008.

Guerra, C., Pinto-Correia, T. and Metzger, M. 2014. Modeling framework mapping soil erosion using an ecosystem land management. *Journal of Ecosystems*, 17(5), pp.878–889. https://doi.org/10.1007/s10021-014-9766-4.

Herzfeld, T., Drescher, L.S. and Grebitus, C. 2011. Cross-national adoption of private food quality standards. *Food Policy*, 36(3), pp. 401–411. https://doi.org/10.1016/j.foodpol.2011.03.006.

Herzon, I., Raatikainen, K.J., Wehn, S., Rūsiņa, S., Helm, A., Cousins, S.A.O. and Rašomavičius, V. 2021. Semi-natural habitats in boreal Europe: A rise of a social-ecological research agenda. *Ecology Society*, 26(2). https://doi.org/10.5751/ES-12313-260213.

Hill, B. 2017. *Farm incomes, Wealth and Agricultural Policy*. 3rd ed. Routledge.

Jones, G., Escalante, C. and Rusiana, H. 2015. Reconciling information gaps in organic farm borrowers' dealings with farm lenders. *Agricultural Finance Review*. https://doi.org/10.1108/AFR-01-2015-0002.

Keenleyside, C., Beaufoy, G., Tucker, G. and Jones, G. 2014. *High nature value farming throughout EU-27 and its financial support under the CAP*. Institute for European Environmental Policy, London, Vol. 10, p. 91086. [online] Available at: <https://ieep.eu/publications/high-nature-value-farming-throughout-eu-27-and-its-financial-support-under-the-cap/> [Acccesed 27 September 2023].

Lee, S. and Kim, H.J. 2010. Agricultural transition and rural tourism in Korea: Experiences of the last forty years. *Agricultural Transition in Asia*, pp. 37–64.

Li, Y. and Beghin, J.C. 2014. Protectionism indices for non-tariff measures: An application to maximum residue levels. *Food Policy,* 45, pp. 57–68.

Lomba, A., Guerra, C., Alonso, J., Honrado J.P., Jongman, R. and McCracken. D. 2014. Mapping and monitoring high nature value farmlands: Challenges in European landscapes. *Journal of Environmental Management,* 143, pp. 140–150. http://dx.doi.org/10.1016/j.jenvman.2014.04.029.

Lomba, A., Moreira, F., Klimek, S., Jongman R.H.G., Sullivan, C., Moran J., Poux, X., Honrado, J.P., Pinto-Correia, T., Plieninger, T. and McCracken D. 2020. Back to the future: Rethinking socioecological systems underlying high nature value farmlands. *Frontiers in Ecology the Environment,* 18(1), pp. 36–42. https://doi.org/10.1002/fee.2116.

MacDonald, D., Crabtree, J.R., Wiesinger, G., Dax, T., Stamou, N., Fleury, P., Gutierrez Lazpita J and Gibon, A. 2000. Agricultural abandonment in mountain areas of Europe: Environmental consequences and policy response. *Journal of Environmental Management,* 59(1), pp. 47–69. https://doi.org/10.1006/jema.1999.0335.

Mäkeläinen, S., Harlio, A., Heikkinen, R.K., Herzon, I., Kuussaari, M., Lepikko, K., Maier, A.,, Seimola, T., Tiainen, J. and Arponen, A. 2019. Coincidence of high nature value farmlands with bird and butterfly diversity. *Agriculture, Ecosystems & Environment,* 269, pp. 224–233. https://doi.org/10.1016/j.agee.2018.09.030.

Melo, O., Engler, A., Nahuehual, L., Cofre, G. and Barrena, J. 2014. Do sanitary, phytosanitary, and quality-related standards affect international trade? Evidence from Chilean fruit exports. *World Development,* 54, pp. 350–359. https://doi.org/10.1016/j.worlddev.2013.10.005.

Moreno, G., Gonzalez-Bornay, G., Pulido, F., Lopez-Diaz, M.L., Bertomeu, M., Juarez, E. and Diaz, M. 2016. Exploring the causes of high biodiversity of Iberian dehesas: the importance of wood pastures and marginal habitats. *Agroforestry Systems,* 90, pp. 87–105. https://doi.org/10.1007/s10457-015-9817-7.

Moreno, G., Aviron, S., Berg, S., CrousDuran, J., Franca, A., Garcia de Jalon, S., Hartel, T., Mirck, J., Pantera, A., Palma, J.H.N., Paulo, J.A., Re G.A., Sanna, F., Thenail, C., Varga, A., Viaud, V. and Burgess, P.J. 2018. Agroforestry systems of high nature and cultural value in Europe: Provision of commercial goods and other ecosystem services. *Agroforestry Systems,* 92, pp. 877–891. https://doi.org/10.1007/s10457-017-0126-1.

Musa, S.F.P.D. and Basir, K.H. 2021. Smart farming: Towards a sustainable agri-food system. *British Food Journal,* 123(9), pp. 3085–3099. https://doi.org/10.1108/BFJ-03-2021-0325.

Musafili, I., Ngabitsinze, J.C., Niyitanga, F. and Weatherspoon, D. 2019. Farmers' usage preferences for Rwanda's volcanoes national park. *Journal of Agribusiness in Developing Emerging Economies*, 9(1), pp. 63–77. https://doi.org/10.1108/JADEE-01-2018-0004.

Norderhaug, A. and Johansen, L. 2011. Semi-natural sites and boreal heaths. In: A. Lindgaard and S. Henriksen, eds. *The 2011 Norwegian Red List for ecosystems and habitat types*. Norway: Norwegian Biodiversity Information Centre, pp. 87–93.

O'rourke, E., Charbonneau, M. and Poinsot, Y. 2016. High nature value mountain farming systems in Europe: Case studies from the Atlantic Pyrenees, France and the Kerry Uplands, Ireland. *Journal of Rural Studies*, 46, pp. 47–59. http://dx.doi.org/10.1016/j.jrurstud.2016.05.010.

Olper, A., Curzi, D. and Pacca, L. 2014. Do food standards affect the quality of EU imports?. *Economics Letters*, 122(2), pp. 233–237. https://doi.org/10.1016/j.econlet.2013.11.031.

Park, J. and Lee, S. 2019. Smart village projects in Korea: Rural tourism, 6th industrialization, and smart farming. In: A. Visvizi, M.D. Lytras and G. Mudri, eds. *Smart Villages in the EU and Beyond*. Leeds: Emerald Publishing Limited, pp. 139–153. https://doi.org/10.1108/978-1-78769-845-120191011.

Pereira, P.F., Lourenco, R., Lopes, C., Oliveira, A., Ribeiro-Silva, J., Rabaca, J.E., Pinto-Correia, T., Figueiredo, D., Mira, A. and Tiago Marques, J. 2019. The influence of management and environmental factors on insect attack on cork oak canopy. *Forest Ecology Management*, 453, p. 117582. https://doi.org/10.1016/j.foreco.2019.117582.

Pinto-Correia, T., Ferraz-de-Oliveira, I., Guimaraes, M.H., Sales-Baptista, E., Pinto-Cruz, C., Godinho, C. and Santos, R.V. 2022. Result-based payments as a tool to preserve the High Nature Value of complex silvo-pastoral systems: progress toward farm-based indicators. *Ecology Society*, 27(1). https://doi.org/10.5751/ES-12973-270139.

Plieninger, T. and Bieling, C. 2013. Resilience-based perspectives to guiding high-nature-value farmland through socioeconomic change. *Ecology Society*, 18(4). http://dx.doi.org/10.5751/ES-05877-180420.

Power, A.G. 2010. Ecosystem services and agriculture: Tradeoffs and synergies. *Philosophical Transactions of the Royal Society B: Biological Sciences*, 365(1554), pp. 2959–2971.

Rockström, J., Williams, J., Daily, G., Noble, A., Matthews, N., Gordon, L., Wetterstrand, H. ... and Smith, J. 2017. Sustainable intensification of agriculture for human prosperity and global sustainability. *Ambio*, 46, pp. 4–17. https://doi.org/10.1007/s13280-016-0793-6.

Sandhu, N. 2021. Dynamics of banks' lending practices to farmers in India. *Journal of Small Business Enterprise Development,* 28(1), pp. 102–120. https://doi.org/10.1108/JSBED-05-2019-0161.

Šatalová, B., Špulerova, J., Štefunkova, D., Dobrovodska, M., Vlachovicova, M. and Kozelova, I. 2021. Monitoring and evaluating the contribution of the rural development program to high nature value farmland dominated by traditional mosaic landscape in Slovakia. *Ecological Indicators,* 126, p. 107661. https://doi.org/10.1016/j.ecolind.2021.107661.

Schermer, M., Darnhofer, I., Daugstad, K., Gabillet, M., Lavorel, S. and Steinbacher, M. 2016. Institutional impacts on the resilience of mountain grasslands: An analysis based on three European case studies. *Land Use Policy,* 52, pp. 382–391. https://doi.org/10.1016/j.landusepol.2015.12.009.

Sheoran, P., Sharma, R., Kumar, A., Singh R.J., Barman, A., Prajapat, K., Kumar S. and Sharma P.C. 2022. Climate resilient integrated soil–crop management (CRISCM) for salt affected wheat agri–food production systems. *Science of The Total Environment,* 837, p. 155843. https://doi.org/10.1016/j.scitotenv.2022.155843.

Sippel, S.R., Lawrence, G. and Burch, D. 2017. The financialization of farming: The hancock company of Canada and its embedding in rural Australia. In: *Transforming the Rural* (Research in Rural Sociology and Development, vol. 24). Leeds: Emerald Publishing Limited, pp. 3–23.

Soussana, J.-F. 2014. Research priorities for sustainable agri-food systems and life cycle assessment. *Journal of Cleaner Production,* 73, pp. 19–23. http://dx.doi.org/10.1016/j.jclepro.2014.02.061.

Stotten, R. 2021. The role of farm diversification and peasant habitus for farm resilience in mountain areas: the case of the Ötztal valley, Austria. *International Journal of Social Economics,* 48(7), pp. 947–964. https://doi.org/10.1108/IJSE-12-2019-0756.

Swinton, S.M., Lupi, F., Robertson, G.P. and Hamilton, S.K. 2007. Ecosystem services and agriculture: Cultivating agricultural ecosystems for diverse benefits. *Ecological Economics,* 64(2), pp. 245–252. https://doi.org/10.1016/j.ecolecon.2007.09.020.

Tew, C. and Barbieri, C. 2012. The perceived benefits of agritourism: The provider's perspective. *Tourism Management,* 33(1), pp. 215–224. https://doi.org/10.1016/j.tourman.2011.02.005.

Torres-Miralles, M., Sarkela K., Koppelmaki, K., Lamminen, H.L. and Herzon, I. 2022. Contribution of High Nature Value farming systems to sustainable livestock production: A case from Finland. *Science of The Total Environment,* 839, p. 156267. https://doi.org/10.1016/j.scitotenv.2022.156267.

Verburg, P.H., van Berkel, D.B., van Doorn, A.M., van Eupen, M. and van den Heiligenberg H.A. 2010. Trajectories of land use change in Europe: a model-based exploration of rural futures. *Landscape Ecology*, 25, pp. 217–232. https://doi.org/10.1007/s10980-009-9347-7.

Waheed, R., Chang, D., Sarwar, S. and Chen, W. 2018. Forest, agriculture, renewable energy, and CO2 emission. *Journal of Cleaner Production*, 172, pp. 4231–4238. https://doi.org/10.1016/j.jclepro.2017.10.287.

Xiong, B. and Beghin, J. 2014. Disentangling demand-enhancing and trade-cost effects of maximum residue regulations. *Economic Inquiry*, 52(3), pp. 1190–1203. https://doi.org/10.1111/ecin.12082.

Karoly Bodnar, Zoltan Istvan Privoczki, Ioan Brad

Development of the price of agricultural land in the South-Eastern region of Hungary

Abstract: *In the South-Eastern region of Hungary, the price of agricultural land in general has risen steadily over the past decade. In order to investigate the reasons, data and information from the Central Statistical Office, the National Chamber of Agriculture and the OTP were processed. The increase in prices was 5–15 % on average and annually. The change in prices depended significantly on the location of the area, the direction of utilization, the type and quality of the soil, the labor situation, the possibility of irrigation, etc. The currently available agricultural subsidies also affected the supply and demand relationship. The frequent changes in the legal regulation of land sales and purchases require special attention from those concerned. In order to speed up the generational change, the legislation tries to put young farmers in a more favorable situation by changing the pre-purchase order. The established situation mainly favors large enterprises with strong capital and makes it increasingly difficult for farms belonging to the SME category to get access to land.*

Keywords: land price, area differences, land quality, economic factors, legal regulation.

Introduction

Land (as the basis and source of agricultural production) has a prominent role among all means of production. The ownership, use, value and price of land as a natural resource has become one of the key problems of political, social, legal and economic decisions. In the European Union, there is no unified theory or method that would be a generally accepted basis for practical land valuation. There are places where an overseas assessment method is used, and there are places where a method based on national traditions is used. Thus, the price and value of land varies from country to country, influenced by many factors. Romania, Germany, Italy, Poland and Hungary have the largest share of agriculturally usable land in the EU in relation to the size of the country (Stoica and Dumitru, 2021).

In Hungary, the price of agricultural land in all fields of cultivation showed a significant and continuous increase in the previous decades (Stoica and Dumitru (2021) showed a similar process in Romania), and it is expected that if production continues on them, it will retain its value. A significant increase in the number of sales is expected thanks to the new state land sale campaign in 2023. The purchase of such farmland may be a one-time opportunity. The increase in the value of farmland continued last year, and in 2022, agricultural land in Hungary changed hands at an average price of 11% higher. The prices of vineyards and orchards raised the most; the average price per hectare was around HUF four million. Those interested in buying land remained the most expensive in the central Hungarian region, while the prices were lowest in the northern part of the country.

During the research, the answer was sought to the question of how agricultural land prices develop in the Southern Great Plain region of Hungary, and which factors typically influence them.

Literature review

In the last decade and a half, the fastest increase in agricultural land prices (13.8 % on average per year) occurred in Poland, followed by Slovakia and Hungary with an annual average value increase of 7.3–7.4 %, and then the Czech Republic with an average increase of 4.6 % per year. Convergence of prices was also observed within the country group. The smallest increase occurred in the Czech Republic, where around 2000 prices were significantly higher than in the other three Visegrad Four countries. Over the course of fifteen years, the relative deviation between the price levels of agricultural land in the four districts of Visegrad decreased from 60 to 80 % to 40 %, and the difference between the lowest and highest price levels decreased from four times to three times.

The differences in the legal regulation of land transactions within the group of countries can be said to be significant. While all the Visegrád Group countries (V4) try to limit the purchase of land by foreign private individuals, there are significant differences in the regulation of the land purchase rights of legal entities and foreign-owned companies with legal personality. In the latter field, the Slovak state regulation is the most permissive, while the Hungarian land transfer law is the least, which, apart from the specially mentioned exceptions, does not allow legal entities to acquire the ownership of agricultural land (Sőreg, Naár and Naárné Tóth, 2017). In addition to ownership, the use of agricultural land is also of particular importance, since who can exercise the rights of use over a piece of land – depending on the will of the owner – also primarily affects people's quality of life, since both their own land use and land use granted to others (leases) means economic advantages (or disadvantages) for people (Posta et al., 2022).

The basis of the value of agricultural land is that it is a limited, conditionally renewable natural and agricultural production resource. Since its supply is inelastic, its price is basically determined by the demand for land. At the same time, the demand for agricultural land is derivative, and its value is determined by the benefit of the produced products. The basis for the calculation of the land capital is that it can be defined as the present value of future land rents, which is modified by the effect of additional factors affecting the land price (Tavares, Tavares and Santos, 2022). These are utilization intensity and alternative utilization options, in addition to the location, location and quality of the land, general economic development and the state of the national economy, government interventions (subsidies and deductions, land market regulation), as well as changes in population and technology; and other factors (speculation, mortgage interest, environmental protection, rural development, etc.). In addition to the production function, agricultural land also enables the accumulation of capital through its ability to preserve assets. If the owners rent out the agricultural land, they are still entitled to income through land rent (Bíró, 2018).

On the agricultural land market, ownership of agricultural land is traded, and the concentration of holdings is influenced by regulating the allocation of agricultural land. In the case of a perfect market, the distribution of land ownership would economically only affect the welfare of households, there would be no difference between land purchase and lease, the land price would give the present value of land income and land rents. The operation of land markets in reality differs significantly from this. Imperfections in land markets and the economic environment in terms of land ownership and use reduce social well-being.

In Romania (as well as in Hungary), agricultural land was often purchased for non-farming purposes. The demand stimulated by area-based direct subsidies and the increase in prices also led to speculative purchases by persons/companies with strong capital (Burja, Tamas-Szora and Dobra, 2020). The laws governing land acquisition in Romania facilitated this, especially among international interested parties, while, for example, in Hungary and Serbia (Baturan, 2013; Vincze-Lendvai, 2009), the acquisition of land by foreigners was hindered or made difficult by law for a long time, although there was also an effort to circumvent them.

The value of arable land is basically determined by what can be produced on it and with what results. Production is significantly influenced by the geographical, physico-chemical and biological characteristics of the soils, as well as the topography and climate of the area. In the case of land use for agricultural purposes, the subsidy system should not be ignored either, since they provide a significant part of the income of arable crop cultivation, and direct subsidies are capitalized in land prices.

The price of land can also be significantly influenced by long-term development ideas, expectations, potential opportunities, and the lack of them. In field crop production, such a prominent factor is, for example, in addition to transport infrastructure, the creation and expansion of the possibility of irrigation in drought areas (Figure 1) in connection with climate change, as well as estate planning in relation to economies of scale.

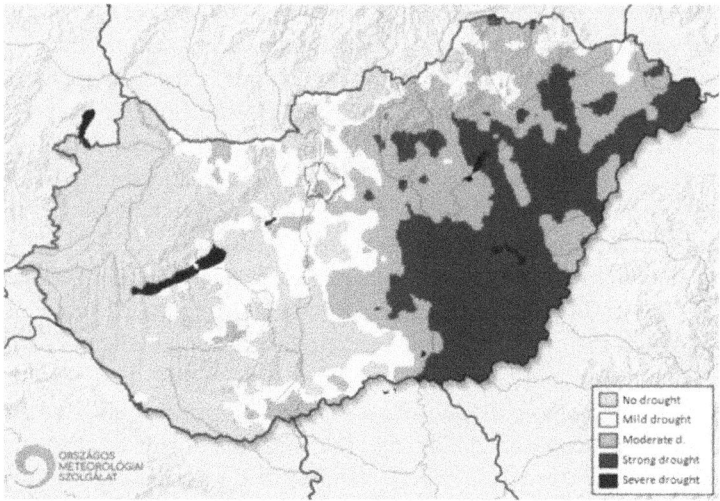

Figure 1. Drought level for summer crops June 2022
Source: OMSZ, 2022

Unfortunately the Southern Great Plain is constantly and year after year struggling with drought. The historical drought of 2022 is predicated on the fact that already in the summer and autumn of the previous year, less precipitation fell than usual. Thus, the entire vegetation period of the plants sown in the fall of 2021 and the spring of 2022 was drier than the long-term average. The summer period with little precipitation was preceded by winter and early spring drought months, when the deeper layers of the soil were not filled with moisture unlike usual, so the effect of the summer drought was even more severe. The effects of the already extraordinary lack of precipitation were also exacerbated by the heat waves that occurred during the summer. According to some studies, the extreme drought in our country in 2022 is not a regional phenomenon, but has a global background (Erdődiné Molnár and Kovács, 2023).

Arable land is a key asset in agriculture, which often changes hands in Hungary due to changes in life situations, so determining its value and market value is essential. Land price and land value are not categories with the same content, because land price means the current market value, while land value is an economic concept and examines land as a production factor (Naárné, 2018).

By joining the EU, Hungary adopted a number of regulations and legal norms into its legal system and practice that are not mandatory, but are based on generally accepted procedures or rules at the international level or within the community. This category includes the standards and recommendations related to asset valuation, regulations for experts, of which Buzás et al. (2020) highlighted the most important ones below:

- IVS – International Valuation Standards
- EVS – European Valuation Standards
- TEGoVA -The European Group of Valuers' Associations
- RICS-Royal Institution of Chartered Surveyors (GB)
- USPAP – Uniform Standards of Professional Appraisal Practice (US).

Main factors influencing the value of agricultural land (Kapronczai, 2018):

- soil quality
- climate
- topography
- size and shape of the plot
- type of cultivation
- accessibility of the area
- subsidies and tax system
- land market regulation
- distance from markets
- landmarks that hinder cultivation
- possibility of irrigation
- the labor supply
- income-producing ability of the land
- the environmental classification of the area
- the (growing) non-agricultural demand.

Methodology

The choice fell on the Southern Great Plain region because agriculture still plays a significant role in the region's economy. Compared to other regions of the country, its agricultural character has always been stronger and more decisive; the agricultural potential in the region is above average.

Secondary research was carried out using data from the Central Statistical Office, OTP Mortgage Bank and MBH Mortgage Bank. The changes in land prices in recent years and the factors influencing the valuation and the land turnover were examined. In the examined period, the average exchange rate was: 386 HUF = 1 EUR. It was also concerned with the opportunities for young farmers to access land.

Results and discussions

Hungary is divided into seven regions and the capital, Budapest, one of which is the Southern Great Plain Region. The Southern Great Plain Region is located in the south-east and south of the country and borders Romania and Serbia. It includes Bács-Kiskun, Békés and Csongrád-Csanád counties (Figure 2). A significant part of its territory is a plain rich in diverse natural and landscape values. Its area is 18,339 km^2, which is 19.7 % of the country's territory. 85 % of the land area is productive land, of which 84 % is agricultural. The importance of agriculture in the economic structure of the region is decreasing, but it is still the most important sector in almost half of the districts.

Figure 2. Counties of the South Great Plain Region
Source: Mokrán, 2015

Profits from agricultural income strongly influence land prices. Despite the uncertain geopolitical situation and economic downturn, agricultural lands have an extraordinary stability of value and income level, which also means that they largely retain their value, the invested capital, even during periods of economic downturn. The fundamentals that determine farmland yields are driven by long-term structural trends, such as population growth, that do not correlate with broader economic cycles. Both the 2008 financial crisis and the coronavirus pandemic highlighted that agricultural land as a means of production can generate strong producer income even in recessionary periods, which results in a stable agricultural land value.

At the same time, a period loaded with drought has a double effect. On the one hand, the decreasing amount of crops increases crop prices, especially if it is a global phenomenon and a commodity exchange product. On the other hand, the lost crop volume reduces the available sales revenue and the effective utilization of the used input materials; therefore the profit achieved may decrease or even turns into a loss. Another effect may be that the rate of increase in the price of soils with lower productivity will slow down, while the price of good-quality farmland with favorable water balance will rise even more. Furthermore, the value of the areas where the rainfall supply has not decreased significantly due to climate change will also increase in value.

Yields on agricultural land are strongly correlated with inflation, which highlights their good value-preserving properties during inflationary periods. Rising producer prices result in higher farm income, which ultimately leads to an increase in the value of farmland. The value of land ownership is significant, which is further increased by EU funds, especially area-based subsidies. The concentration of the structure of land use after the Land Compensation and Cooperatives Act became significant from the beginning of the nineties until the accession to the European Union, according to both historical and international comparisons (Kovách, 2018).

In Hungary in the county breakdown, Tolna had the highest prices per hectare 2021 in, the average amount to be paid was (Hungarian Forint) HUF 2.35 million, and in the counties of Békés and Hajdú-Bihar HUF 2.27 and 2.24 million, respectively. Prices exceeding the national average were also recorded in the counties of Győr-Moson-Sopron and Fejér (HUF 2.09 and 1.97 million respectively), while they were around the national average in the counties of Bács-Kiskun, Csongrád-Csanád and Szabolcs-Szatmár (HUF 82-1.83 million). Just like in 2020, before last year they asked for the least in Nógrád, just over one million HUF for one hectare of arable land. It was somewhat more expensive to buy in Heves (HUF 1.19 million), as well as in Zala and Borsod-Abaúj-Zemplén (HUF 1.27 million per hectare).

As in 2020, Hajdúszoboszlói had the highest specific price of agricultural land among the districts, exceeding HUF 3 million. In the districts of Mezőkovácsháza and Hajdúböszörmény, you had to pay almost as much, on average HUF 2.94 million per hectare (Nyemcsok, 2023). In the districts of Bélapátfalva, Pétervására, Putnok and Ózd, the land was sold the cheapest in 2021, on average below HUF 700,000. However, there are often few transactions in these areas, so prices can fluctuate strongly.

With the exception of arable land, the growth of the amounts to be paid per hectare accelerated in 2021, the most significant, with an increase of 15.5 %, in the sale of forests and wooded areas. In the same period, the increase in the price of vineyards exceeded 10% (10.3 %), even though the price per hectare decreased in the previous two years. The price of orchards rose by 9 %, and the price of lawns, meadows and pastures by 8.8% compared to the year before last.

Compared to ten years ago, the price of orchards and fields increased by 300 %, the price of grasslands, meadows and pastures by 180 %, and the price of forests and wooded areas by 250 % per hectare. The smallest price increase occurred in the case of grapes: prices more than doubled in 11 years (Table 1).

In 2021, there were also significant differences in the various branches of cultivation in the individual regions. The amount to be paid for a hectare of grapes in Central Transdanubia, for example, reached HUF 3.08 million, but in the Southern Great Plain it was only HUF 1.62 million. In the Southern and Western Transdanubia parts of the country, one hectare of vineyard could be purchased at an average price of over two million HUF (2.76 and 2.39 million), while in the Northern Great Plain and Northern Hungary the average price was close to HUF 1.7 million.

The average price of agricultural land in 2022 increased by 7.4 % compared to the values calculated in 2021. The growth rate was higher than the 6–7 % experienced between 2019 and 2021. However, due to high inflation, real prices fell. The average price of arable land per hectare was HUF 1.94 million in 2022, while HUF 1.81 million had to be paid for it in 2021. The price increase was nationwide. Prices rose the most in Northern Hungary, by about 18 %. The greatest price increase (more than HUF 2 million/hectare) was observed in Southern Transdanubia, Central Hungary, the Northern Great Plain and the Southern Great Plain. Among the counties, Hajdú-Bihar, Tolna and Békés led the list, where the price of farmland in all three places (Figure 3) was agreed at around HUF 2.5 million.

Development of the price of agricultural land

Table 1. The price of the arable land and the land rent according to the counties and the region

	Average land price (HUF/ha)				Average land rent (HUF/ha)			
	Bács-Kiskun	Békés	Csongrád-Csanád	Southern Great Plain	Bács-Kiskun	Békés	Csongrád-Csanád	Southern Great Plain
Year	county	county	county	region	county	county	county	region
2008	–	–	–	–	28,200	27,800	32,500	29,000
2009	410,500	643,000	468,300	520,000	28,500	29,200	29,300	28,900
2010	455,500	638,900	510,000	541,200	31,600	32,200	32,800	32,200
2011	529,500	720,400	582,700	609,000	35,800	37,800	39,900	37,700
2012	675,500	848,300	675,000	736,900	41,900	43,000	48,100	43,900
2013	745,800	990,100	798,400	835,200	45,200	45,200	45,000	45,100
2014	924,600	1,195,400	952,800	1,031,400	48,200	48,800	46,200	47,900
2015	1,035,900	1,323,800	1,059,800	1,161,600	51,100	52,200	48,100	50,900
2016	1,377,100	1,448,200	1,371,200	1,396,200	55,500	57,000	51,200	55,100
2017	1,439,600	1,745,300	1,382,700	1,530,400	57,900	61,200	55,600	58,800
2018	1,530,500	1,930,600	1,528,500	1,695,900	61,200	64,500	59,100	62,100
2019	1,627,700	2,130,600	1,669,200	1,831,000	67,700	71,300	64,100	68,400
2020	1,693,300	2,239,700	1,739,500	1,916,700	71,000	73,600	70,000	71,900
2021	1,986,600	2,460,900	1,781,000	2,100,000	78,600	83,700	82,200	81,700
2022	2,080,500	2,646,500	2,020,100	2,263,300	88,700	95,700	88,100	91,500

Source: Authors' own research based on KSH, 2023

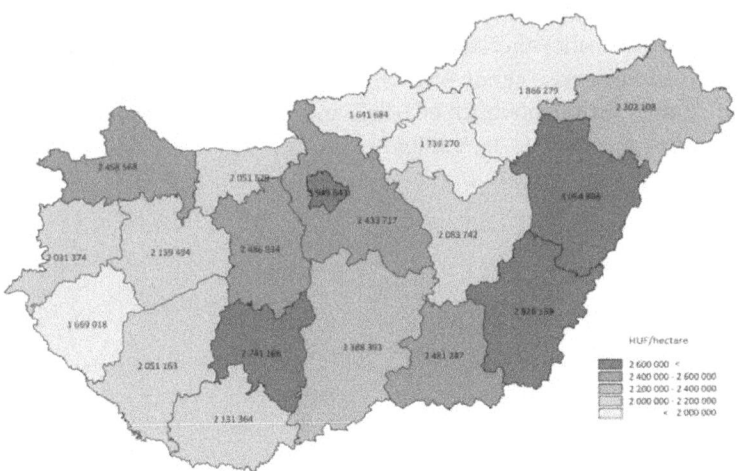

Figure 3. Average sales prices of arable land in 2022 per county
Source: Agrotax Kft, 2022.

Among Hungary's counties (19 counties, not counting Budapest), the arable fields of Békés county are ranked 2nd, Csongrád-Csanád county 5th, and Bács-Kiskun county 8th in the ranking according to the average sales price. For the sake of comparability, the data of the previous 15 years and the land rents for the given period are presented in Table 1.

The demand for agricultural land, and this is especially true for arable land, typically grows faster than the supply. Due to this and the simple reason that we are talking about a limited means of production, prices tend to rise. Of course, this does not mean that land prices cannot move downwards. A correction is natural when the market, often under emotional or other influence, exceeds the level where those on the buy side are willing to enter. For this reason, the graph of the land price is not straight, but typically stepped. As for the outlook, according to the forecast, domestic arable land prices will not show a significant increase this year. This is related to the fact that subsidies are not increasing and that the state land program has been closed. At current price levels, the payback period is so long that it makes investors think twice.

Today, we typically evaluate land areas based on traffic data, including our cropland classified as arable land. The method based on yield calculation, which focuses on the productivity of the land, in practice only partially affects the value of the land, precisely in relation to the investment capital that appears in the system as a result of the continuously increasing value. Last year, a total of 14,742 parcels (22,943 hectares) changed hands in the country, this number exceeded 20,000 a year earlier. While in 2016 the turnover exceeded ten thousand hectares in seven counties, since then no other county has had such a large annual sales volume. The largest area, about 4.5 thousand hectares, was sold last year in Bács-Kiskun County. In the counties of Békés and Csongrád-Csanád, however, the rate of decline was close to 40 % (Figure 4).

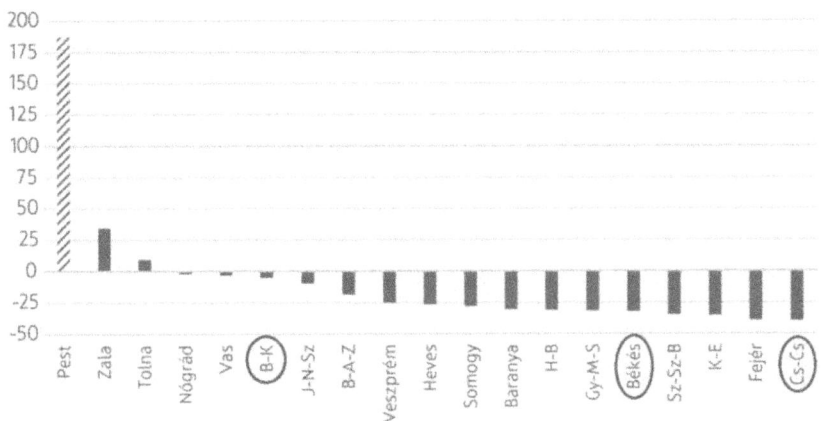

Figure 4. Change in county farmland turnover (2022, based on hectare area, %)
(Legend: B-K = Bács-Kiskun, Cs-Cs = Csongrád Csanád)
Source: Authors' own research based on OTP Mortgage Bank, 2023

In the opinion of farmers, the investments made with subsidies in irrigation were a significant value and price-increasing factor. Although the areas made irrigable are in demand, they are sold less often, because their owners also want to take advantage of this competitive factor (Bodnar and Privoczki, 2022). According to this, market turnover fell by 27 % last year compared to the previous year, and the average size of land plots sold also decreased by about 25 %. The database of the analysis did not include the transactions carried out during the termination of undivided joint ownership, or the number of mergers, as they are not tradable in the market sense, except for the narrow circle of owners, and are not required to be published.

The highest average price was still in Central Hungary, where sellers asked for around HUF 2.8 million for one hectare. In addition, the Great Plain and Central Transdanubia were also characterized by average prices above HUF two million. Plots were still sold the cheapest in the northern part of the country, the average price in northern Hungary was HUF 1.68 million per hectare (Agrotrend, 2023).

In settlements with more than a hundred transactions, the average price is significantly higher, by 24%, than in settlements with fewer than ten transactions. Furthermore, the average prices in some areas of the counties of Hajdú-Bihar, Csongrád, Bács-Kiskun, Tolna, Győr-Moson-Sopron are permanently higher than in other areas of the country.

One of the biggest problems of the agricultural economy of the EU and Hungary is the unfavorable age structure of producers, the low share of young farmers. In the Southern Great Plain region, the proportion of young farmers is the 2nd highest in the country (Kőszegi, 2014), so agriculture here can still be an income supplement or a source of livelihood for many young people in the future. That is why it would be important for young farmers to be in an advantageous position when determining the order of pre-purchase of the land. The Hungarian Land Act has undergone numerous amendments. Based on the current Land Transfer Act, the groups of holders are entitled to the right of pre-emption in the following order (Table 2).

In the first place, the state has the right of pre-emption in order to enforce the land policy guidelines defined in the Act on the National Land Fund, as well as to achieve public employment and other public interest goals.

Secondly in the case of jointly owned land, the farmer co-owner follows the Hungarian State in the case of the sale of the co-owner's share in favor of a third party.

In third place is the farmer who uses the land, who is considered a neighbor living locally, or who is considered a local resident, or whose place of residence or agricultural center has been in the settlement for at least 3 years, the administrative border of which is the administrative boundary of the settlement according to the location of the land that is the subject of the sale and purchase it is no more than 20 km away from its border on a public road or a private road not closed to public traffic.

In fourth place is a farmer who qualifies as a neighbor living locally.

Table 2. Order of pre-purchasers in land transactions from 1st July 2023

1.	Status:		
2.	Farmer who has been a co-owner for at least 3 years		
3.	Land user for 3 years, local farmer neighbor	A farmer who has been using the land for 3 years and lives locally	A farmer whose place of residential use as a lifestyle or the center of his agricultural business has been in a settlement whose administrative border is no more than 20 km away for at least 3 years.
4.	Local resident, farmer next door		

Table 2. Continued

5.	Livestock farmer who has lived locally for 3 years	A local farmer who produces propagating material	A local farmer who works in horticulture	The producers of the product under protection and the organic farmer
6.	Local resident farmer			
7.	Farmers who have been at a distance of no more than 20 km from their place of residential use or their agricultural center for at least 3 years.			

Source: Authors' own editing based on Farkas, 2023

In fifth place, the local farmer neighbor is followed by the local farmer who operates the animal farm, whose purpose of acquiring property is to ensure the necessary and proportional feed requirements for animal husbandry and has the specified animal density.

In sixth place, a farmer who is considered a local resident has the right of pre-emption.

Finally, the right of first refusal belongs to those farmers whose place of residence or farm center has been in the settlement for at least 3 years, the administrative boundary of which is no more than the administrative boundary of the settlement in which the land that is the subject of the sale is located, on a public road or a private road not closed to public traffic, at most It is 20 km away.

If several persons within a given group of rights holders, using their right of pre-emption, make a statement of acceptance regarding the contract, the Land Transfer Act establishes the following sequence:

(a) a member of a family agricultural company or a member of a family farm of primary producers;
(b) young farmer;
(c) novice farmer.

Finally, the current change was reached, according to which, with effect from July 1, 2023, the right of pre-emption cannot be exercised against those farmers who acquire the first ten hectares of land, who live locally, whose land is locally, or 20 kilometers from it they have an agricultural center inside (Zima, 2023). In practice, according to the new legislation, sales contracts that meet the conditions do not have to be published, which means that the licensing procedure can be completed earlier. This can not only significantly speed up the buying and selling process, but also greatly reduce the administrative burden on farmers.

The regulation also prevents abuses, because the first 10 hectares must be understood as all land previously acquired by the buyer under any legal title, including those areas that were later removed from his ownership, rather than the actual land ownership at the time of purchase. This is how to avoid abusing the purchase discount. Privoczki stated in 2020 (Privoczki, 2020) that "It would significantly improve the land supply indicators of young farmers if, similar to another 'Land for Farmers' program, they submit a first priority claim for undivided common land." However, young farmers applying within the framework of the Rural Development Program are still not in a sufficiently favored position in the pre-purchase ranking, and their access to land is only possible up to 10 hectares, which is insufficient for the realization of economies of scale in most sectors.

Conclusions

Land is not only a means of production, but also an asset. In addition to the fact that its price is rising all over the country, including in the examined region, it is also an excellent investment in the short term. In addition, the rise in the price of land usually approaches, and in some cases even exceeds, inflation. Perhaps there is hardly a better investment in our country today. Based on past international examples, despite the expected economic difficulties – the even more direct effects of the coronavirus pandemic and the Russian-Ukrainian war – sustainably cultivated and technologically maintained agricultural land will certainly continue to hold its value. Due to population growth and, consequently, food supply, as well as in light of climate change, the importance of farmland is increasing even further.

Whether we are on the right track or not? The questions can be answered in a complex way: From whose point of view, from the producer or from the investor? Is the producer owner, investor and tenant at the same time? The Hungarian arable land market is constantly changing, so the future development of price, value and turnover and the determination of the root causes require further investigations.

References

Agrotax Kft. 2022. *Internal documents.* Hungary.

Agrotrend. 2023. *Átlépte a lélektani határt a termőföldek ára [The price of agricultural land has crossed the psychological limit].* [online] Available at: <https://agrotrend.hu/gazdalkodas/szantofold/atlepte-a-lelektani-hatart-a-termofoldek-ara/> [Acccesed 4 October 2023], [in Hungarian].

Baturan, L. 2013. Economic analysis of the ban on foreigners acquiring property rights on agricultural land in Serbia. *Economics of Agriculture,* 60(3), pp. 479–491.

Bíró, Sz. 2018. Milyen tényezők határozzák meg a termőföld értékét a szántóföldi művelésben? [What factors determine the value of farmland in field cultivation?], *Agronapló,* 2018. 04.10. [online] Available at: <https://www.agronaplo.hu/agrofokusz/20180410/milyen-tenyezok-hatarozzak-meg-a-termofold-erteket-a-szantofoldi-muvelesben-38263> [Acccesed 8 October 2023], [in Hungarian].

Bodnar, K. and Privoczki, Z.I. 2022. Factors influencing the price of arable land in Hungary. *Research Journal of Agricultural Science,* 54(1), pp. 30–36.

Burja, V., Tamas-Szora, A. and Dobra, I.B. 2020. Land concentration, land grabbing and sustainable development of agriculture in Romania. *Sustainability,* 12(5), p. 2137. https://doi.org/10.3390/su12052137.

Buzás, F., Cehla, B., Kiss, S., Mező, I. and Tóth, Cs. 2020. Recommendations for actualization of land valuation legal frameworks and directives. *Régiókutatás Szemle,* 5(1), pp. 104–116. https://doi.org/10.30716/RSZ/20/1/9.

Erdődiné Molnár, Zs. and Kovács, A. 2023. A 2022-es aszály agrometeorológiai elemzése. [Agrometeorological analysis of the 2022 drought.] *Légkör: Az Országos Meteorológiai Intézet Szakmai Tájékoztatója,* 68(1), pp. 20–27. https://doi.org/10.56474/legkor.2023.1.3 [in Hungarian].

Farkas, D. 2023. *Elővásárlási sorrend a földforgalomban (érvényes: 2023. Július 1-től) [Order of pre-emption in land transactions (valid from July 1, 2023)].* Nemzeti Agrárgazdasági Kamara. [online] Available at: <https://agrotrend.hu/hirek/elovasarlasi-sorrend-a-foldforgalomban-ervenyes-2023-julius-1-tol/> [Acccesed 8 October], [in Hungarian].

Kapronczai, I. 2018. Milyen tényezők határozzák meg a termőföld értékét a szántóföldi művelésben? [*What factors determine the value of farmland in field cultivation?*], *Agronapló,* 4, pp. 27–32. [online] Available at:<https://dev.agronaplo.hu/szakfolyoirat/2018/04/gazdasag/milyen-tenyezok-hatarozzak-meg-a-termofold-erteket-a-szantofoldi-muvelesben> [Acccesed 4 October 2023], [in Hungarian].

Kovách, I. 2018. *Földhasználat és földtulajdon szerkezet.* [*Land use and land ownership structure*], pp. 248–263. [online] Available at: <http://real.mtak.hu/180440/1/10.61501_TRIP.2018.14.pdf> [Acccesed 4 October 2023], [in Hungarian].

Kőszegi, I.R. 2014. *Az üzemszerkezet átalakulása a Dél-alföldi Régióban rendszerváltás után Magyarországon* [The transformation of the plant structure in the Southern Great Plain Region after the regime change in Hungary]. In: J.T. Karlovitz, ed. *Kulturális és társadalmi sokszínűség a változó gazdasági környezetben.* Komárno: International Research Institute, pp. 26–34, [in Hungarian].

KSH. 2023. *A szántó ára és földbérleti díja vármegye és régió szerint (forint/hektár)* [*Arable land price and land rent by county and region (HUF/hectare)*]. [online] Available at: <https://www.ksh.hu/stadat_files/mez/hu/mez0069.html> [Acccesed 4 October 2023], [in Hungarian].

Mokrán, S. 2015. *Dél-Alföld és megyéinek gazdasági fejlettsége a GDP alapján* [*Economic development of the Southern Great Plain and its counties based on GDP*]. [online] Available at: <https://prezi.com/vbzdjftcnfq0/del-alfold-es-megyeinek-gazdasagi-fejlettsege-a-gdp-alapjan/> [Acccesed 4 October 2023], [in Hungarian].

Naárné Tóth, Z. 2018. *A piaci és hozadékelvű földértékelés elmélete és gyakorlata nemzetközi kitekintésben* [*The theory and practice of market and yield-based land valuation in an international perspective*]. Budapest: Szaktudáskiadó Ház, [in Hungarian].

Nyemcsok, L. 2023. *A szántók átlagára Békésben volt a második legmagasabb az országban* [*The average price of arable land in Békés was the second highest in the country*]. BEOL, 2023. 01.04. [online] Available at: <https://www.beol.hu/helyi-gazdasag/2023/01/a-szantok-atlagara-bekesben-volt-a-masodik-legmagasabb-az-orszagban> [Acccesed 8 October], [in Hungarian].

OMSZ. 2022. *Aszálytérkép* [*Drought map*]. Országos Meteorológiai Szolgálat. [online] Available at: <https://pestisracok.hu/wp-content/uploads/2022/07/aszalyterkep-OMSZ.jpg> [Acccesed 8 October], in Hungarian].

OTP Mortgage Bank. 2023. *Termőföld Értéktérkép* [*Farmland Value Map*]. [online] Available at: <https://www.otpbank.hu/static/otpjelzalogbank/sw/file/OTP_Termofold_Ertekterkep_2023.pdf> [Acccesed 8 October], [in Hungarian].

Posta, L., Szentesi, I., Túróczi, I. and Tóth, R. 2022. The role of factors affecting the value of Hungarian farmland in the Hungarian economy. *Polgári Szemle*, 18(1–3), pp. 218–234. https://doi.org/10.24307/psz.2022.1116.

Privoczki, Z.I. 2020. *The socio-economic situation of young farmers in the Southern Great Plain, in particular with utilization of requested supports*. Ph.D. Szent Istvan University, Kaposvár Campus, Faculty of Economic Sciences, pp. 35.

Sőreg, Á.P., Naár, A.T. and Naárné Tóth, Z. 2017. Regionális különbségek és árkonvergencia a visegrádi országok termőföldpiacán [Regional differences and price convergence in the agricultural land market of the Visegrád countries]. *Statisztikai Szemle*, 95(4), pp. 349–381. https://real.mtak.hu/51351/1/2017_04_349.pdf, [in Hungarian]

Stoica, D.G. and Dumitru, E.A. 2021. Aspects that contributed to changes in the price of agricultural land in Romania and other countries in the European Union. *Scientific Papers Series Management, Economic Engineering in Agriculture and Rural Development*, 21(4), pp. 565–570.

Tavares, V.C., Tavares, F. and Santos, E. 2022. The value of farmland and its determinants – The current state of the art. *Land*, 11, p. 1908. https://doi.org/10.3390/land11111908.

Vincze-Lendvai, E. 2009. A földvásárlás és –birtoklás főbb kérdései Magyarországon [The main issues of land purchase and ownership in Hungary]. *Jelenkori Társadalmi és Gazdasági Folyamatok*, 4(1), pp. 62–66, [in Hungarian].

Zima, Sz. 2023. *A fiatalok könnyebben hozzájuthatnak az első 10 hektár földjükhöz [A fiatalok könnyebben hozzájuthatnak az első 10 hektár földjükhöz]*. Agrofórum, Press release, 13.07.2023. [online] Available at: <https://agroforum.hu/szakcikkek/egyeb/fontos-valtozas-mostantol-a-fiatalok-konnyebben-hozzajuthatnak-az-elso-10-hektar-foldjukhoz/> [Acccesed 8 October].

Sorinel Ionel Bucur

The sustainability of rural area – An approach from the perspective of biocapacity and ecological footprint

Abstract: *As a measure of human impact on the Earth's ecosystem, the ecological footprint indicates the dependence of the human economy on natural capital. At the opposite pole is the concept of biocapacity, which represents the total of productive surfaces. The difference between ecological footprint and biocapacity basically shows whether a country is an ecological debtor or creditor. Globally, footprint assessments show how high human demands on natural resources are compared to the planet's renewable resources. Biocapacity and ecological footprint influence the sustainability of each individual area. From this point of view, the present approach aims to carry out an analysis of the biocapacity and the ecological footprint at the level of the Romanian rural space, including in a regional profile.*

Keywords: ecological footprint, biocapacity, sustainability, rural area, development regions, gaps.

Introduction

The issue of biocapacity and the ecological footprint has recently been the subject of many discussions both among the authorities, but also among the academic environment and civil society. Knowing the surplus or deficit of biocapacity is an important element in measuring the degree to which existing, limited resources will be able to sustain the pressure exerted by the economy on them. However, such an approach at the local (regional) level runs into the limited informational stock that would allow, on the one hand, the measurement of the ecological footprint, and on the other, the biocapacity reserve.

Literature review

In the initial attempts to define the ecological footprint (Constantinescu, 2015), its purpose was to identify and propose new definitions and measurement units for development, favouring the reorientation of the economy towards a superior valorisation of man and nature, aiming to consider the effects social and economic aspects of development, effects that do not appear in traditional economic calculations.

At an international level, the first relevant academic work on the ecological footprint was however published by William Rees in 1992 (Rees, 1992). According to him, in order to develop a framework for evaluating the ecological footprint specific to an area, it is necessary to know the human transport capacity and natural capital (Rees, 1992). The concept and calculation method were later developed in the period 1990–1994 by Wackernagel (Wackernagel and Rees, 1996). Wackernagel and Rees (1996) initially defined this concept as "adequate transport capacity" because later, to be better understood, the term ecological footprint was arrived at.

Other critics (Lenzen, Borgstrom and Bond, 2006) argue that this is an accurate characterization because rural farmers with agricultural machinery in developed countries can easily consume more resources than urban dwellers due to transport requirements and the unavailability of economies of scale. Therefore, critics argue that the footprint can only be applied globally. This method seems to compensate for the replacement of original ecosystems with high-productivity agricultural monocultures by allocating such higher biocapacity to these regions.

In 2007, other specialists (Grazi, Van den Bergh and Rietveld, 2007) conducted a systematic comparison between the ecological footprint method and the analysis of spatial well-being that includes environmental externalities, crowding effects and commercial advantages. They found that the two methods can lead to distinct and even opposite classification of different spatial patterns of economic activity.

According to other specialists (Gordon and Richardson, 2012), the ecological footprint concept expands the indicator of land use per capita both spatially (to cover the globe) and functionally (land requirements to support all types of consumption). Global aggregates imply that demands are greater than available land worldwide, suggesting that current consumption patterns are "unsustainable".

The ecological footprint concept has also established itself as an important environmental indicator, being constantly subjected to a severe evaluation. Practically, this analysis system initially started from indicators of "sustainable development" that were redefined, in the sense of applicability, by using a large database made available by computer systems (Constantinescu, 2015).

In other words, the notion of "ecological footprint" is connected to the biologically productive land needed to satisfy the consumption of a population and absorb all its waste (Ding and Li, 2011; Wiedmann, Minx and Barrett, 2006).

Starting 2003, GFN has been calculating the ecological footprint based on UN statistical data, for the whole world and for more than 200 individual countries. Although these calculations are updated with the latest statistical data, the trends for each country tend to remain broadly constant (Lin et al., 2018).

Ecological footprint analysis is widely used around the world, and in support of sustainability assessments, which enable the measurement and management of resource use in all economic sectors, as well as exploring the sustainability of individual lifestyles, goods and services, organizations, industry sectors, neighbourhoods, cities, regions and nations.

The ecological footprint can be calculated on any scale, but specialists appreciate that, as a rule, due to population concentration, cities tend to have large ecological footprints, and measures to reduce this indicator are necessary (Tyszczuk, 2014).

The bearing capacity of natural ecosystems is exceeded by the high level of consumption recorded, for example, at the level of European states. Even in conditions where people's behavior would be comparable worldwide, it is estimated that approx. 3 planets to ensure, on the one hand, the necessary resources, and on the other hand, to neutralize the generated waste (Bucur, 2020).

Practically, the increase of the ecological footprint and the reduction of biocapacity lead to the depletion of resources, which are already limited, against the background of the increase in the amount of waste (European Commission, 2016).

Ecological footprint ideas first appeared in a program report on sustainable development indicators for a US organization (Hayes, 2019). The objective of this institution was to identify and propose new definitions and units of measure for development, favouring the reorientation of the economy towards a superior valorisation of the capabilities of man and nature, seeking to follow the social and economic effects of development methods, effects that do not figure in traditional economic calculations.

At the national level, the issue of the ecological footprint and its impact on the sustainability of rural areas has been little addressed, given the unavailability of adequate informational support for its measurement. It is, on the one hand, the difficulties of identifying relevant primary or aggregated indicators at the national level, but mainly those at the regional level.

In the national specialized literature, a series of studies can be found regarding the indicators that are the basis of the quantification of the ecological footprint at the national or local level (Bucur 2020; Stanciu, 2009; Toderoiu, 2010).

However, lowering the analysis to the level of the rural space and especially to the local level, whether regional or local, the scientific "fund" is extremely reduced. From this point of view, we consider that any scientific approach in the sense of identifying some indicators that can form the basis of determining the ecological footprint at the local level represents a starting point in building any alternative for rural development, based, along with other indicators, on the effect of pressure exerted by the population on the biosphere.

Methodology

From a methodological point of view, the present approach is based on public information provided by the Global Footprint Network (GFN), regarding the determination of ecological footprint, biocapacity and biocapacity reserve/deficit, as well as information from European and national statistics, regarding certain indicators at the local (regional) level.

The approach uses established statistical methods, such as comparisons, dynamics, the time period analyzed being 2012–2022. In terms of ecological footprint and biocapacity, for comparison with other states, the time period considered is 1961–2022.

It should also be noted that the information on ecological footprint and biocapacity is expressed in global hectares per person. The global hectare (GHA) is a unit of measurement for the ecological footprint of people or activities and the biocapacity of the earth or its regions (wikipedia, 2020). A global hectare represents the world's annual amount of biological production for human use and assimilation of human waste, per hectare of biologically productive land and fisheries. The global hectare per inhabitant thus represents the amount of waste production and assimilation per person on the planet.

Results and discussions

As stated previously, the GFN calculates the ecological footprint of a country both by reference to biocapacity (expressed in global hectares/per inhabitant), but also by comparison with the number of planets like Earth in size, necessary to ensure the requirements of the population. Thus, over the course of 61 years, Romania has continuously registered a deficit of biocapacity caused, mainly, by the high level of the ecological footprint compared to the stock of biocapacity.

Since 1961, the ecological footprint has registered a significant increase compared to the level of biocapacity, sustained growth maintained until 1980, the year after which the trend of the footprint begins to decrease, so that, after 2018, a biocapacity reserve of national level (Table 1).

Table 1. The evolution of the biocapacity stock and reserve, as well as the ecological footprint in Romania in the period 1961–2022 (global hectares/person)

Year	Biocapacity	Ecological footprint	Biocapacity reserve/deficit
1961	2.30	2.50	−0.20
1965	2.30	3.00	−0.70
1970	2.10	3.60	−1.50
1975	2.20	4.10	−1.90
1980	2.20	4.80	−2.60
1985	2.30	4.60	−2.30
1990	2.20	4.40	−2.20
1995	2.20	3.20	−1.00
2000	2.10	2.40	−0.30
2005	2.40	3.30	−0.90
2010	2.40	2.90	−0.50
2015	2.70	2.80	−0.10
2017	3.10	3.40	−0.30
2018	3.47	3.02	0.45
2019	3.33	2.94	0.39
2020	2.69	2.48	0.21
2021	2.84	2.50	0.34
2022	2.74	2.45	0.29
2022/1961	19.10	−2.00	

Source: Author own research based on the GFN, 2023.

Calculated as the number of planets necessary to ensure the population's requirements, Romania's ecological footprint increased in the period 1961–2022, by no less than 245.5 %, respectively from 0.79 planets (relative to the year 1961) to no less than 2.74 planets (2022), as a result of the changes made in the level of the main indicators considered, respectively: the area of built-up land, carbon emissions, the area with agricultural crops, the area of fishing grounds, forest products and the area with grazing lands (Table 2).

Table 2. The evolution of the ecological footprint in Romania in the period 1961–2017 (no. of planets needed to ensure the population's requirements)

Year	Built-up Land	Carbon	Cropland	Fishing Grounds	Forest Products	Grazing Land	Total
1961	0.03	0.32	0.22	0.01	0.17	0.04	0.79
1965	0.03	0.50	0.26	0.01	0.20	0.04	1.03
1970	0.03	0.81	0.23	0.01	0.21	0.04	1.33
1975	0.04	1.04	0.31	0.02	0.21	0.04	1.66
1980	0.05	1.32	0.42	0.03	0.2	0.07	2.10
1985	0.05	1.26	0.42	0.03	0.28	0.05	2.10
1990	0.05	1.35	0.47	0.02	0.17	0.09	2.14
1995	0.07	0.94	0.46	0.01	0.17	0.05	1.70
2000	0.04	0.71	0.32	0.01	0.16	0.04	1.28
2005	0.07	0.94	0.55	0.02	0.22	0.06	1.87
2010	0.07	0.9	0.43	0.02	0.22	0.05	1.69
2015	0.08	0.94	0.42	0.02	0.31	0.05	1.82
2016	0.12	1.29	0.74	0.04	0.54	0.16	2.90
2017	0.12	1.00	0.59	0.03	0.34	0.05	2.13
2018	0.16	1.43	1.20	0.04	0.55	0.08	3.47
2019	0.15	1.42	1.07	0.04	0.52	0.13	3.33
2020	0.10	1.30	0.60	0.0	0.50	0.10	2.70
2021	0.11	1.43	0.62	0.04	0.54	0.10	2.84
2022	0.10	1.35	0.61	0.04	0.53	0.10	2.74
2022/1961	245.5	322.5	177.6	306.9	211.4	153.4	246.5

Source: Author own research based on the GFN, 2023.

Compared to the other states of the European Union, for which relevant information has been identified, in 2022, Romania occupies the last place in terms of biocapacity (expressed in global hectares/inhabitant), respectively with the lowest level (2.7 gha/place). Regarding the ecological footprint (expressed both in global hectares/inhabitant), Romania occupies the 13th place, after in 2017 it was in the first position with the lowest level of the footprint (Table 3). However, the first position is maintained by Romania in terms of the ecological footprint expressed in the number of planets (1.8 planets in 2022).

Table 3. Evolution of the biocapacity stock and reserve. as well as the ecological footprint in EU member states and worldwide in 2022

	Biocapacity (global hectares/inhabitant)	Ecological footprint global (hectares/inhabitant)	Biocapacity reserve/deficit (global hectares/inhabitant)	Ecological footprint (no. of planets)
Austria	5.6	2.9	2.7	3.7
Belgium	6.7	1.2	5.5	4.4
Bulgaria	3.6	3.1	0.5	2.4
Czechia	5.1	2.5	2.6	3.4
Croatia	3.7	2.6	1.1	2.5
Danmark	7.3	4.2	3.1	4.8
Estonia	8.1	9.8	−1.7	5.4
Finland	5.4	11.8	−6.4	3.6
France	4.3	2.5	1.8	2.9
Germany	4.5	1.6	2.9	3
Greece	3.8	1.6	2.2	2.5
Ireland	4.5	3.1	1.4	3
Italy	4	1	3	2.6
Letonia	7.7	9.9	−2.2	5.1
Lithuania	6.4	6	0.4	4.3
Luxembourg	11	1.2	9.8	7.3
Poland	4.6	2	2.6	3.1
Portugal	3.7	1.5	2.2	2.5
Slovakia	4.2	2.7	1.5	2.8
Slovenia	4.8	2.5	2.3	3.2
Spain	3.9	1.7	2.2	2.6
Sweden	4.9	8.5	−3.6	3.3
Netherland	6	1.1	4.9	4
Hungary	3.8	2.6	1.2	2.5
Romania	**2.7**	**2.5**	**0.2**	**1.8**
World	**2.6**	**1.5**	**1.1**	**1.7**

Source: Author own research based on the GFN, 2023.

The high degree of economic development in certain EU member states has led to high population pressure on the biosphere to meet the growing demands of the population. However, for example, in Luxembourg, the biocapacity stock in 2022 was 9.8 global hectares/capita, while the ecological footprint was only 1.2 global hectares/capita.

The values recorded by the ecological footprint and biocapacity are the result, among others, of some primary and derived indicators, respectively: life expectancy, human development index, GDP/capita and population. From the point of view of life expectancy, it should be mentioned that Romania occupies the penultimate place among European states, with a life expectancy of 73 years, Spain, Sweden, Italy and Luxembourg occupying the first four positions (83 years). Calculated as a measure of life expectancy, literacy, education and standard of living, the human development index (HDI) allows comparison of a country's level of development in a more detailed manner than GDP/capita.

Regarding this indicator, Romania is also in the penultimate position, the first three places being occupied by Denmark, Sweden and Ireland, with an HDI of 0.95. Even in terms of GDP/inhabitant, Romania is not in a leading position, occupying the 22nd place. Thus, in 2022, 10.1 % of the total GDP per inhabitant was achieved by Luxembourg, followed by Ireland (Table 4).

Table 4. Relevant demo-economic indicators in EU member states, 2022

EU States	Life expectancy (years)	Human development index	GDP/inhabitant ($/inhabitant)	Population (million inhabitants)
Austria	81	0.92	55,460	9.1
Belgium	82	0.94	52,749	11.7
Bulgaria	72	0.80	24,490	6.8
Czechia	77	0.89	40,707	10.7
Croatia	76	0.86	31,007	4.1
Danmark	81	0.95	59,333	5.8
Estonia	77	0.89	38,353	1.3
Finland	82	0.94	49,686	5.6
France	82	0.90	47,995	65.6
Germany	81	0.94	54,192	83.9
Greece	80	0.89	30,488	10.3
Ireland	82	0.95	106,717	5.0
Italy	83	0.90	43,010	60.3
Letonia	73	0.86	31,973	1.8
Lithuania	74	0.88	39,810	2.7
Luxembourg	83	0.93	120,505	0.6
Poland	76	0.88	35,703	37.7
Portugal	81	0.87	34,950	10.1
Slovakia	75	0.85	33,079	5.5

Table 4. Continued

EU States	Life expectancy (years)	Human development index	GDP/inhabitant ($/inhabitant)	Population (million inhabitants)
Slovenia	81	0.92	41,570	2.1
Spain	83	0.91	39,753	46.7
Sweden	83	0.95	53,897	10.2
Netherland	81	0.94	58,732	17.2
Hungary	74	0.85	35,069	9.6
Romania	73	0.82	31367	19.0

Source: Author own research based on the GFN, 2023.

Deepening the analysis at the regional level, similar to those previously detailed, several important aspects should be mentioned, respectively (NIS, 2023):

- in the period 2012–2022, the population registered a reduction trend, including by residential areas; thus, if for the whole country, the population decline was 5 %, in the rural area, the population decreased by only 1.5 %; it should be mentioned that out of the eight development regions, in five of them the rural population increased by percentages that oscillated between 0.4 % (North-East) and 33.8 % (Bucharest-Ilfov);
- in 2021, the last year for which HDI information is available, the Bucharest-Ilfov region has the highest level of HDI (0.926), at the opposite pole is the North-East region (0.775); viewed dynamically, the HDI registered an increasing trend in 2021 compared to 2012, with percentages oscillating between 0.89 % (South-East region) and 3.77 % (North-West region);
- in terms of GDP, in 2021, the Bucharest-Ilfov region achieved 27.4 % of the total GDP, being followed by the North-West (12.4 %), the Center (11.4 %) and the South-Muntenia (11.4 %). Practically, the three regions achieved, cumulatively, 62.6 % of the total GDP;
- The average life expectancy tends to increase especially in rural areas, with percentages varying between 2.5 % (South-East) and 5.8 % (Bucharest-Ilfov);
- Regarding the ecological footprint, at the regional level, it should be mentioned that in 2022, the West region registers the highest level (3.888 global hectares/inhabitant), while at the opposite pole is the Bucharest-Ilfov region (0.145 global hectares/inhabitant). The same positions are occupied by the two regions also in terms of biocapacity. However, all eight development

regions record a biocapacity deficit that varies between -0.042 global hectares/inhabitant (North-West) and -0.408 global hectares/inhabitant (West);
- In the rural area, the ecological footprint and biocapacity gaps are much higher compared to the national average. Thus, for example, the West region records the highest values (10.168 global hectares/inhabitant – ecological footprint, respectively 9.102 global hectares/inhabitant – biocapacity), generating a biocapacity deficit of -1.066 global hectares/inhabitant (Table 5).

Table 5. Ecological footprint and biocapacity in rural area, on development regions, in 2022 (global hectars/inhabitant)

	Ecological footprint	Biocapacity	Biocapacity reserve/deficit
North-Vest	6.138	5.494	−0.644
Centre	7.666	6.862	−0.804
North-East	4.319	3.866	−0.453
South-East	7.059	6.319	−0.740
South – Muntenia	4.470	4.001	−0.469
Bucharest – Ilfov	1.352	1.210	−0.142
South -Vest Oltenia	6.487	5.807	−0.680
Vest	10.168	9.102	−1.066

Source: Author own research based on the GFN, 2023 and NIS, 2023.

Conclusions

At the national level, in the period 1961–2022, Romania registered an oscillating evolution regarding the biocapacity reserve/deficit caused by the fluctuation of the ecological footprint compared to the biocapacity stock. The most recent data reveal the fact that in 2022, Romania's ecological footprint, in equivalent planets, was 1.8 planets.

Compared to the other states of the European Union, Romania ranks last in terms of ecological footprint and biocapacity.

The estimation of the ecological footprint and biocapacity at the regional level highlighted the fact that all regions register a deficit of biocapacity/inhabitant, both in total and in rural areas. Thus, the pressure exerted by the region's population on the environment exceeded the sum of the productive surfaces and their productivity level.

Perceiving and knowing the size of the ecological footprint and the reserve/deficit of biocapacity, together with other indispensable components (demography, economic potential, etc.), constitute important actions for ensuring a sustainable development of the rural space.

References

Bucur, S. 2020. *Dezvoltarea durabilă complexă în spațiul rural românesc.* Bucharest: Ed. Universitară.

Constantinescu, D. 2015. *Amprenta ecologică – Metode de evaluare și analiză.* [online] Available at: <https://www.researchgate.net/publication/301602561_AMPRENTA_ECOLOGICA_-_Metode_de_Evaluare_si_Analiza> [Accessed 3 October 2023].

Ding, Z. and Li, J. 2011. Ecological footprint and reflections on green development of Hangzhou. *Energy Procedia*, 5, pp. 118–124. https://doi.org/10.1016/j.egypro.2011.03.022.

European Commission. 2016. *Ecological footprint and biocapacity. The world's ability to regenerate resources and absorb waste in a limited time period.* Working Papers and Studies. [online] Available at: <https://ec.europa.eu/eurostat/documents/3888793/5835641/KS-AU-06-001-EN.PDF.pdf/d17ffc94-bf5a-404f-ac90-cc64891a2b67?t=1414779227000> [Accessed 3 October 2023].

Global Footprint Network (GFN). 2023. *Ecological footprint.* [online] Available at:<https://www.footprintnetwork.org/our-work/ecological-footprint/>[Accessed 3 October 2023].

Gordon, P. and Richardson, H. 2012. *Farmland preservation and ecological footprints: A critique.* [online] Available at: <https://web.archive.org/web/20100627044350/http:/www-pam.usc.edu/volume1/v1i1a2print.html> [Accessed 3 October 2023].

Grazi, F., Van den Bergh, J.C.J.M. and Rietveld, P. 2007. Welfare economics versus ecological footprint: Modeling agglomeration, externalities and trade. *Environmental and Resource Economics*, 38(1), pp.135–153. https://doi.org/10.1007/s10640-006-9067-2.

Hayes, A. 2019. *What is Genuine Progress Indicator (GPI)?.* [online] Available at: <https://www.investopedia.com/terms/g/gpi.asp> [Accessed 3 October 2023].

Lenzen, M.C., Borgstrom, H. and Bond, S. 2006. On the bioproductivity and land-disturbance metrics of the ecological footprint. *Ecological Economics*, 61(1), pp. 6–10. https://doi.org/10.1016/j.ecolecon.2006.11.010.

Lin, D., Hanscom, L., Murthy, A., Galli, A., Evans, M., Neill, E., Mancini, M.S, Martindill, J., Medouar, F-Z., Huang, S. and Wackernagel, M. 2018. Ecological footprint accounting for countries: Updates and results of the national footprint accounts, 2012–2018. *Resources,* 7(3), p. 58. https://doi.org/10.3390/resources7030058.

National Institute of Statistics of Romania (NIS). 2023. Tempo-Online database. [online] Available at: <http://statistici.insse.ro:8077/tempo-online/#/pages/tables/insse-table> [Accessed 4 October 2023].

Rees, W. E. 1992. *Ecological footprints and appropriated carrying capacity: What urban economics leaves out. Environment & Urbanization,* 4(2), pp. 121–130. https://doi.org/10.1177/095624789200400212.

Stanciu, M. 2009. Amprenta ecologică a României – o nouă perspectivă asupra dezvoltării. *Revista Calitatea vieții,* XX(3–4), pp.271–288.

Toderoiu, F. 2010. Amprenta ecologică și biocapacitatea în zona Țara Hațegului-Retezat. In: P.I. Otiman, V. Florian and C. Ionescu, eds. *Conservarea geo- și biodiversității și dezvoltarea durabilă în Țara Hațegului-Retezat.* Bucharest: Ed. Academiei Române.

Tyszczuk, R. 2014. Architecture of the *Anthropocene: The crisis of agency. Scroope 23: The Cambridge Journal of Architecture.* [online] Available at: <https://eprints.whiterose.ac.uk/99142/1/Renata_Scroope23_20140604.pdf> [Accessed 3 October 2023].

Wackernagel, M. and Rees, W. 1996. *Our ecological footprint: Reducing human impact on the earth.* Philadelphia: New Society.

Wiedmann, T., Minx, J. and Barrett, J. 2006. *Allocating ecological footprints to final consumption categories with input-output. Analysis. Ecological Economics,* 56(1), pp. 28–48.

Wikipedia. 2020. *Global hectare.* [online] Available at: <https://en.wikipedia.org/wiki/Global_hectare> [Accessed 3 October 2023].

Lavinia Denisia Cuc, Dana Rad, Silviu Gabriel Szentesi,
Gabriel Croitoru, Gavril Rad

Evaluation of safety and hygiene measures in Romanian hospitality industry in the context of the COVID-19 pandemic and customer profile

Abstract: *In the wake of the unprecedented disruptions caused by the COVID-19 pandemic, the global hospitality industry finds itself facing a unique challenge – the need to seamlessly integrate safety measures with the tenets of sustainable development. This challenge is particularly salient in the Romanian hospitality sector, which, like its global counterparts, is navigating the complexities of a post-pandemic landscape. As the Romanian hospitality industry adapts to the evolving dynamics of the pandemic, it must delicately balance the imperative of safeguarding public health with a broader commitment to ecological, economic, and social sustainability. The stakes are high, as the decisions made during this pivotal period will have lasting consequences not just for the industry but for the well-being of the communities it serves. This research represents a concerted effort to gain a comprehensive understanding of the intricate dynamics within the Romanian hospitality sector during these unparalleled times. To achieve this, an extensive survey engaged 323 customers who have been directly exposed to evolving safety and hygiene measures in hotels and restaurants. These customers, as integral stakeholders, offer unique insights that underpin this study, shedding light on the numerous challenges, successes, and potential areas for improvement. A significant finding of this research is the intricate interplay between the perceived effectiveness of safety and hygiene measures and customers' perspectives on sustainable practices within hospitality establishments. The perception of sustainable development emerges as a crucial mediator in this complex relationship. In essence, customers' impressions of the industry's commitment to sustainability significantly influence their overall satisfaction with services provided, and subsequently, their perception of the safety of their experiences. The results of this survey offer valuable guidance for the Romanian hospitality industry as it navigates these turbulent waters. Insights from customers not only reflect their experiences but also provide direction to industry decision-makers as they navigate the uncharted territory of the "new normal." This guidance is instrumental in making informed choices that balance safety, sustainability, and the evolving expectations of a diverse clientele.*

Keywords: hospitality industry, COVID-19 pandemic, safety measures, sustainable development, customer perception.

Introduction

The global hospitality industry, like many other sectors, found itself thrust into a new era of unprecedented challenges and transformations in the wake of the COVID-19 pandemic (Davahli et al., 2020; Gursoy and Chi, 2020; Pillai et al., 2021). In the aftermath of this profound disruption, the industry stands at a unique crossroads, where safety measures and sustainable development are inextricably intertwined, demanding a delicate balancing act (Jones and Comfort, 2020; Alonso et al., 2020; Canhoto and Wei, 2021; Ntounis et al., 2022). Within the specific context of the Romanian hospitality industry, a microcosm representative of global challenges, the post-pandemic landscape presents both formidable hurdles and compelling opportunities (Tigu et al., 2021; Volkmann et al., 2021).

As the Romanian hospitality sector grapples with the ever-evolving dynamics of the pandemic, it faces a pivotal conundrum. It must adeptly navigate the intricate and evolving terrain of safety and hygiene measures, seeking to safeguard public health, while simultaneously upholding a broader commitment to practices that are ecologically, economically, and socially sustainable. The stakes are undeniably high, as the choices made during this pivotal period will reverberate not only throughout the industry itself but also have far-reaching implications for the well-being of the communities it serves.

This research embarks on a journey to provide a comprehensive understanding of the multifaceted dynamics within the Romanian hospitality sector during these unprecedented times. To achieve this, an extensive survey was conducted, engaging 323 customers who have been directly exposed to the evolving landscape of safety and hygiene measures in hotels and restaurants. These customers, positioned as key stakeholders in the industry, offer unique insights that serve as the bedrock of this study, shedding light on the myriad challenges, successes, and potential areas for improvement.

However, within the context of the current landscape, a research gap emerges. While numerous studies have explored safety measures and their impact on customer satisfaction, few have delved into the complex interplay between safety and sustainability within the hospitality industry. This study bridges this gap by unraveling the intricate connections between safety measures, sustainable development, and overall customer satisfaction. It explores how sustainable development perceptions mediate the impact of safety measures on overall customer satisfaction. By addressing this research gap, our study contributes to a more holistic understanding of the factors influencing customer perceptions in the post-pandemic hospitality industry.

One of the most intriguing findings of this research is the intricate interplay between the perceived effectiveness of safety and hygiene measures and customers' perspectives on sustainable practices within hospitality establishments. Sustainable development, long recognized as an essential component of responsible business practices, emerges as a pivotal mediator in this complex relationship. Customers' impressions of the industry's commitment to sustainability significantly influence their overall satisfaction with the services provided, and in turn, their perception of the safety of their experiences. In essence, the perception of sustainable development shapes the overall customer experience within the industry, rendering it not only safer but more ecologically and socially responsible.

The results of this survey, therefore, present valuable guidance for the Romanian hospitality industry as it charts its course through these turbulent waters. The insights garnered from the customers do not merely serve as a reflection of their experiences but also act as a compass to guide decision-makers within the sector as they navigate the uncharted territory of the "new normal." This guidance is instrumental in making informed choices that balance safety, sustainability, and the ever-evolving expectations of the industry's diverse clientele.

Literature review

The COVID-19 pandemic has catalyzed a profound transformation within the Romanian hospitality industry, propelling it into a new era where safety and sustainability are inextricably intertwined. Understanding the subtleties of customer perceptions in this context is paramount in upholding and enhancing the resilience and viability of the industry. By effectively harmonizing safety measures with sustainability principles, the industry can evolve to meet the ever-changing demands and preferences of its clientele, ensuring its enduring growth and prosperity while contributing positively to the well-being of both society and the environment.

Understanding the intricate relationship between safety measures, sustainable development, and customer satisfaction is essential in the context of the hospitality industry. Each of these factors plays a pivotal role in shaping the customer experience, and their interplay is a subject of growing interest in both academic research and industry practice.

Safety measures within the hospitality industry have traditionally been associated with ensuring the physical well-being of customers. Numerous studies have emphasized the importance of safety measures in influencing customer satisfaction. For instance, a study by Ali et al. (2021) found a strong positive correlation between the perceived safety of a hotel environment and customer satisfaction. Their research highlighted that customers who feel safe and secure during their stay are more likely to express high levels of satisfaction (Ali et al., 2021, 2019).

Sustainable development practices within the hospitality industry encompass a range of initiatives, from energy-efficient operations to community engagement. These practices have garnered increasing attention for their potential to positively impact customer satisfaction. Research by Koch, Gerdt, and Schewe (2020) explored the influence of sustainable practices in hotels on customer satisfaction. They discovered that customers tend to have more positive perceptions of hotels that engage in sustainable practices, leading to higher levels of satisfaction (Koch, Gerdt and Schewe, 2020).

A fascinating dimension emerges when investigating the interplay between safety measures and sustainable development in the context of customer satisfaction. A study by Pozo, do Amaral Moretti and Tachizawa (2016) revealed that sustainable development perceptions mediate the impact of safety measures on customer satisfaction. Their research showed that customers not only value safety but also perceive the commitment to sustainable development as an added value. In essence, safety is necessary but not sufficient for full customer satisfaction; it is the inclusion of sustainable development practices that enhances overall satisfaction (Pozo, do Amaral Moretti and Tachizawa, 2016).

These studies collectively underline the significance of these three factors in shaping the customer experience in the hospitality industry. Safety measures are foundational for customer well-being, sustainable development practices are an added value, and the combination of both provides a comprehensive and enriched customer experience. The relationship between these factors is not only of theoretical interest but also of practical importance. It holds profound implications for the operational strategies of hospitality establishments, as aligning safety and sustainability practices can lead to improved customer satisfaction and loyalty.

The central research question that underpins this study delves into the intricacies of customer perceptions within the hospitality industry. It specifically focuses on the concept of mediation, seeking to unravel the extent to which perceptions of sustainable development act as a mediating factor in the relationship between two critical aspects of the customer experience: satisfaction with safety and hygiene measures and overall customer satisfaction.

This research question is pivotal in unveiling the underlying mechanisms that shape customer perceptions and preferences in the context of the hospitality industry. By exploring the mediating role of sustainable development perceptions, the study aims to shed light on the nuanced interplay between safety and hygiene satisfaction and the broader satisfaction of customers. It seeks to ascertain whether and to what degree the perception of sustainable practices serves as a bridge that influences the impact of safety and hygiene satisfaction on the overall satisfaction of customers.

The inquiry into this mediation mechanism is not only relevant but also timely, especially in the evolving landscape of the post-pandemic hospitality industry, where issues of safety, sustainability, and customer satisfaction have gained significant prominence. Thus, this research question serves as a focal point for a comprehensive understanding of the complex dynamics within the hospitality sector. It provides a lens through which to explore the multifaceted relationships and interdependencies between safety measures, sustainable development initiatives, and the ultimate satisfaction of customers. Ultimately, the insights derived from this research question can guide decision-makers within the industry as they navigate this ever-changing terrain, making informed choices that harmonize safety, sustainability, and customer expectations to enhance the well-being of both patrons and the environment.

Methodology

This study employed a cross-sectional research design, aimed at capturing a snapshot of customer perceptions within the hospitality industry during the year 2022. Data collection was carried out through an online questionnaire distributed to a sample of 323 customers who had direct experience with the evolving safety and hygiene measures in hotels and restaurants. The online questionnaire facilitated efficient data collection.

The data collected for this study was subjected to several statistical analyzes to explore various dimensions of customer perceptions within the hospitality industry. The following analytical methods were employed. Cronbach's alpha was computed to assess the internal consistency and reliability of the scale used to measure customer perceptions. This analysis helped ensure that the survey items effectively captured the intended construct. Factor analysis was conducted to explore the underlying structure of the survey items and to identify he latent factors or dimensions within customer perceptions referring to safety and hygiene satisfaction, overall customer satisfaction and sustainable development perceptions. This technique allowed for a more in-depth understanding of the interrelationships among the survey items and the potential grouping of related items into factors. Mediation analysis was further employed to examine the mediating role of sustainable development perceptions in the relationship between safety and hygiene satisfaction and overall customer satisfaction within the hospitality industry. This analysis aimed to elucidate the extent to which the perception of sustainable practices mediates the impact of safety and hygiene satisfaction on overall satisfaction.

Research question: To what extent does the perception of sustainable development mediate the relationship between safety and hygiene satisfaction and overall customer satisfaction within the hospitality industry?

These data analysis methods were employed to comprehensively assess and understand the complex dynamics of customer perceptions within the hospitality industry, including the interplay between safety measures, sustainable development, and overall satisfaction.

Participants

A total of 323 customers actively participated in this study, offering their invaluable insights into the customer perceptions within the hospitality industry. Participants were selected using a convenience sampling method, which involves individuals chosen based on their availability and accessibility. This method allowed for a practical and efficient approach to gather a diverse set of customer perspectives. Of the participants, 59 % identified as female, while 41 % identified as males, providing a balanced gender representation within the sample. This diversity in gender representation enables a more comprehensive understanding of how different customer demographics perceive safety measures, sustainability, and overall satisfaction within the hospitality context. The study also considered the average mean age of the respondents, which was calculated at 32 years. This age distribution reflects a range of generational perspectives and experiences, making the findings more reflective of a broader customer demographic.

The study adhered to ethical principles, including the protection of participants' rights and their right to anonymity. Prior to participating in the study, all respondents provided informed consent, acknowledging their willingness to contribute to the research by responding to the online questionnaire. Participants were assured that their responses would remain confidential and be used solely for research purposes. Any potentially sensitive information was treated with the utmost discretion, safeguarding the privacy of all individuals who offered their time and insights.

Instrument

An 11 items scale was developed to measure hospitality industry customer perceptions. Each of the 11 items were inspired by relevant research on the topic

(Zibarzani et al., 2022; Szentesi et al., 2021; Srivastava and Kumar, 2021; Cuc et al., 2022; Sun et al., 2022; Jiménez-Medina et al., 2022).

The first item assesses the level of satisfaction regarding safety and hygiene measures implemented by hotels and restaurants, with respondents providing ratings on a 1 to 5 scale: Item 1 – "I am completely satisfied with the safety and hygiene measures implemented by the hotel/restaurant.". The fifth item focuses on customers' expectations and satisfaction concerning services provided by hotels, particularly in terms of safety and protection against the Sars-Cov-2 virus: Item 5 – "I believe that the services provided by the hotel are fully in line with my expectations for safety and protection against the Sars-Cov-2 virus.". Respondents were asked to rate their agreement on a 1 to 5 scale. The tenth item delves into the extent to which respondents believe that the hospitality establishment is involved in the local community or supports social and sustainable development projects, with ratings also provided on a 1 to 5 scale: Item 10 – "To what extent do you believe the hospitality establishment engages with the local community or supports social and sustainable development projects?". These items collectively contributed to a comprehensive understanding of customer perceptions and are essential for assessing the industry's alignment with safety, sustainability, and community engagement.

The scale's reliability was evaluated using Cronbach's alpha and individual item reliability statistics, offering insights into the consistency and dependability of the scale and the contributions of each item to its overall reliability. The scale designed to assess customer perceptions within the hospitality industry demonstrates a commendable level of internal consistency, as evidenced by a Cronbach's alpha of 0.871. This high alpha value suggests that the items within the scale consistently gauge the same underlying construct, ensuring a robust level of reliability. The scale's mean score is 38.452, with a standard deviation of 5.905. The analysis of individual item reliability reveals that most items exhibit good to excellent reliability, contributing positively to the overall reliability of the scale. These findings signify that the scale is a dependable instrument for evaluating customer perceptions in the context of the hospitality industry.

Results

In this study, exploratory factor analysis (EFA) was conducted to uncover the underlying structure of the survey items assessing customer perceptions within the hospitality industry. The chi-squared test revealed a significant association between the model and the observed data ($\chi^2=37.675$, $df=12$, $p<.001$), indicating a valid model fit.

The EFA results are presented in the Table 1. A promax rotation method was applied to enhance interpretability. Three factors emerged from the analysis. Factor 1 had high loadings for items 7, 3, 6, 5, and 9, demonstrating that these items are strongly related to this factor. Factor 2 exhibited a strong loading for items 11 and 10, whereas Factor 3 was primarily associated with items 2 and 1. Factor 1, Factor 2 and Factor 3 appeared to have no shared variance, indicating their orthogonality.

Table 1. Factor loadings

	Factor 1	Factor 2	Factor 3	Uniqueness
item7	0.873			0.389
item3	0.812			0.422
item6	0.728			0.484
item5	0.567			0.387
item9	0.414			0.422
item11		1.085		0.004
item10		1.020		0.103
item2			0.725	0.549
item1			0.441	0.442

Note: Applied rotation method is promax.
Source: Authors' own research.

The chi-squared test statistic (χ^2) assessed the goodness-of-fit of the model to the data. The obtained value of 37.675, with 12 degrees of freedom, demonstrated a statistically significant association between the model and the observed data ($p<.001$). This suggests that the identified factor structure fits the data significantly better than a model with no factors, providing support for the validity of the model in explaining the relationships among the survey items measuring customer perceptions in the hospitality industry.

The factor loadings, as indicated in the table, highlight the strength of the associations between the items and the identified factors. The high factor loadings for specific items on each factor signify their close relationship to that factor, supporting the validity of the factor structure.

Descriptive statistics were calculated to provide an overview of the central tendencies and variability of the three identified factors within the context of customer perceptions in the hospitality industry. Descriptive statistics are depicted in Table 2.

For Factor 1, representing safety perceptions, the mean score was 4.128 (M= 4.128, SE=0.043). The 95 % confidence interval ranged from 4.044 to 4.213, with a standard deviation of 0.775. The coefficient of variation for this factor was 0.188. The mean score for Factor 2, which represents sustainable development perception, was 4.452 (M=4.452, SE=0.041). The 95 % confidence interval ranged from 4.372 to 4.532, with a standard deviation of 0.736. The coefficient of variation for this factor was 0.165. Factor 3, associated with overall services satisfaction, had a mean score of 4.443 (M=4.443, SE=0.039). The 95 % confidence interval ranged from 4.367 to 4.519, with a standard deviation of 0.695. The coefficient of variation for this factor was 0.157.

Table 2. Descriptive statistics

	Valid	Missing	Mean	Std. Error of Mean	95 % Confidence Interval Mean Upper	95 % Confidence Interval Mean Lower	Std. Deviation	Coefficient of variation
Factor 2 Sustainable development	323	0	4.452	0.041	4.532	4.372	0.736	0.165
Factor 3 Overall services satisfaction	323	0	4.443	0.039	4.519	4.367	0.695	0.157
Factor 1 Safety perceptions	323	0	4.128	0.043	4.213	4.044	0.775	0.188

Source: Authors' own research.

Table 3, the factor characteristics provides insights into the unrotated and rotated factor solutions, elucidating the underlying structure of customer perceptions.

In the unrotated solution, Factor 1 (safety perceptions) exhibited an eigenvalue of 4.844 and a sum of squared loadings of 4.543, accounting for 50.5 % of the variance. In the rotated solution, the proportion of explained variance was 29.0 %, with an eigenvalue of 2.609. The unrotated solution for Factor 2 (sustainable development perception) yielded an eigenvalue of 1.219 and a sum of squared loadings of 0.718, explaining 8.0 % of the variance. In the rotated solution, this factor accounted for 27.3 % of the variance, with an eigenvalue of 2.456. Factor 3 (overall services satisfaction) had an unrotated solution with an eigenvalue of 0.782 and a sum of squared loadings of 0.538, explaining 6.0 % of the variance. In the rotated solution, this factor accounted for 8.1 % of the variance, with an eigenvalue of 0.733.

Table 3. Factor characteristics

		Unrotated solution			Rotated solution		
	Eigenvalues	SumSq. Loadings	Proportion var.	Cumulative	SumSq. Loadings	Proportion var.	Cumulative
Factor 1 – safety perceptions	4.844	4.543	0.505	0.505	2.609	0.290	0.290
Factor 2 – sustainable development perception	1.219	0.718	0.080	0.585	2.456	0.273	0.563
Factor 3 – overall services satisfaction	0.782	0.538	0.060	0.644	0.733	0.081	0.644

Source: Authors' own research.

These results provide a deeper understanding of the factor characteristics, shedding light on the relative importance of each factor in explaining the variance in customer perceptions in the hospitality industry. The rotated factor solution suggests a more robust interpretation of the relationships between safety perceptions, sustainable development perceptions, and overall services satisfaction, allowing for a more comprehensive understanding of their interplay.

Next, a correlation analysis was conducted to examine the relationships between the three identified factors: Factor 1 – Safety Perceptions, Factor 2 – Sustainable Development, and Factor 3 – Overall Services Satisfaction, within the context of customer perceptions in the hospitality industry. Results are presented in Table 4.

Table 4. Correlation matrix

Variable	1	2	3
1. Factor 2 Sustainable development	–		
2. Factor 3 Overall services satisfaction	0.389***	–	
3. Factor 1 Safety perceptions	0.648***	0.320***	–

* $p < .05$, ** $p < .01$, *** $p < .001$
Source: Authors' own research.

Factor 2 – Sustainable Development demonstrated a statistically significant positive correlation with Factor 3 – Overall Services Satisfaction (r=0.389, p<.001). This finding indicates that as customers' perceptions of sustainable development within the establishment increase, their overall satisfaction with services also tends to increase. Factor 1 – Safety Perceptions exhibited a strong and statistically significant positive correlation with both Factor 2 – Sustainable Development (r=0.648, $p<.001$) and Factor 3 – Overall Services Satisfaction (r= 0.320, $p<.001$). These results suggest that as customers' perceptions of safety within the establishment improve, their perceptions of sustainable development and overall service satisfaction are likely to improve as well. The significance levels for these correlations are denoted as follows: $p<.05$, $p<.01$, and $p<.001$. The statistically significant correlations underscore the importance of safety perceptions, sustainable development perceptions, and overall service satisfaction in shaping customer perceptions in the hospitality industry.

Further, the mediation analysis in this study aimed to explore the mediating role of sustainable development perceptions (Factor 2) in the relationship between safety perceptions (Factor 1) and overall services satisfaction (Factor 3) within the hospitality industry. The analysis sought to unravel the extent to which sustainable development perceptions act as a mediator in shaping customer satisfaction. The mediation analysis was performed in JASP software.

The direct effect of safety perceptions (Factor 1) on overall services satisfaction (Factor 3) was estimated at 0.150 (SE=0.080, z=1.868, p=0.062). Although this direct effect was marginally significant, it suggested that safety perceptions had a positive influence on overall services satisfaction.

The analysis revealed a statistically significant indirect effect, indicating that the influence of safety perceptions on overall services satisfaction was partially mediated by sustainable development perceptions (Factor 2). The indirect effect was estimated at 0.262 (SE=0.048, z=5.422, $p<.001$), signifying a substantial mediating role of sustainable development perceptions in this relationship.

The total effect of safety perceptions (Factor 1) on overall services satisfaction (Factor 3) was calculated to be 0.412 (SE=0.077, z=5.344, $p<.001$), reflecting the combined direct and indirect influences of safety perceptions on overall satisfaction.

Table 5. Path coefficients

		Estimate	Std. Error	z-value	p	95 % Confidence Interval	
						Lower	Upper
Factor2_Sustainable_development	→ Factor3_Overall_services_satisfaction	0.314	0.055	5.698	<.001	0.206	0.421
Factor1_Safety_perceptions	→ Factor3_Overall_services_satisfaction	0.150	0.080	1.868	0.062	-0.007	0.308
Factor1_Safety_perceptions	→ Factor2_Sustainable_development	0.836	0.073	11.458	<.001	0.693	0.979

Note. Robust standard errors, robust confidence intervals, ML estimator.
Source: Authors' own research.

The path coefficients depicted in Table 5 displayed robust relationships in the mediation model. Factor 2 (Sustainable Development) had a direct effect on Factor 3 (Overall Services Satisfaction) with an estimate of 0.314 (SE=0.055, z=5.698, $p < .001$), emphasizing its role in directly influencing overall satisfaction. Factor 1 (Safety Perceptions) also exhibited a positive but marginally significant direct effect on Factor 3 (Estimate=0.150, SE=0.080, z=1.868, p= 0.062). Furthermore, Factor 1 had a strong direct effect on Factor 2 (Sustainable Development) with an estimate of 0.836 (SE=0.073, z=11.458, $p<.001$), signifying its significant influence on sustainable development perceptions.

The R-squared values indicated that Factor 2 (Sustainable Development) explained approximately 42.0 % of the variance, while Factor 3 (Overall Services Satisfaction) explained about 15.9 % of the variance within the mediation model, further underlining their importance in understanding customer perceptions in the hospitality industry.

These results demonstrate the mediation role of sustainable development perceptions in the relationship between safety perceptions and overall services satisfaction. While safety perceptions had a direct positive influence on overall satisfaction, the mediation analysis highlighted the significant indirect effect mediated by sustainable development perceptions, emphasizing their substantial role in shaping customer satisfaction in the hospitality industry.

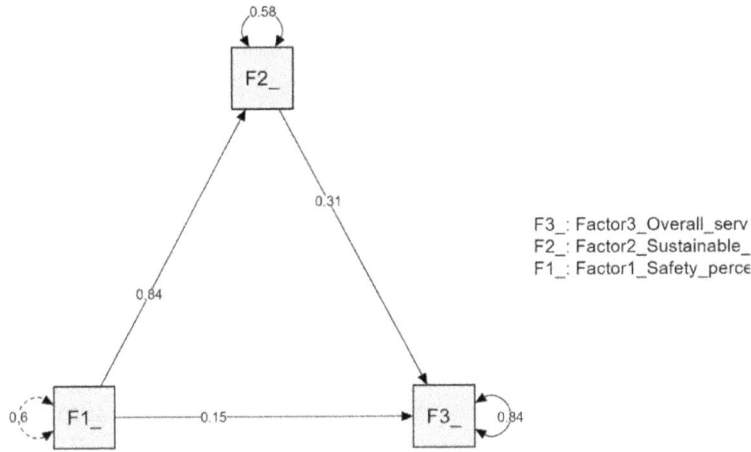

Figure 1. Path Plot.
Source: Authors' own research.

Overall, the path plot depicted in Figure 1 provides a clear visual representation of how safety perceptions influence sustainable development perceptions, how both safety and sustainable development perceptions impact overall services satisfaction, and the mediating role of sustainable development in this relationship. It offers a concise and interpretable summary of the mediation model's pathways and associations, helping researchers and stakeholders understand the complex dynamics at play within the context of customer perceptions in the hospitality industry.

Discussions

The findings of this study offer valuable insights into the multifaceted nature of customer perceptions within the hospitality industry, specifically in the context of safety measures, sustainable development, and overall service satisfaction. This discussion section delves into the key results, their implications, and the broader significance of the study.

The exploratory factor analysis (EFA) revealed a three-factor structure that effectively captured the underlying dimensions of customer perceptions. Factor 1, labeled "Safety Perceptions," encompassed aspects related to safety and hygiene measures implemented by hotels and restaurants. Factor 2, termed "Sustainable Development Perception," centered on customers' perceptions of the establishment's engagement in sustainable development initiatives. Factor 3, denoted as "Overall Services Satisfaction," represented the broader satisfaction with the services provided. The robustness of the rotated factor solution highlighted the distinct and significant relationships between these factors, contributing to a deeper understanding of how they interrelate and collectively influence customer perceptions.

The correlation analysis unveiled several relationships within the model. The positive correlation between Sustainable Development Perception (Factor 2) and Overall Services Satisfaction (Factor 3) suggests that as customers' perceptions of sustainable development practices within the establishment increase, their overall satisfaction with services also tends to increase. This finding emphasizes the importance of sustainable development initiatives in enhancing customer satisfaction, similar to other research (Prud'homme and Raymond, 2013; Pozo, do Amaral Moretti and Tachizawa, 2016; Chen, Chen and Lee, 2011; Teng and Wu, 2019; Koch, Gerdt and Schewe, 2020).

Furthermore, Safety Perceptions (Factor 1) demonstrated a strong positive correlation with both Sustainable Development Perception (Factor 2) and Overall Services Satisfaction (Factor 3). This suggests that as customers' perceptions of safety improve, their perceptions of sustainable development engagement and overall service satisfaction also tend to improve. Safety is a fundamental component that underpins customer experiences, influencing their perceptions in various facets, as found in other research (Jang and Lee, 2020; Ghorbani et al., 2023; Wang, Ma and Yu, 2021; Quan, Al-Ansi and Han, 2022).

The mediation analysis provided substantial evidence of the mediating role of Sustainable Development Perception (Factor 2) in the relationship between Safety Perceptions (Factor 1) and Overall Services Satisfaction (Factor 3). While Safety Perceptions had a direct positive effect on Overall Services Satisfaction, the indirect effect mediated by Sustainable Development Perception was notably significant. This result underscores the importance of sustainable development perceptions as a mediator that amplifies the positive influence of safety perceptions on overall satisfaction, as concluded by other relevant research (Modica et al., 2020; Malik et al., 2020; Hu, Liu and Zhang, 2020).

The total effect of safety perceptions on overall services satisfaction was substantial, including both the direct and indirect influences. These findings suggest that fostering a perception of safety among customers is a vital aspect of enhancing overall satisfaction. Moreover, the mediating role of sustainable development perception highlights the value of sustainable practices in amplifying the positive impact of safety measures on customer satisfaction.

While this study has yielded valuable insights into customer perceptions in the hospitality industry, it is important to acknowledge several limitations that may affect the interpretation and generalizability of the findings. First and foremost, the study relied on convenience sampling, which may introduce selection bias. The use of an online questionnaire could have excluded individuals who are less inclined to participate in such surveys, potentially limiting the diversity of the sample and the generalizability of the results. Another limitation is the cross-sectional design of the study. The data were collected at a specific point in time, providing a snapshot of customer perceptions. A longitudinal approach, tracking these perceptions over an extended period, would offer a more comprehensive understanding of how they evolve in response to changing circumstances and external events. Additionally, the study relied on self-reported data, which comes with inherent limitations. Respondents may have provided socially desirable responses, impacting the accuracy of the findings. Future research could benefit from a combination of self-report data and objective measures to mitigate this limitation. Lastly, this study primarily focused on the Romanian hospitality industry. While this allowed for in-depth exploration, the findings may not be universally applicable to other cultural or regional contexts. To address this limitation, future research should consider cross-cultural comparisons to explore potential variations in customer perceptions.

To build upon the findings of this study and address the limitations outlined, there are several promising avenues for future research. Longitudinal studies could provide valuable insights into the dynamics of customer perceptions over time. By tracking changes in safety, sustainability, and satisfaction as they evolve in response to changing circumstances, researchers can gain a more nuanced understanding of these relationships. Diverse and representative sampling is essential for robust research. Future studies should aim to include a broader range of demographics and customer profiles to enhance the generalizability of findings and ensure a more comprehensive understanding of the factors at play. In addition to quantitative research, qualitative investigations can offer in-depth insights into the factors influencing customer perceptions. Exploring individual experiences and expectations can complement the quantitative data, providing a richer context for the findings. Intervention studies can assess the impact of safety and sustainability initiatives on customer perceptions and satisfaction. By conducting experiments or interventions, researchers can provide practical recommendations for the hospitality industry, helping to shape its practices and policies.

To gain a more comprehensive understanding of the impact of sustainability, future research could explore additional sustainability metrics. This may include environmental practices and community engagement, shedding light on the broader implications of sustainability efforts on customer perceptions. Finally, researchers may want to examine the influence of external factors, such as government regulations and global events, on customer perceptions. Understanding how these factors interact with safety and sustainability initiatives can provide a more holistic view of the forces shaping customer perceptions in the hospitality industry.

Conclusions

The implications of these findings are pertinent for the post-pandemic hospitality industry. Safety measures are foundational, directly influencing customer satisfaction. However, the study highlights that sustainable development initiatives play a mediating role, enhancing the impact of safety on overall satisfaction. This suggests that a comprehensive approach, combining safety and sustainability, is essential for achieving high levels of customer satisfaction.

The implications of the study's findings are highly pertinent for the post-pandemic hospitality industry, which has undergone profound changes and faces new challenges. The results emphasize the critical role of safety measures in directly influencing customer satisfaction. In times of heightened health concerns, safety is a fundamental prerequisite for any customer interaction within the industry. However, the study goes further by revealing that sustainable development initiatives play a pivotal mediating role, significantly amplifying the positive impact of safety on overall satisfaction.

This nuanced relationship can be interpreted through the lens of Social Exchange Theory (Cook et al., 2013; Cropanzano and Mitchell, 2005; Cropanzano et al., 2017). In the context of the hospitality industry, customers engage in social exchanges with service providers. Safety measures represent a fundamental offering in this exchange – customers expect and demand a safe and hygienic environment. In this sense, safety measures represent a basic social contract within the exchange. Sustainable development initiatives, on the other hand, represent an additional, value-adding element in the exchange. They go beyond the basic contract, signaling to customers that the establishment is not only meeting their basic needs but also contributing to broader social and environmental goals. This additional value creates a positive imbalance in the exchange, leading to enhanced customer satisfaction.

The study's findings underscore the importance of maintaining and even exceeding the basic social contract of safety while actively engaging in sustainable development initiatives to create this positive imbalance. This can result in a more satisfied and loyal customer base, which is critical for the long-term success and sustainability of hospitality establishments.

References

Ali, B.J., Gardi, B., Othman, B.J., Ahmed, S.A., Ismael, N.B., Hamza, P.A., Aziz, H.M., Sabir, B.Y. and Anwar, G. 2021. Hotel service quality: The Impact of service quality on customer satisfaction in hospitality. *International Journal of Engineering, Business and Management*, 5(3), pp. 14–28.

Alonso, A.D., Kok, S.K., Bressan, A., O'Shea, M., Sakellarios, N., Koresis, A., ...and Santoni, L.J. 2020. COVID-19, aftermath, impacts, and hospitality firms: An international perspective. *International Journal of Hospitality Management*, 91, p. 102654.

Canhoto, A. I. and Wei, L. 2021. Stakeholders of the world, unite!: Hospitality in the time of COVID-19. *International Journal of Hospitality Management*, 95, p. 102922.

Chen, C.M., Chen, S.H. and Lee, H.T. 2011. The destination competitiveness of Kinmen's tourism industry: Exploring the interrelationships between tourist perceptions, service performance, customer satisfaction and sustainable tourism. *Journal of Sustainable Tourism*, 19(2), pp. 247–264.

Cook, K.S., Cheshire, C., Rice, E.R. and Nakagawa, S. 2013. Social exchange theory. *Handbook of Social Psychology*, pp. 6–88.

Cropanzano, R. and Mitchell, M.S. 2005. Social exchange theory: An interdisciplinary review. *Journal of Management*, 31(6), pp. 874–900.

Cropanzano, R., Anthony, E.L., Daniels, S.R. and Hall, A.V. 2017. Social exchange theory: A critical review with theoretical remedies. *Academy of Management Annals*, 11(1), pp. 479–516.

Cuc, L.D., Feher, A., Cuc, P.N., Szentesi, S.G., Rad, D., Rad, G., Pantea, M.F. and Joldes, C.S.R. 2022. A parallel mediation analysis on the effects of pandemic accentuated occupational stress on hospitality industry staff turnover intentions in COVID-19 context. *International Journal of Environmental Research and Public Health*, 19(19), p. 12050.

Davahli, M.R., Karwowski, W., Sonmez, S. and Apostolopoulos, Y. 2020. The hospitality industry in the face of the COVID-19 pandemic: Current topics and research methods. *International Journal of Environmental Research and Public Health*, 17(20), p. 7366.

Ghorbani, A., Mousazadeh, H., Akbarzadeh Almani, F., Lajevardi, M., Hamidizadeh, M. R., Orouei, M., ... and Dávid, L.D. 2023. Reconceptualizing customer perceived value in hotel management in turbulent times: A case study of Isfahan metropolis five-star hotels during the COVID-19 Pandemic. *Sustainability*, 15(8), p. 7022.

Gursoy, D. and Chi, C.G. 2020. Effects of COVID-19 pandemic on hospitality industry: Review of the current situations and a research agenda. *Journal of Hospitality Marketing & Management*, 29(5), pp. 527–529.

Hu, B., Liu, J. and Zhang, X. 2020. The impact of employees' perceived CSR on customer orientation: An integrated perspective of generalized exchange and social identity theory. *International Journal of Contemporary Hospitality Management*, 32(7), pp. 2345–2364.

Jang, H.W. and Lee, S.B. 2020. Serving robots: Management and applications for restaurant business sustainability. *Sustainability*, 12(10), p. 3998.

Jiménez-Medina, P., Navarro-Azorín, J.M., Cubillas-Para, C. and Artal-Tur, A. 2022. What safety and security measures really matter in the post-COVID recovery of the hospitality industry? An analysis of the visitor's intention to return in Spain. *Tourism and Hospitality*, 3(3), pp. 606–617.

Jones, P. and Comfort, D. 2020. The COVID-19 crisis and sustainability in the hospitality industry. *International Journal of Contemporary Hospitality Management*, 32(10), pp. 3037–3050.

Koch, J., Gerdt, S.O. and Schewe, G. 2020. Determinants of sustainable behavior of firms and the consequences for customer satisfaction in hospitality. *International Journal of Hospitality Management*, 89, p. 102515.

Malik, S.A., Akhtar, F., Raziq, M.M. and Ahmad, M. 2020. Measuring service quality perceptions of customers in the hotel industry of Pakistan. *Total Quality Management & Business Excellence*, 31(3–4), pp. 263–278.

Modica, P.D., Altinay, L., Farmaki, A., Gursoy, D. and Zenga, M. 2020. Consumer perceptions towards sustainable supply chain practices in the hospitality industry. *Current Issues in Tourism*, 23(3), pp. 358–375.

Ntounis, N., Parker, C., Skinner, H., Steadman, C. and Warnaby, G. 2022. Tourism and hospitality industry resilience during the Covid-19 pandemic: Evidence from England. *Current Issues in Tourism*, 25(1), pp. 46–59.

Pillai, S.G., Haldorai, K., Seo, W.S. and Kim, W.G. 2021. COVID-19 and hospitality 5.0: Redefining hospitality operations. *International Journal of Hospitality Management*, 94, 102869.

Pozo, H., do Amaral Moretti, S.L. and Tachizawa, T. 2016. Hospitality practices as sustainable development: An empirical study of their impact on hotel customer satisfaction. *Tourism & Management Studies*, 12(1), pp. 153–163.

Prud'homme, B. and Raymond, L. 2013. Sustainable development practices in the hospitality industry: An empirical study of their impact on customer satisfaction and intentions. *International Journal of Hospitality Management*, 34, pp. 116–126.

Quan, L., Al-Ansi, A. and Han, H. 2022. Assessing customer financial risk perception and attitude in the hotel industry: Exploring the role of protective measures against COVID-19. *International Journal of Hospitality Management*, 101, p. 103123.

Srivastava, A. and Kumar, V. 2021. Hotel attributes and overall customer satisfaction: What did COVID-19 change?. *Tourism Management Perspectives*, 40, p. 100867.

Sun, S., Jiang, F., Feng, G., Wang, S. and Zhang, C. 2022. The impact of COVID-19 on hotel customer satisfaction: Evidence from Beijing and Shanghai in China. *International Journal of Contemporary Hospitality Management*, 34(1), pp. 382–406.

Szentesi, S.G., Cuc, L.D., Feher, A. and Cuc, P.N. 2021. Does Covid-19 affect safety and security perception in the hospitality industry? A Romanian case study. *Sustainability*, 13(20), p. 11388.

Teng, Y.M. and Wu, K.S. 2019. Sustainability development in hospitality: The effect of perceived value on customers' green restaurant behavioral intention. *Sustainability*, 11(7), 1987.

Tigu, G., Ciora, C., Petcu, M.A., Boboc, D., Crismariu, O.D. and Curteanu, A.B. 2021. Restart the hotel, restaurant, and travel industry in Romania after the COVID-19 pandemic. In: A.M. Dima, I. Anghel and R.C. Dobrea, eds. *Economic recovery after COVID-19*. Springer.

Volkmann, C., Tokarski, K.O., Dincă, V.M. and Bogdan, A. 2021. The impact of COVID-19 on Romanian tourism. An explorative case study on Prahova County, Romania. *Amfiteatru Economic*, 23(56), pp. 196–205.

Wang, K.Y., Ma, M.L. and Yu, J. 2021. Understanding the perceived satisfaction and revisiting intentions of lodgers in a restricted service scenario: Evidence from the hotel industry in quarantine. *Service Business*, 15(2), pp. 335–368.

Zibarzani, M., Abumalloh, R.A., Nilashi, M., Samad, S., Alghamdi, O.A., Nayer, F.K.,Ismail M.Y., Mohd S. and Akib, N.A.M. 2022. Customer satisfaction with restaurants service quality during COVID-19 outbreak: A two-stage methodology. *Technology in Society*, 70, p.101977.

Nadiia Davydenko, Oleksandra Smirnova,
Zoia Titenko, Alina Buriak

The state of formation of the state financial control system in the field of land relations in Ukraine

Abstract: *Agricultural land (apart from human potential) is the most important asset for Ukraine. Almost 28 million hectares of Ukrainian chernozems are No. 1 in Europe and No. 4 in the world (8.7 % of world reserves). In order to use this potential and ensure the sustainable development of agriculture. In the context of state regulation of land relations, state financial control over reproduction of soil fertility occupies a special place in view of its multifacetedness and complexity as a natural phenomenon with many forms of manifestation, which is a necessary condition for the development of the country's economy. The purpose of the article is to study the current state of land tax payment, identify the problem of organizing and conducting state financial control, and develop measures to improve the effectiveness of this control. The article defines that state financial control in the system of land relations is aimed at ensuring the effective and legal use of financial resources related to land issues. The main task of this control is to ensure the transparency, efficiency and legality of economic activities related to land resources. The reserves of land resources of Ukraine and the state of their use were analyzed. It was established that there is an imperfect system of state financial control in Ukraine. Characteristic schemes of committing crimes in the field of land relations are: abuse of officials during registration of land plots; arbitrary seizure of land plots; illegal transfer of agricultural lands to other categories: urbanization of agricultural lands; reduction of rental payment rates, illegal leasing of land; alienation of agricultural land plots by peasants at a price that is much lower than economically justified; inefficient and non-targeted land use, which does not contribute to the regeneration of agricultural land.*

Keywords: financial control, land resources, environmental tax, land management, land relations.

Introduction

In the face of global challenges, in particular the fight against land degradation, the implementation of an important stage of land reform, namely the full opening of the agricultural land market, the implementation of conceptual changes in the decentralization of power, Ukraine faces the important task of substantiating and implementing strategic directions for the development of land policy based on the principles recommended by the UN. In addition, the question of attracting the potential of land and other natural resources to effective and efficient economic processes (turnover) while preserving the functionality of land resources as an important element of ensuring the well-being and ecological security of the population does not lose its relevance, since this contributes to access to food and water, stable employment and life support, resistance to climate change and extreme weather events, and overall safety of human life (Igor Chugunov et al., 2020; Sergij Pirozhkova and Mykhailo Khvesika, 2015).

The current land policy in Ukraine is not aimed at the protection and preservation of land, does not defend the interests and rights of private owners, and also does not ignore the capitalization (increase in value) of land use. Growing demand for food, feed, fuel and raw materials leads to increased pressure on land and competition for land and other natural resources located on them. At the same time, the area of available productive lands is decreasing as a result of their degradation. Driving forces of degradation are external factors that directly or indirectly affect the health and productivity of the land, as well as land-related resources – soil, water, and biological diversity (Maria Ilyina and Yuliia Shpylyova, 2020). In the conditions of global negative changes, land policy needs significant institutional changes in the system of state financial control and land use and reform into a multifunctional (multi-purpose) management system "from top to bottom" and "from bottom to top" based on the principles recommended by the UN to overcome poverty, ensuring well-being and environmental security of people.

Literature review

Land relations occupy a special place in the system of fundamental socio-economic categories. In general, the problem of regulating land relations is quite widely presented in the works of domestic scientists. Thus, Leonbd Novakovsky (2015), made a significant contribution to the study of the problems of regulating land relations having proposed ways to improve its legislative support and directions for completing the land reform in Ukraine (Nadiia Davydenko anf Yuliia Porohivnyk, 2019).

An equally important contribution to the theory of regulation of land relations belongs to Mykola Fedorov (1998), who substantiated the theoretical-methodological and methodological foundations of its development, developed the methodological foundations of the normative monetary valuation of land, proposed methodological approaches to the economic stimulation of rational land use.

A significant contribution to the development of the conceptual foundations of the formation and functioning of market land relations in agriculture and the methodological foundations of land management, as a fundamental state management mechanism in the field of land use and protection, belongs to Dmytro Dobryak (2007).

As scientists Svitlana Antonova (2021), Shamil Ibatullin et al. (2012), Anatoliy Moskalenko (2011), Hanna Shust (2019) state financial control in the field of land relations is an independent type of activity that ensures compliance with current legislation during state supervision of the use and protection of land and the mechanism of effective management of land resources.

Methodology

The methodological basis of the study is the fundamental provisions of the finance of nature use and environmental protection, the theory of innovation, anti-crisis management, state financial control, research by domestic and foreign authors in the field of sustainable development and sustainable agricultural land use, data from the State Treasury Service of Ukraine, the State Service of Ukraine for Geodesy, cartography and cadastre, regulatory and legislative acts of state authorities, periodical scientific publications.

Results and discussions

Control over land resources and land use is concentrated in the governing bodies themselves (state committees for land resources, water and forestry). State control is aimed mainly at responding to abuses in the field of land use, but not at preventing them and, even more so, not at adjusting state land policy.

Reports on inspections of the activities of land resources management bodies are published in a generalized form and irregularly. The results of the response to specific violations are not always made public, mostly without analyzing the causes and identifying the guilty parties. Control over activities in the field of land use has not become an effective mechanism for fighting corruption (Oleksandr Gutorov 2006; Anya Ivanova, 2017; Aurel Lup, Liliana Miron and Indira D. Alim, 2016).

As practice shows, the object of financial control is not limited to checking only monetary funds, but also covers material, natural, labor and other resources of the state, since their use is carried out in monetary form. In other words, financial control extends not only directly to financial, but also to related economic relations (Sviatoslav Baliuk and Anatolii Kucher, 2015; Liubov Melnychuk,

2015). Therefore, this gives us the opportunity to assert that land relations are also included in the sphere of influence of state financial control.

The legislative support for the implementation of land reform, which was formed and improved during the period of its implementation (1991–2018), in general contributed to the change of land relations in Ukraine and their improvement.

Considering the conceptual nature of most of the changes to both the Land Code and a number of other related legislative acts, the process of formation and implementation of state policy in this area, including in connection with the presence of high risks of untimely and incomplete adoption appropriate management decisions, was and remains complicated.

Repeated organizational changes in the system of central bodies, whose powers include, in particular, the implementation of state policy in the field of land relations, land use and protection, as well as the reorganization of the central body that forms such policy and coordinates its implementation, did not contribute to the proper implementation of land reform and ensuring appropriate internal control (Maria Mendonca and Fabio Pitta 2022; Sijia Liu et al., 2021).

Analyzing the reserves of land resources in Ukraine and the state of their use, it should be noted that our country is among the top five countries with more than 50 hectares of arable land per 100 inhabitants. The fact that Ukraine is among the five countries with the highest arable land per capita indicates the importance of the agricultural sector in the country. This can be an advantage, but also requires a proper balanced approach and financial control over the use of these resources. The land fund of Ukraine is characterized by the presence of high bioproductive potential. Its structure is dominated by lands with fertile soils. 7 % of the world's black soil reserves are concentrated in Ukraine.

The land fund of Ukraine, as of January 1, 2020, amounted to 603,549 million hectares (Table 1).

Table 1. Land areas of Ukraine and their changes from 2010 to 2020

Types of main land plots	Land area as of 12/31/2010		Land area as of 12/31/2015		Land area as of 12/31/2020	
	total, thousand ha	% to the total area	total, thousand ha	% to the total area	total, thousand ha	% to the total area
Agricultural lands	42,813.7	70.94	42,731.5	70.80	42,682.0	70.7
Forests and other wooded areas	10,591.9	17.55	10,630.3	17.61	10,686.8	17.7
Built-up lands	24,99.1	4.14	2,550.4	4.23	2,480.4	4.1
Earth under water	2,423.2	4.01	2,426.4	4.02	2,399.1	3.9
Open wetlands	979.4	1.62	982.6	1.63	990.1	1.6
Other lands	1,047.5	1.74	1,033.7	1.71	905.9	1.5
Total (territory) of Ukraine	60,354.8	100	60,354.9	100	60,354.9	100

Source: Authors' own research based on the data of the State Service of Ukraine for Geodesy, cartography and cadastre, 2023.

The area of built-up land decreased: from 2.49 million hectares as of 12/31/2010 to 2.48 million hectares as of 12/31/2020, i.e. by 18.6 thousand hectares. The decrease in the share of built-up land is associated with a decrease in the scale of private construction near large cities and urban agglomerations, with the intensification of housing construction in many cities of Ukraine and the expansion of the boundaries of large cities.

The total area of forested land increased from 10.59 million hectares to 10.69 million hectares, i.e. by 10 thousand hectares. Taking into account that in the near future afforestation of unproductive lands will be stimulated, the share of forests and forested areas will increase. This is quite a positive factor, because the assimilation potential of the territories will increase and the functioning of agricultural landscapes will be strengthened.

Land relations in Ukraine have undergone significant changes after the implementation of the land reform. The private form of land ownership, including agricultural land, has developed. The land fee has become one of the main components of local budget revenues (Table 2).

Table 2. The structure of payment for land in Ukraine during 2013–2020, million hryvnias

Indicator	2013	2014	2015	2016	2017	2018	2019	2020
Land tax on legal entities	2950	2450	3558	7060	8262	8241	7501	8589
Rent from legal entities	8232	8114	9413	13351	14701	15588	13778	14795
Land tax for individuals	416	414	501	1032	1387	1629	1262	1369
Rent from individuals	1204	1104	1357	1878	2033	2162	1799	1938
Total land fee	12802	12083	14831	23323	26384	27321	24342	26693

Source: Authors' own research based on the data of the State Treasury Service, 2023.

The average amount of the land fee per individual in 2020 was UAH 449, in 2019 it was UAH 403. In the general structure of payment for land by individuals in 2020, 42 % fell on land tax, and 60 % on rent. In 2020, the average amount of the land fee per legal entity was UAH 163,000, in 2019 it was UAH 141,000. Thus, in the structure of the land fee, legal entities paid 37 % of the land tax and 65 % of the rent in 2020.

Today, there is a trend towards an annual increase in the receipts of the land fee, which will continue. Thus, for the period 2013–2020, the receipt of land fees was increased by 2 times.

However, despite the general trend of increasing the amount of land tax, its share in tax revenues of local budgets for the analyzed period is decreasing. The decrease in land tax revenues in the structure of tax revenues is explained primarily by the difficult socio-political situation in the country.

It should be noted that until 2015 the land fee was part of the national taxes, however, from January 1, 2015, the land fee became a component of the property tax and was transferred to the local taxes. It is worth noting that village, settlement, and city councils make decisions on setting local taxes and fees within their powers. Thus, local self-government bodies have been given the authority to establish land payment rates within the limits of the maximum amount, land tax benefits, and normative monetary valuation of land plots located on the territory of settlements. We believe that the transfer of the land fee to local taxes contributed to the improvement of its administration.

The share of land fees in tax revenues of local budgets of Ukraine in 2019 is 12.1 %, which is 4.2 % less than in 2013.

According to the results of the analysis presented in Figure 1 information, it can be noted that the environmental tax is not characterized by a high level of fiscal efficiency, since its share in the revenues of the State Budget of Ukraine does not exceed 1 % (with the exception of 2014, in which this specific weight was 1.01 %).

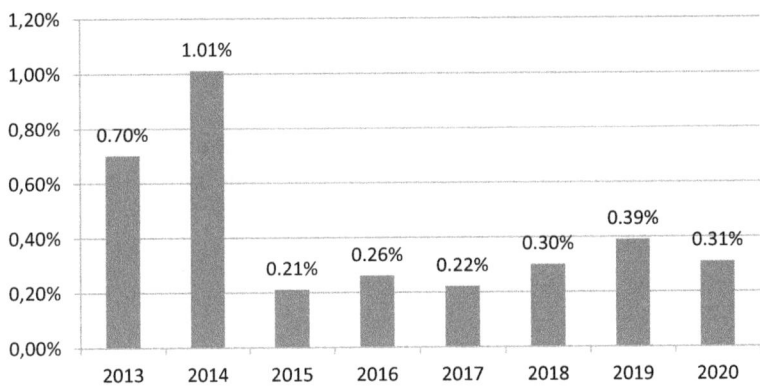

Figure 1. The specific weight of the environmental tax in the revenues of the State Budget of Ukraine
Source: Authors' own research based on the data of the State Treasury Service, 2023.

It is also worth noting that on average during the research period, the value of the indicator is 0.41 %, which indicates the absence of significant fluctuations, the most significant change of the studied parameter was in 2013 and 2014, while this trend leveled off already with the beginning of the implementation of the financial decentralization reform.

In general, throughout the studied period, the specific weight of the land tax is decreasing, except for 2016, this is primarily due to such factors as:

- an increase in the areas for which the normative monetary assessment of land was carried out (Yurii Yaremko, 2018);
- more favorable conditions for leasing land reserves and the reserve fund;
- rent increase as a result of registration of the right of permanent ownership and use of a plot of land of enterprises and organizations of state and communal ownership (Irina Adamenko, 2019);
- carrying out an inventory of land plots provided for use and rent;
- review of concluded lease agreements and the amount of rent;

- increase in rental rates;
- lack of regulatory properties of the land fee, namely the connection between the amount of land tax per hectare of land and the actual amount of rent income from land use, which leads to the application of low land tax rates (Svitlana Antonova, 2021);
- the lack of a rent-based approach in the application of tax rates and the determination of the normative monetary value has led to mismanagement in the use of land resources, a decrease in soil fertility and, as a result, to low efficiency of agricultural production (Lans Bovenberg and Ruud A. de Mooij, 2014);
- insufficient level of implementation of the stimulating and deterrent property of land taxes, which does not allow to settle the issue of development of land relations, as a result of which the fertility of the land decreases and the area of agricultural land decreases, which negatively affects the agricultural sector and creates significant threats to the food security of the country (Dariya Vasilieva et al. 2021).

Thus, the above problems of land taxation require an immediate solution, which is possible by improving the mechanism for paying the land tax, streamlining effective benefits for its payment, which will allow additional tax revenues to local budgets to solve fiscal and regulatory issues and effective state financial control.

The effectiveness of state financial control cannot be measured by the number of audits carried out or the number of objects subject to control, or the number of decisions made based on the results of control. And even the number of imposed sanctions cannot be a basis for recognizing control as effective and efficient. The main criterion is the achieved real results – normal, proper functioning of the controlled object in the future, preventing it from violating financial discipline (Nadiia Davydenko, Olga Klyuchka and Sergey Klyuchka, 2020).

In general, the current state of control institutions is characterized by the fact that the formation of a system of state financial control, which would meet the requirements of economic development and the state system of the country, has not yet been completed. There is no sufficient legal basis for control. State control does not cover all means belonging to it, especially in terms of their use.

The control organization is characterized by lack of regularity, duplication of audits and inspections. The lists of financial violations in the current classifiers are weakly substantiated both economically and legally. Violations that caused real damage to the budget or the budget recipient are not highlighted. In such conditions, institutions of state financial control cease to be an effective tool in the fight against corruption.

Conclusions

State financial control over the use and protection of land is an extremely important activity for the preservation and proper management of land resources. The main purpose of this control is to ensure compliance with land legislation, conservation of natural resources, and prevention of illegal activities and minimization of negative impact on the environment. The main factors in increasing the effectiveness of state control over the use and protection of land are clear legal regulations, solving problems of control organization, increasing the effectiveness of the activities of authorities in the field of control over the use and protection of land, interaction with other law enforcement agencies during the detection of offenses, as well as the implementation of preventive measures in order to avoid violations of land legislation. The way out of the crisis situation that occurred in the state policy in the field of control over the use and protection of land is the balancing of the system of authorities in the field of land relations between the executive power and local self-government, the establishment of a complex balance and mutual control. In turn, to date, the state administration system in the field of state control over the use and protection of land is imperfect. Therefore, in the current conditions of reforming the land policy, the problems of land use, reproduction and protection have acquired national importance, and require double attention from the state authorities and local self-government bodies.

References

Adamenko, I.P. 2019. Financial stabilization in economic transformations. *Ekonomichnyi visnyk universytetu*, 43, pp. 126–133 [in Ukrainian].

Antonova, S.E. 2021. State control over land use and protection: Regional aspect. *Public administration: Improvement and development*. https://doi.org/10.32702/2307-2156-2021.2.32 [in Ukrainian].

Baliuk, S.A. and Kucher, A.V. 2015. *Rational use of ground resources and soil fertility restoration: organizational and economic, environmental and regulatory aspects: collective monograph.* Kharkiv: Smuhasta Typohrafiia [in *Ukrainian*].

Bovenberg, A.L. and De Mooij, R.A. 2014. Environmental levies and distortionary taxation. *American Economic Review*, 94(4), pp. 1085–1088.

Chugunov, I., Makohon, V., Vatulov, A., Markuts, Y. 2020. General government revenue in the system of fiscal regulation. *Investment Management and Financial Innovations*, 17(1), pp. 134–142 [in Ukrainian].

Davydenko, N.M. and Porohivnyk, Yu.O. 2019. Financial support for solving ecological and economic problems of land use in the agrarian sector of the economy of Ukraine. *Business-navigator*, 4(530), pp.146–149 [in Ukrainian].

Davydenko, N.M., Klyuchka, O.V. and Klyuchka, S.S. 2020. Ways of improving state financial control in the system of land relations in Ukraine. *Scholarly Notes of V. I. Vernadsky Tavri National University. Series: Economics and Management*, 31(70), pp. 63–69 [in Ukrainian].

Dobryak, D.S. 2007. Modern land management – the fundamental state management mechanism in the field of land resource use and protection. *Land Management and Cadastre*, 2, pp. 3–8 [in Ukrainian].

Fedorov, M.M. 1998. *Economic problems of land relations in agriculture*. Kyiv: In-t ahrarnoï ekonomiki [in *Ukrainian*].

Gutorov, O. 2006. *Evaluation of land resources and efficiency of investments*. Kharkiv National Agrarian University named after V.V. Dokuchaeva [in Ukrainian].

Ibatullin, Sh.I., Stepenko, O.V. and Sakal, O.V. 2012. *The mechanisms of land relations in the context of sustainable development*. Derzhavna ustanova Kyiv: Instytut ekonomiky pryrodokorystuvannia ta staloho rozvytku Natsionalnoi akademii nauk Ukrainy [in Ukrainian].

Ilyina, M.V. and Shpylyova, Yu.B. 2020. Ecosystem services as a tool for ecologically oriented organization of rural space. *Business Navigator*, 2(58), pp. 54–58 [in Ukrainian].

Ivanova, A. 2017. An evaluation of the domestic financial supervisory authorities in land relations in ukraine. *Baltic Journal of Economic Studies*, 3(4), pp. 96–100. https://doi.org/10.30525/2256-0742/2017-3-4-96-100

Liu, S.J., Wang, X.D., Guo, G.L. and Yan, Z.G. 2021. Status and environmental management of soil mercury pollution in China: A review. *Journal of Environmental Management*. p. 277. https://doi.org/10.1016/j.jenvman.2020.111442.

Lup, A., Miron, L. and Alim, I.D. 2016. Management of land resource, agricultural production and food security. *Scientific Papers-Series Management Economic Engineering in Agriculture and Rural Development*, 16(2), pp. 219–228.

Melnychuk L.S. 2015. Efficiency of use of land resources of agricultural enterprises. *Sustainable Economic Development*, 1. pp. 135–140 [in Ukrainian].

Mendonca, M.L and Pitta, F.T. 2022. Land speculation by international financial capital in Brazil. *Latin American Perspectives*, 45(5), pp. 146–160. https://doi.org/10.1177/0094582X221115693.

Moskalenko, A. 2011. Land taxation: history, methodology, practice. *Regional Economy*, 4, pp. 112–116 [in Ukrainian].

Novakovsky, L.A. 2015. *National report on completion of land reform.* Kyiv: Agrarian [in *Ukrainian*].

Pirozhkov, S.I. and Khvesik, M.A. 2015. *Economic assessment of the natural wealth of Ukraine.* Kyiv: State University of IEPSR of the National Academy of Sciences of Ukraine [in *Ukrainian*].

Shust, G.P. 2019. *Administrative and legal protection of land resources of Ukraine.* Ph.D. Ternopil National Economic University [in Ukrainian].

State Service of Ukraine for Geodesy, cartography and cadastre. 2023. [online] Available at: <https://land.gov.ua/> [Accessed 12 September 2023].

State Treasury Service of Ukraine. 2023. [online] Available at: <https://mof.gov.ua/en/state-treasury> [Accessed 12 September 2023].

Vasilieva, D., Vlasov, A., Parsova, V. and Khasaev, G. 2021. Increase of economic efficiency of agricultural and management on regional level. *International Scientific Conference Engineering for Rural Development,* Jelgava, Latvia, 26–28 May 2021. https://doi.org/10.22616/ERDev.2021.20.TF069.

Yaremko, Yu.I. 2018. *Environmental and economic principles of rational land use within the Southern Steppe zone of Ukraine.* Kherson: PP "Resnik" [in Ukrainian].

Camelia Anişoara Gavrilescu

Romania's international agri-food trade – Why in a permanent deficit? A post-accession analysis

Abstract: *Joining the EU meant free access to the Single Market for Romanian agri-food products, reflected in a spectacular increase in exports. At the same time, imports from the EU increased significantly. After 2010, Romania's entry into the markets of the Near and Middle East led to a major expansion of extra-EU exports, generating a positive balance on these destinations. At the same time, the balance deficit on the intra-Community market increased significantly at a much faster rate, so that the total agri-food trade balance remained negative. This paper analyzes the changes in the value, volume, geographical direction and structure of Romanian trade flows in the post-accession period, compared to the performance in pre-accession period and of neighboring countries, highlighting the main influencing factors. The results show important fluctuations in the values of Romanian imports and exports. Out of 24 groups of agri-food products, only four (cereals, oilseeds, live animals and tobacco products) show continuous positive balances. The very high share of unprocessed products in exports and of processed products in imports is another major factor that contributes to the permanent deficit of the Romanian agri-food trade balance.*

Keywords: trade balance, extra-EU trade, intra-Community trade, processed products, trade barriers.

Introduction

The problem of increasing imports, continuously deficit of the agri-food trade balance and lack of competitiveness on the international markets is frequently debated in academic and political environments. It has been studied in the last two decades, and the present paper is a continuation and development of such studies.

The performances and counter-performances of a country's agri-food trade is determined mainly by the corresponding performance of the background agri-food sector. Other factors with potential strong influence upon the evolutions on the world food markets and trade are: entering in new trade agreements (such as CEFTA, accession to the EU Single Market for the New Member States or DCFTA for Partner Eastern European countries in 2016); enforcement of various trade barriers (such as the Russian ban on imports from EU, USA and other several countries in 2014); occurrence of dangerous and serious animal diseases

(Avian flu, African Swine Fever), or widely spread human diseases such as the COVID-19 pandemic.

All these influencing factors, acting either successively or simultaneously have complex implications on the evolution, performance and geographical orientation of the international agri-food trade flows. Various studies were made, taking different methods and points of view in order to reveal the inward mechanisms of the agri-food trade evolutions.

Literature review

An analysis of the Romanian agri-food foreign trade in transition and pre-accession period was made by Gavrilescu et al. (2005), Rusali (2006), and Gavrilescu (2019b), showing the main directions and composition of the trade flows. For the post-accession period, Rusali and Gavrilescu (2008) concluded that despite the opportunities offered by (then the recent) EU accession, the Romanian food industry, which at the time overall still featured low efficiency, needed restructuring and modernization, in terms of equipment, technologies complying with EU sanitary and quality requirements, as well as in the management and marketing of products. Differences in trade performance and competitiveness in the first decade of membership were analyzed for the New Member States by Csaki and Jambor (2015), and Gavrilescu (2018), while Piglowski (2023) showed that their shares in intra-EU agri-food trade, although continuously increasing in the post-accession period, remained much lower that of the Old Member States.

The competitiveness of the Romanian agri-food trade has been assessed by several indicators and methods, such as: revealed comparative advantage (Alexandri Giurcă and Șerbănescu, 2002; Rusali and Gavrilescu, 2008); trade balances (Gavrilescu 2018, 2023); trade unit values (Gavrilescu and Voicilaș, 2014; Gavrilescu, 2023); intra-industry values and indices (Jambor, 2014).

In time, as agricultural technology advanced, food industry developed and consumers tastes and demand evolved and diversified, the agri-food value chains expanded and became increasingly large and complex. The current supply of agri-food products is extremely diversified, ranging from raw agricultural products to highly processed food products. Consequently, the classification of agricultural and food products into relatively homogenous groups taking into account their processing degree generated many debates in the literature in the last two decades. Some authors, like Regmi et al. (2005), separate two main categories: bulk commodities and high-value products. Jankune et al. (2015)

proposed a three-category classification: basic agricultural goods, primary and secondary processed food products.

Methodology

The present study used agri-food trade data from Eurostat, at 2-digits level (HS 24 chapters) and 4-digits level (HS 196 subgroups) in the Combined Nomenclature (CN). Data were processed using Excel software in order to reveal the evolution of exports, imports and trade balances by main partners and directions. Four time periods were considered and compared: first, T_0=2000–2006 (pre-accession period), and another three for the post-accession period: T_1=2007–2011, T_2= 2012–2016 and T_3=2017–2022.

To analyze the structure of the agri-food trade by the processing degree of the products, the classification and subsequent method used by Jankune et al. (2015) was used in the present paper. The basic agricultural goods include products that can be consumed or used as such as raw material, possibly with minimal conditioning but no processing (e.g.: cereals, oilseeds, fresh fish, eggs, fresh vegetables and fruits, tobacco leaves). Products requiring minimal processing (such as meat, liquid milk, frozen vegetables, dried fruits, edible oils, coffee, tea, etc.) are included in primary processed category. Products with higher added value resulting from more complex industrial transformation (e.g. ice cream, yogurt, cheese, bread, cakes, pasta, sauces, wine, chocolate, cigarettes, etc.) fall into the secondary processed category.

Results and discussions

The successive EU enlargements in 2004, 2007 and 2013 resulted in a significant increase in the overall value and volume of the EU agri-food trade, consolidating its position among the top world players in the world markets. Only ten Member States (MS) show a positive trade balance: Netherlands, Spain, France, Denmark, Belgium and Ireland (of the old MS), and Poland, Hungary, Bulgaria and Lithuania (of the new MS) (Figure 1).

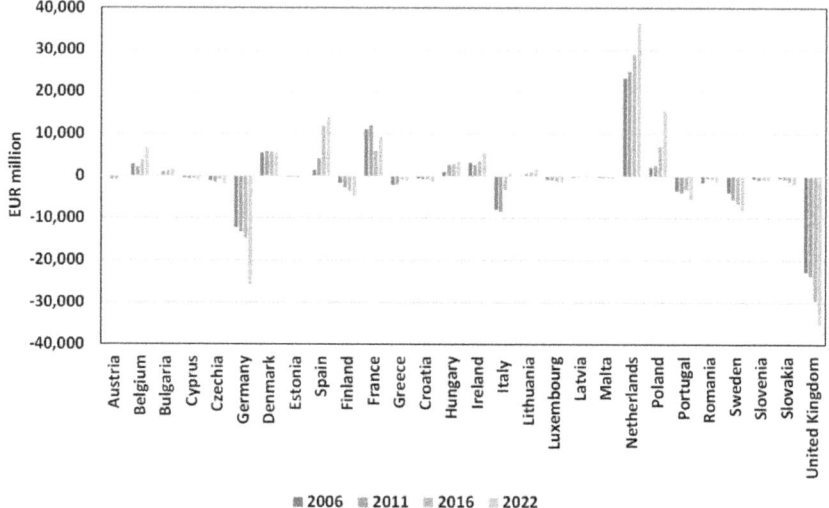

Figure 1. Agri-food trade balance in the European Union by Member States (selected years)

Source: Author's own research based on Eurostat, 2023 and HM Revenue & Customs. UK Trade info, 2023.

The remaining seventeen MS show trade deficits, Romania being among them. The Netherlands had the highest surplus in 2022 (EUR +36 billion), followed by Poland (EUR +16 billion) and Spain (EUR +14 billion), while UK had the highest deficit (EUR -35 billion), followed by Germany (EUR -26 billion) and Sweden (EUR -8 billion). In 2022, Romania ranked 10-th, with a deficit of EUR -1.3 billion.

When Romania joined the EU in January 2007, it had already a history of agri-food trade deficits. In the period of centrally planned economy, Romania has been continuously a net exporter of agri-food products (Gavrilescu, 2019a). Large farms (collective farms or state-owned) were supplying agricultural raw materials for large processing units in the food industry, and the resulting products were sold on both domestic and external markets by state-controlled distribution enterprises.

That changed since 1990, when an export ban was imposed on all agri-food products, in an attempt to increase the food availability for the population. Although the export ban relaxed gradually in the subsequent years, down to a few staple products (such as meat, milk, cereals, oil), imports increased, and the agri-food trade balance turned to deficit.

In the 1990's, the agricultural sector went through major transformations all along the supply chains, starting with land ownership and farming systems. The former soviet-type cooperative farms were dismantled; the arable land, orchards and vineyards were returned to former owners or their successors within a limit of 10 ha. The result was a huge number – more than 4.2 million very small private farms (NIS, 2005). The ban on the land market, in force until 1997, prevented land consolidation and the emergence of medium-size family farms, commercially oriented.

At the same time, massive reforms were implemented in the economy – restructuring and privatization. Most large food industry enterprises were privatized, divided or went bankrupt. A new private sector of mostly small and medium-size units emerged, with domestic and foreign capital. Nevertheless, the former value chains were broken, but not replaced with others, adapted to the new structure. The fragility of the new economic units, the lack of supply concentration, together with various export barriers resulted in slow upward trend in exports, while increasing domestic demand pushed to a stronger upward trend of imports. The result was a continuous agri-food negative trade balance.

Although still a member of CMEA (Council of Mutual Economic Assistance), led by USSR, already in 1989 (last year of the centrally planned economy), half of the Romanian agri-food exports were oriented to the EEC (European Economic Community). After 1990, Romania entered in various free trade agreements: with EFTA (1991), Republic of Moldova (1994), Turkey (1997), CEFTA (1997), Israel (2001). The implementation of the Interim Agreement with the European Union (1993), the EU Association Agreement (1995) and the Double-Zero Agreement (2000) contributed to the further liberalization and expansion of the agri-food trade with these partners (Gavrilescu, 2019b) (Figure 2). Five CEFTA member countries (Czech Republic, Hungary, Poland, Slovakia and Slovenia) took a rather important share of the Romanian agri-food trade; so, when they left CEFTA and joined the EU in May 2004, the trade values added to those of the EU-15.

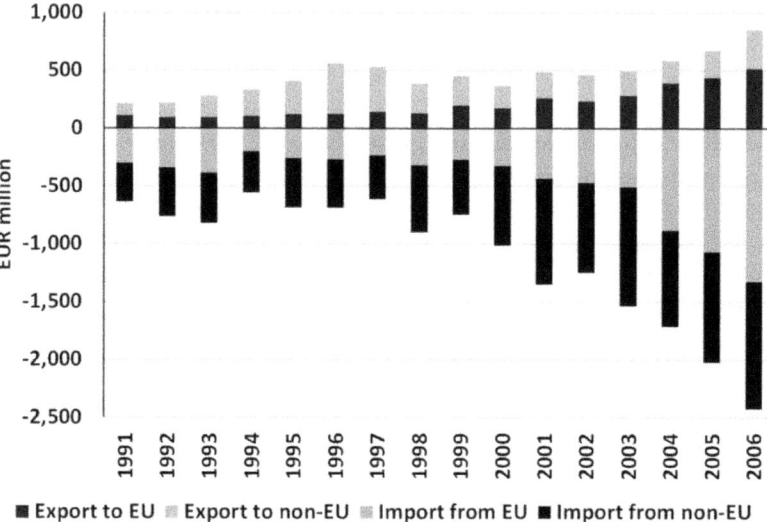

Figure 2. Structure of agri-food trade by main destinations/origins during the transition and pre-accession period (1991–2006)

Note: exports are represented as positive values, and imports as negative values.

Source: Author's own research based on NIS, 2023a, 2023b.

CEFTA was supposed to be a "training field" for the candidate countries to function in the terms of a free trade agreement, before joining the Single Market. Romania joining CEFTA in 1997 realized its lack of competitiveness on international markets in the absence or even low import tariff barriers. While exports increased almost continuously (albeit at a slow pace), imports increased sharply after 1997, generating a rapid increase of the trade deficit (from EUR -88 million in 1997 to EUR -1,037 million in 2003).

Immediately after joining the EU in 2007, the Romanian agri-food trade increased considerably, both in exports (2.5 times in 2008 as compared to 2006) and in imports (1.8 times). The European economic crisis of 2008–2009 impacted negatively the Romanian agri-food trade as well, resulting in a 12 % contraction in 2009 as compared to previous year. Subsequently, the upward trend of both exports and imports resumed and continued as such to the present day; exports multiplied in value terms by a factor of 14 (2022/2006), reaching a maximum of EUR 12 billion, while imports multiplied by a factor of 5.5, up to EUR 13.3 billion in 2022 (Figure 3). The resulting trade balance was mostly negative, varying in a wide range (from EUR -2.22 billion in 2007 to EUR -0.13 billion in 2015). A notable exception from the continuous deficit occurred in 2013–2014, the only years with agri-food trade surplus in the last three decades (Figure 5).

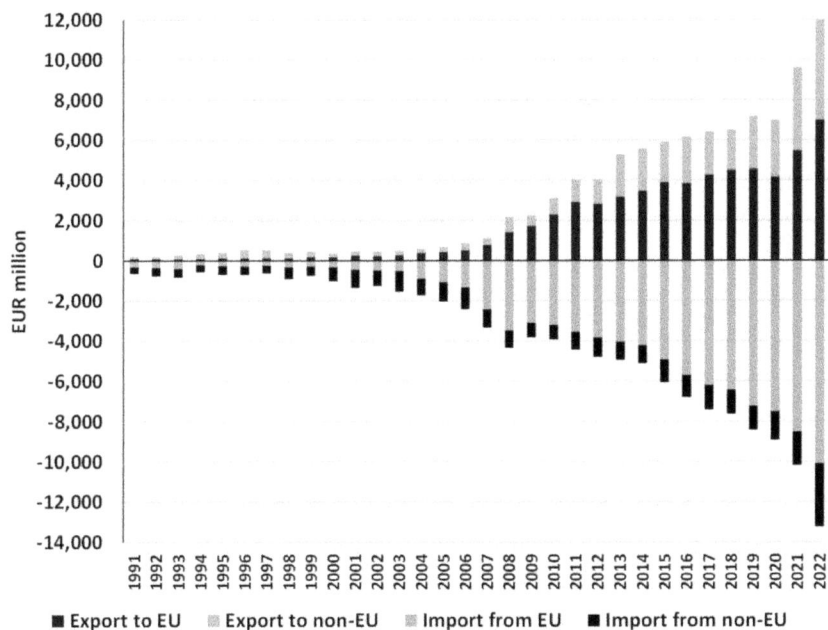

Figure 3. Structure of agri-food trade by main destinations/origins (1991–2022)
Note: imports are represented as negative values.
Source: Author's own research based on NIS, 2023a, 2023b.

The major orientation of the Romanian exports to the EU in the pre-accession period continued in the post-accession years as well, with an inflexion point in 2009. The share of exports to the EU, already over half since the implementation of the Double-Zero Agreement in 2000, increased slowly, to reach a maximum value (79 %) in 2009 (Figure 4). Since 2010, Romania penetrated the Middle-Eastern cereal markets and expanded its exports there, on the expense of other EU destinations, their share varying in a 17 % range (between 57 % and 74 % of the total agri-food exports). On the other hand, given the Community preference principle, the share of EU-originating imports increased after accession and varied in a 13 % range (between 73 and 86 % of the total agri-food imports) (Figure 4).

The evolution of trade with non-EU countries was very much different: export values started increasing after 2010, doubled from 2012 to 2013 (to reach EUR 2.1 billion in 2013), varied slightly between EUR 2.1 and 2.8 billion in 2014–2020 and almost doubled again in 2021 and 2022, to reach EUR 5 billion in 2022.

Imports from non-EU countries showed an upward trend in the post-accession period as well, but far lower than exports (ranging from EUR 0.7 to 3.1 billion EUR in 2007–2022).

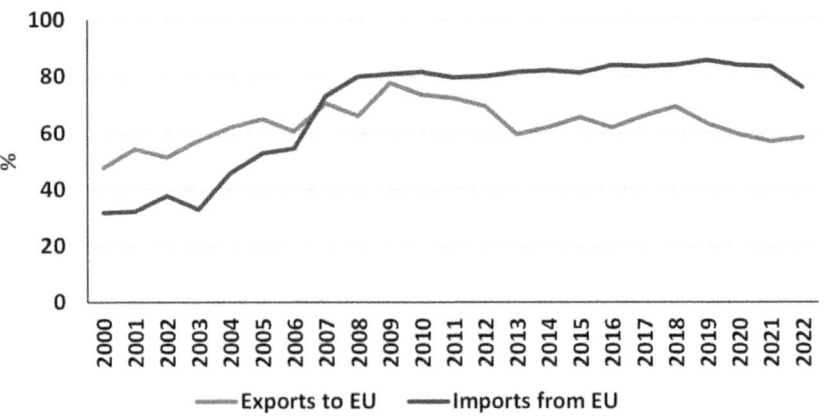

Figure 4. Share of EU in the Romanian agri-food trade
Source: Author's own research based on Eurostat, 2023.

The resulting trade balance with extra-EU countries changed from negative until 2009 to increasingly positive every year since 2010 (Figure 5). The surplus generated by the non-EU trade counterbalanced partially the large deficit generated by the exchanges with the EU countries.

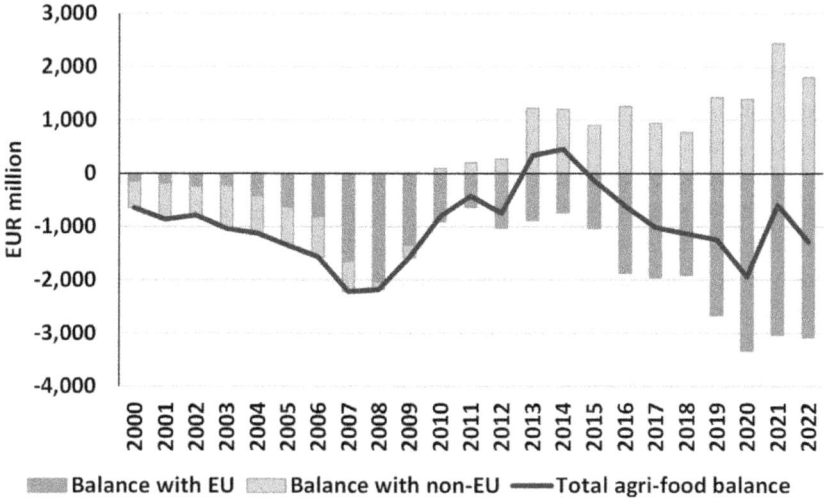

Figure 5. Romanian agri-food trade balance by main destinations
Source: Author's own research based on Eurostat, 2023.

The total result was a variable – but overall negative agri-food trade balance, except for two years, 2013 and 2014. Subsequently, the total trade balance reversed to deficit again. Since the trade deficit has been generated primarily in the exchanges with the other EU countries, the analysis is focused on that part of the agri-food trade.

The agri-food exports to the EU in the post-accession period increased by a factor of 13.5 (2022/2006), up to EUR 12 billion in 2022. At the same time, imports went up by a factor of 7.6 (2022/2006), with the highest value in 2022 (EUR 13.3 billion). Consequently, the trade deficit with the EU has been permanent, albeit varying also in a large range: between EUR -0.63 billion in 2011 and EUR -3.33 billion in 2000.

The analysis of exports to the EU show that after accession, there were three main groups of products that amounted for more than half the total value: oilseeds (HS12), mostly sunflower seeds, cereals (HS10), mostly wheat, and tobacco products (HS24) (Figure 6a). Edible oils (HS15) were ranking fourth in exports, but with shares less than 8 %.

While in the first period (T_1=2007–2011) the fifth product group exported was live animals (HS01), in the subsequent years, its value diminished due the reorientation of live sheep exports to Arabic countries, as well as due to the enforcement of recent regulations on animal welfare during transportation. In the next periods (T_2=2012–2016; T_3=2017–2022), meat exports ranked fifth (mostly chicken). Milk and dairy products (HS04), worth EUR 183 million were exported in T3 period. Despite their continuous increasing values in exports, the various groups of processed products have low shares (less than 4 % each).

The increasing demand for higher quality and more diversified products resulted in increasing imports from EU. The Romanian food industry is not yet able to cover the domestic demand. On the other hand, the fruit and vegetables supply is scattered due to the poor organization of domestic producers, therefore the large transnational retailers in food products prefer to import products rather than dealing with a multitude of domestic producers able to deliver only small amounts of products.

Meat (mostly pork) has been all along the post-accession period the major imported product from the EU (Figure 6b). Its value increased continuously, from EUR 559 million in T1 to EUR 94 million in T3, its share in all imports varying between 12–18 %, and is reflecting the inability of the domestic production sector to cover the internal demand, as well as the low competitiveness of the sector, given that the prices for the domestic products are higher than their counterparts coming from Hungary and Poland. Nevertheless, one should remember also that Romania was among the most affected countries by the African Swine Fever, that resulted in killing more than 1 million pigs in both commercial and private households since late 2018 when the disease occurred in several locations in the country.

Milk and dairy products (HS04) is the second important product group in imports coming from EU. Its value more than tripled between T_1 and T_3 (from EUR 211 million to EUR 654 million), taking 8 % as average share in total imports. Although the domestic production is rather large, the main processors use important imported quantities of raw milk (from Hungary and Poland), due to their lower price, concentrated supply and higher quality.

Miscellaneous edible preparation (HS21) is another important product group in imports from the EU. Value of imports more than tripled between T1 and T3 (from EUR 157 to 513 million). Various sauces, soups, other ready to bake food produced by the food industry entered in the Romanian consumption model, pushing up the demand for such products.

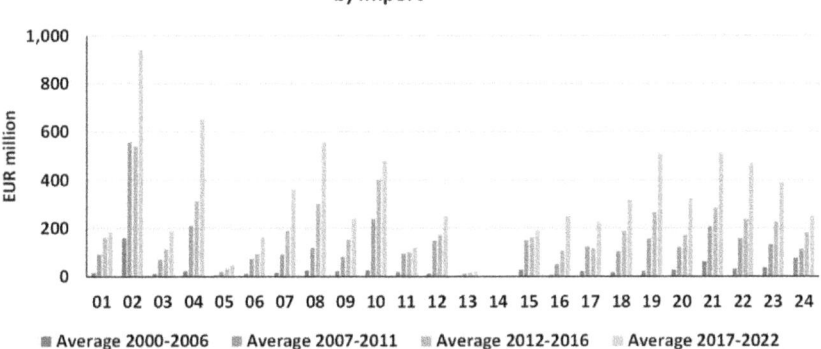

Figure 6. Structure of agri-food balance with EU by product groups (HS chapters): (a) Export; (b) Import

Note: HS chapters 01–24 include all agri-food products (see list of HS chapters in Table 1).
Source: Author's own research based on Eurostat, 2023.

Preparations of cereals, bakery and pastry products (HS19), worth between EUR 157 to 513 million, were imported from the EU between T1 and T3. Given the established export markets for cereals (mostly wheat and maize) to the Middle Eastern markets (worth EUR 2.9 billion in 2022) Romanian producers prefer to export the raw material instead of processing it locally. That is favored also by the fact that most of the cereals are produced (and exported) from very large farms by international traders. Most of the production that is processed in the country comes from small farms; as a result, large quantities of frozen dough and other bakery and pastry products are imported to help covering the domestic demand.

Table 1. Structure of the trade balance with EU, by product groups (HS chapters)

HS chapter	Product group	T_0 Average 2000–2006	T_1 Average 2007–2011	T_2 Average 2012–2016	T_3 Average 2017–2022
01	Live animals	58.53	43.76	−19.39	−24.74
02	Meat and offal	−150.08	−467.75	−324.51	−732.48
03	Fish and seafood	−10.58	−64.09	−101.15	−173.85
04	Dairy products, eggs and honey	−3.62	−156.96	−176.68	−472.50
05	Other animal products	−0.24	−6.70	−4.62	−21.42
06	Live plants	−11.86	−72.91	−91.24	−160.41
07	Vegetables	15.81	−44.59	−118.12	−263.41
08	Fruit	−8.91	−85.06	−239.19	−488.49
09	Coffee, tea and spices	−21.74	−76.43	−136.76	−215.27
10	Cereals	3.55	69.25	214.15	511.13
11	Products of the milling industry	−19.79	−87.95	−89.72	−99.59
12	Oilseeds	38.12	218.40	453.22	886.75
13	Lacs, gums and resins	−3.63	−13.10	−17.76	−23.40
14	Other vegetable products	1.36	0.41	0.52	0.10
15	Oils and fats	−17.82	−28.48	7.18	46.46
16	Preparations of meat and fish	5.76	−6.86	2.28	−79.56
17	Sugars and sugar confectionery	−18.20	−67.72	−49.12	−182.12
18	Cocoa and cocoa preparations	−17.20	−88.50	−147.22	−229.48
19	Preparations of cereals, pastry products	−18.49	−118.57	−173.49	−340.20
20	Preparations of vegetables, fruits and nuts	−15.84	−97.40	−131.09	−255.81
21	Miscellaneous edible preparations	−57.93	−163.70	−187.07	−354.11
22	Beverages	−18.99	−104.55	−146.62	−339.28
23	Animal feed	−31.51	−83.89	−114.43	−257.78
24	Tobacco and tobacco products	−72.06	187.60	476.31	604.59
	Total intra-EU	−375.39	−1,315.77	−1,114.52	−2,664.87

Note: cells in yellow highlight product groups with negative trade balance.
Source: Author's own research based on Eurostat, 2023.

The majority of product groups showed negative trade balances in the post-accession period (21 groups in T_1, 19 groups in T_2 and 18 groups in T_3) (Table 1). Therefore, the permanent deficit of the Romanian trade balance (both with the EU and the total agri-food balance) is not given by a few important products, but rather by the cumulation of many product groups, which show lack of competitiveness, poor domestic production and processing infrastructure, as well as poorly functional (if not fractured) value chains.

The analysis of the agri-food trade using the methodology of Jankune et al. (2015) showed that in the post-accession period, despite the fluctuations in the total agri-food balance, the exports of agricultural commodities (basically cereals and oilseeds) surpassed by far the respective imports, generating a significant increasing surplus of trade (Table 2).

Table 2. Romanian exports, imports and balance – Processed products vs. agricultural commodities (EUR million)

	2006	2011	2016	2022
EXPORTS				
Agricultural commodities	537	2,292	3,743	7,104
Total processed products, *of which:*	317	1,728	2,429	4,863
- primary processed products	155	823	851	1,695
- secondary processed products	162	905	1,578	3,168
Total agri-food exports	854	4,020	6,172	11,966
IMPORTS				
Agricultural commodities	499	1,301	2,267	4,072
Total processed products, *of which:*	1,926	3,144	4,523	9,182
primary processed products	1,084	1,770	2,097	3,964
secondary processed products	842	1,374	2,426	5,218
Total agri-food imports	2,425	4,445	6,789	13,253
BALANCE				
Agricultural commodities	38	990	1,476	3,032
Total processed products, *of which:*	−1,609	−1,416	−2,094	−4,319
primary processed products	−929	−947	−1,246	−2,269
secondary processed products	−680	−469	−848	−2,051
Total agri-food balance	−1,571	−425	−617	−1,287

Source: Author's own research, based on Eurostat, 2023.

On the other hand, despite the upward trend of both exports and imports of processed products in the post-accession years, import values were almost double than export values, resulting in sharp increase of the deficit. Among the processed products, both primary and secondary products showed increasing export and import values, and the trade balance followed a similar pattern. For all categories, the year 2022 recorded the highest values. The two main factors associated with the record values in 2022 were the unexpected rise of energy prices (the more processed the products are, more energy is needed in their production technologies), and the high inflation.

A similar analysis was performed for several Central and Eastern European countries (CEEC-s): Bulgaria (BG), Czech Republic (CZ), Hungary (HU) and Poland (PL) in 2022. The results are illustrated in Figure 7 and Table 3. The share of processed products in total exports, respective imports, is shown in Table 3. Among the compared countries, Romania and Czech Republic have continuous negative total agri-food trade balance in the last decade, year 2022 included. Hungary and Poland became net agri-food products exporters shortly after their EU accession in 2004.

Table 3. Share of processed products in total agri-food exports and imports – A comparison (2022)

	Romania	Bulgaria	Czech R.	Hungary	Poland
Processed products (% in total exports)	41	60	77	73	85
Processed products (% in total imports)	71	71	79	70	75
Balance – agricultural commodities (EUR million)	3,025	1,334	−200	638	−1,134
Balance – processed products (EUR million)	−4,308	257	−1,713	2,813	16,640

Source: Author's own research based on Eurostat, 2023.

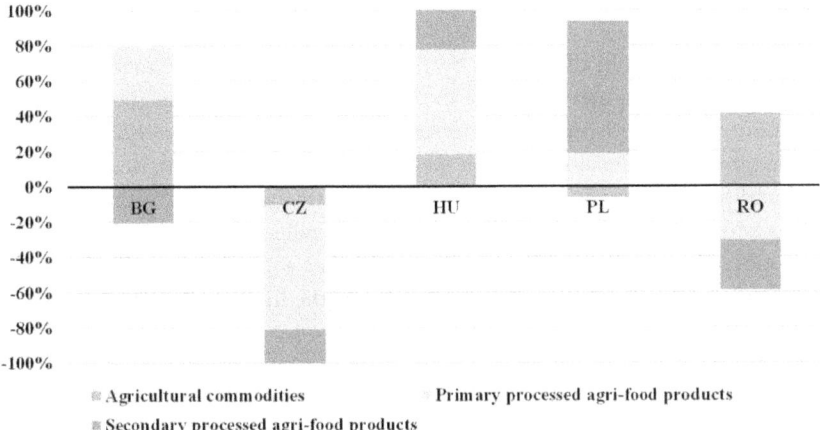

Figure 7. Processed products and agricultural commodities – A comparison of balances in selected CEEC-s (2022)
Source: Author's own research based on Eurostat, 2023.

Bulgaria, which has a similar trade pattern to the Romanian one (relying heavily on cereals and oilseeds exports), experiences also a trade deficit with secondary processed products. Just as in Romania, Bulgarian agricultural companies acquired large plots of land (property or leased), put them together in very large farms (more than 1,000 ha) and specialized in cereals and oilseeds production that goes mainly to export, while paying far less attention to animal production. Yet, they developed the vegetables sector (production and processing), hence the positive trade balance for the category primary processed products. Hungary shows trade surplus for all products categories, while as Czech Republic shows a complete opposite situation – trade deficits for all product categories.

During its pre-accession period, and also in the post-accession times, Poland developed significantly its food industry, by means of large investments, mostly of foreign capital (from Germany, Sweden and Denmark) and from CAP funds (Jambor, 2014; Piglowski, 2023). The result can be observed in the share of processed products in exports, as well as in the positive trade balance with processed products, which value is far higher than all the other selected New Member States. By far Poland has the best structure of processed products (85 % in the total exports), and the highest value of the trade surplus comes from secondary processed products which yield the highest added value.

Conclusions

Romania's accession to the EU has been for the last 15 years the main engine of its economic development. Despite the numerous problems that the Romanian agri-food sector has been and is currently facing, access to the Single market fostered important developments in the sector. Many new enterprises in agricultural and food production emerged with domestic and foreign investments, as well as with funds coming from the second CAP pillar. That imposed using new and modern machinery and technology, as well as achieving and observing the stricter sanitary and veterinary regulations, and the high-quality levels required for exporting on the Single Market.

At the same time, on the less positive side, the country's inclusion in the Single Market allowed unrestricted access of EU products on Romanian markets, thus putting pressure on less developed and less competitive domestic markets, so significant competition came in form of lower prices and diversified quality. One should mention as well as an important negative factor, the easy access of very cheap, low quality, counterfeit products, which are sought by the poorest segments of consumers, but are harmful for the domestic producers that are trying to develop their businesses by providing good quality products, made of locally grown products.

There are a few factors that contributed heavily to the lesser development of the Romanian agri-food sector, as compared to the other Member States:

(i) The inadequate structure of the agricultural production sector (small holdings and lack of association, providing low possibilities for concentration and capitalization);
(ii) Underdeveloped or fractured agri-food chains, that do not allow for supply concentration in subsectors dominated by small family farms (vegetables, fruit, dairy and meat production).
(iii) The unfavorable structure of the agricultural production (crop/animal production ratio), heavily favoring the cereal and oilseeds production for export, while neglecting other important sub-sectors able to provide higher value added (such as animal husbandry and processing);
(iv) Inadequate structure of the food industry, which shows some very large processing units (generally transnational companies), and very small units, operating locally. The middle-size processing units, able to use more local raw materials and able to produce enough to penetrate the large retail (supermarkets and hypermarkets), and also to concentrate enough supply fit for export, are sadly lacking.

There are a number of social and demographic factors that impede the development of a supple and efficient agri-food sector, such as: ageing rural population; farmers' and agricultural workers' poor professional skills and knowledge; perceived lack of opportunities in the rural area; current poorly developed infrastructure in rural area, and possibilities for young people to emigrate.

All these factors contributed to a lesser development of the Romanian agri-food sector, as compared to other new member states such as Hungary and Poland, resulting also in lower competitiveness and low availability of diversified and high value-added products for exports.

The analysis of the Romanian agri-food trade highlighted several very positive developments in the post-accession period: spectacular increase of the trade flows (mostly in exports); increased market shares on international markets; significant improvement in quality, quantity and diversification of processed products (albeit not enough for reversing the ratio between the shares of processed products in exports and imports), and quite important, the maintained positive trade balance with non-EU countries, since 2010.

On the negative side, there are developments that maintain Romania in the area of high imports resulting in a continuous agri-food trade deficit and prevent it from becoming a net agrifood exporter. It relies heavily on cereals and oilseeds exports, and favors their outward shipment instead of increasing the share of products that are transformed and processed in the country, thus reducing the need for imports and increasing export availabilities and diversification. As a consequence, it increased the excessive compositional and directional range of exports, which may prove dangerous when major disruptions occur on the international markets, caused by diseases outbursts (human and animal) or wars such as that in Ukraine and more recently, Israel.

Therefore, future targets of the Romanian agri-food production and trade sector should be: (i) expansion of the exported products range, (ii) medium and long-term investment programs in food industry in order to make better use of the domestically produced agricultural raw materials and increase the share of processed products in exports; (iii) increasing production, concentrating the supply and thus diminishing imports of staples (meat, fruit, vegetables).

Complementary targets for the Romanian agri-food trade should be on the short and medium term regaining the domestic markets for local consumers, and contributing to the UN Sustainable Development Goal SDG 12 – Ensure sustainable consumption and production patterns.

The present study aimed to continue a thorough analysis of the Romanian agri-food trade after 1990, by investigating the main trends and deficiencies that resulted in a permanent deficit. The authors envisage to further continue and deepen the study with more details and new approaches in investigating inner mechanisms of the lower Romanian agri-food trade competitiveness and identify possible solutions and policies that should make a far better use of the natural, economic and human resources, allowing the country to become a net agri-food exporter in the medium and long term.

References

Alexandri, C., Giurcă, D. and Șerbănescu, C. 2002. Food security and competitiveness under the incidence of transition. In: V. Florian, D. Gavrilescu, D. Giurcă, M. Rusali and C. Șerbănescu, eds. *Restructuring and transition of agrifood sector and rural areas in Romania*. Bucharest: Expert Printing House, pp. 189–206.

Csaki, C. and Jambor, A. 2015. Ten years of EU membership: How agricultural performance differs in the new member states. *EuroChoices*, 15(2), pp. 35–41.

Eurostat. 2023. *EU trade since 1988 by HS2-4-6 and CN8*. [online] Available at: <https://ec.europa.eu/eurostat/databrowser/view/ds-045409/default/table?lang=en> [Accessed 20 September 2023].

Gavrilescu, C. 2018. Romanian agri-food trade performance and competitiveness in its first decade of EU membership. *Lucrări Științifice Management Agricol*, 20(2), pp. 46–55.

Gavrilescu, C. 2019a. Historical evolutions of the Romanian international agri-food trade – the period 1950–1989. *Agricultural Economics and Rural Development*, 16(1), pp. 17–38.

Gavrilescu, C. 2019b. The Romanian international agri-food trade in the transition period (1990–2000) – A historical perspective. *Lucrări Științifice Management Agricol*, 21(2), pp. 23–32.

Gavrilescu, C. 2023. Romanian agri-food products competitiveness – An analysis using trade balances and unit values: Did the COVID-19 crisis influenced it?. In: L. Chivu, I. De Los Ríos Carmenado, and J.V. Andrei, eds. *Crisis after the crisis: Economic development in the new normal. ESPERA 2021*. Springer International Publishing, pp. 119–134, https://doi.org/10.1007/978-3-031-30996-0_9.

Gavrilescu, C. and Voicilaș, D.M. 2014. Changes in the Romanian agri-food trade competitiveness in the post-accession period. *Management Theory and Studies for Rural Business and Infrastructure Development*, 36(4), pp. 823–834.

Gavrilescu, D., Giurcă, D., Rusali, M. and Șerbănescu, C. 2005. The analysis of Romania's agrifood foreign trade in the period 1998–2003. *IAE Quarterly Journal Agricultural Economics and Rural Development*, 3–4, pp. 166–174.

HM Revenue & Customs. UK Trade info. 2023. *Summary of import and export trade with EU and non-EU countries – Annual 2015-2022.* [online] Available at: <https://view.officeapps.live.com/op/view.aspx?src= https%3A%2F%2Fwww.uktradeinfo.com%2Fmedia%2F1uvo5mjj%2Fwebtbl s_final2022.xlsx&wdOrigin=BROWSELINK> [Accessed 22 September 2023].

Jambor, A. 2014. Country-specific determinants of horizontal and vertical intra-industry agri-food trade: The case of the EU new member states. *Journal of Agricultural Economics*, 65(3), pp. 663–682. https://doi.org/10.1111/ 1477-9552.12059.

Jankune, K.G, Wagner, H., Gavrilescu, C., Rusali, M., Toderoiu, F., Bucur, S. and Bucur, C. 2015. Competitiveness of Hungarian and Romanian agri-food products in intra and extra EU trade. In: V. Vasary and M.D. Voicilas, eds. *Agricultural Economics and Rural Development at the beginning of the programming period 2014-2020 in EU member states – Comparative analysis for Romania and Hungary.* Budapest: Primerate KFT for Research Institute of Agricultural Economics, pp. 9–30.

National Institute of Statistics (NIS). 2005. *Farm structure survey.* Bucharest: National Institute of Statistics.

National Institute of Statistics (NIS), TEMPO-Online. 2023a. *EXP101E – Export value (FOB) according to CN sections and chapters, total, of which European Union.* [online] Available at: <http://statistici.insse.ro:8077/tempo-online/#/ pages/tables/insse-table> [Accessed 22 September 2023].

National Institute of Statistics (NIS), TEMPO-Online. 2023b. *EXP102D – Import value (CIF) according to CN sections and chapters, total, of which European Union.* [online] Available at: <http://statistici.insse.ro:8077/tempo-online/#/ pages/tables/insse-table> [Accessed 22 September 2023].

Piglowski, M. 2023. Eastern European union countries in the intra-eu food trade in 1999–2019. *Scientific Journal of Gdynia Maritime University*, 126, pp.19–32. https://doi.org/10.26408/126.02.

Regmi, A., Gehlhar, M., Wainio, J., Vollrath, T., Johnston, P. and Kathuria, N. 2005. Market access for high-value foods, USDA, economic research service, agricultural economic report AER-840. [online] Available at <https://www. ers.usda.gov/webdocs/publications/41724/30138_aer840_002.pdf?v=0> [Accessed 20 September 2023].

Rusali, M. 2006. Performance indicators of external competitiveness of the Romanian agri-food products. *IAE Quarterly Journal – Agricultural Economics and Rural Development,* 9–10, pp. 117–123.

Rusali, M. and Gavrilescu, C. 2008. Competitive advantages and disadvantages in Romania's agri-food trade – trends and challenges. *12th Congress of the European Association of Agricultural Economists – EAAE 2018*, Ghent, Belgium 26–29 August 2008. [online] Available at <https://doi.org/10.22004/ag.econ.44118> [Accessed 20 September 2023].

Marko Jeločnik, Lana Nastić, Boris Kuzman

Improving the vegetable growing by the use of new technologies

Abstract: *Vegetable growing represents the significant segment of the overall agricultural activity, as the vegetables are essential component of the human nutrition. In line to population growth trends, climate change and ecological issues, rapid urbanization and development of other economy's activities to the detriment of agriculture, frequent energy shocks, etc., there are constant confrontation of increase in demand for vegetable crops and available supply at global or national markets. In this circumstances, the stable and high yields in vegetable growing are not conceivable without the use of modern technologies, i.e. without the application of contemporary science. The primary goal of the paper is to consider some technical features and economic effects derived from the investment in appliance of few hi-tech solutions mainly based on the use of renewables in vegetable growing, or even in their processing (such are solar mobile robotized electro-generator, "Smart farm" concept, "Agrokapilaris" irrigation system, or solar dryer). Economic effects are reconsidered in accordance the use of general static and dynamic indicators for the assessment of investment economic efficiency. Gained results demonstrate that the investment in observed production alternatives are economically highly recommended, and above all ecologically and socially very welcomed.*

Keywords: vegetable growing, new technologies, renewables, science, Serbia.

Introduction

Vegetable production has profiled its importance in line to human necessity for involving the vegetables in daily nutrition (Hazra, Chattopadhyay and Karmakar, 2011). Besides nutritional, vegetable usually carries strong medicinal benefits (Jena et al., 2018), as they represent large reservoir of various vitamins, minerals, valuable proteins, fiber and carbohydrates, antioxidants, and many other phyto-chemical compounds (Rodriguez-Casado, 2016; Zapucioiu, Sterie and Dumitru, 2023). Vegetables are part of complete, healthy and well-balanced meals in human diet (Liu, 2013), while their consumption affects the good psychophysical state of organism (Finch, Cummings and Tomiyama, 2019). Besides, they prevent, reduce or even eliminate some human disease and health disorders, such are intestinal, cardiovascular, blood and skin, cancer, endocrine and other disorders (Guan et al., 2021; Rao and Rao, 2007; Oguntibeju, Truter and Esterhuyse, 2013). In general, vegetable could be consumed as fresh (raw) or processed, mostly as frozen, pasteurized and canned or dried (Rickman, Bruhn and Barrett, 2007), while it could be eaten as single or as a part of several salads and dishes (Achikanu et al., 2013). Besides in human nutrition, they are irreplaceable element in many industries, such are food and feed industry, textile industry and light chemistry, medicine and pharmacy, cosmetology, etc. (Morin-Crini et al., 2019).

The main goal of this research is to present economic effectiveness of investment in few technological innovations (use of renewable energy sources and subsurface irrigation system) implemented in vegetables production. Innovations are developed in Serbian academic community and locally tested, while they are ready for the further use in vegetable growing and processing on a regional or even global level.

Literature review

Globally, in line to population growth and more intensive popularization of healthy nutrition, there are coming to constant increase in demand for vegetables at worldwide markets (Mason D'Croz et al., 2019). The rise in demand represents the actual issue for vegetable producers, while shaping their strivings to adapt, solve or adjust many technological, organizational, production, environmental, logistic, marketing and other problems and requests (Lumpkin, Weinberger and Moore, 2005; Rolle, 2006; Villalobos et al., 2019).

In 2021, worldwide vegetable production counted to 1,156 million metric tons of fresh agro-products, while over the 78 % of production was settled in Asia. As the top producers are marked China (52 % of global production), India, USA, and Turkey, while the most produced fresh crops were tomatoes (over the 16 % of overall vegetable production), onion, cucumbers, and cabbage (Shahbandeh, 2023). At worldwide level the top vegetable suppliers are set within the "global vegetable belt", including Eastern and Central Asia, Mediterranean countries, or mainly the countries from Southern Europe and North Africa (Dong et al., 2022).

Currently, vegetable is producing in almost all available crop production systems (conventional, integrated or organic production, hydroponic production, etc.), in protected areas or in open field (Ozkan, Kurklu and Akcaoz, 2004; Sabir and Singh, 2013; Sharma et al., 2018). Related to generally short vegetation period and high requirement in water, it usually considers irrigation (Locascio, 2005; Shock et al., 2007). Vegetable production is highly intensive and very profitable sector of agro-food production (Joshi, Joshi and Birthal, 2006; Mariyono, 2018). Pronounced production seasonality, expressed perishability of fresh products and adjustment to demand at distant markets, usually requires organization and maintaining of adequate logistic (storing, packaging and transportation)

(Onwude et al., 2020; Surucu-Balci and Tuna, 2021). Production is the most often organized at the small farms (Dinham, 2003). Realization of fresh vegetables intended for direct consumption is usually throughout the short supply chains, i.e. retail and green markets, while the large quantities are directed to processing industry too (Vojkovska et al., 2017).

Despite the health benefits and authorized recommendations (per capita consumption has to reach at least 240 g/day) (Kalmpourtzidou, Eilander and Talsma, 2020), in average the consumption of vegetable at global level is under the proposed quantity, around 185 g/day, and differing from 55 g/day in Central America to almost 350 g/day in East Asia. Pronounced differences in vegetable consumption in certain countries largely depends to general income level, culture and tradition in nutrition, etc. (Kalmpourtzidou, Eilander and Talsma, 2020).

Contrary to demand, on offer side there are permanent expectations turned to increase in yields and quality of produced vegetables. Faced to effects of global climate changes, increasingly sharp requests of environmental protection, stronger pressure of plant diseases and pests, energy, economy and logistics shocks, certain social and health tendencies, etc. (Ayyogari, Sidhya and Pandit, 2014; Barnwal and Sharma, 2005; Richards and Rickard, 2020; Robačer et al., 2016), producers are forced to constantly flirt with core science, i.e. wide range of new technologies and innovations, in order to reach overall sustainability of preformed production (Dias and Ortiz, 2021; König et al., 2018; Razin, Taktarova and Semenov, 2018). It has to be noted that in current times followed by almost constant economy and energy crises exposed in certain level, facing with climate change consequences and rise in environmental issues, there are a question should we or not generally skip to the use of renewable energy sources (RES) in agriculture, specifically plant production, instead the fossil fuels. Transition to RES exploitation will surely affects the level of emitted GHGs, enabling the better maintaining of available natural ecosystems and environmentally much comfortable agricultural production (Stoian, 2021). Besides, usually implementation of new technologies and innovations in agriculture is the subject of national subsiding (Feder and Umali, 1993; Wu et al., 2022).

Methodology

Methodology framework implies the use of descriptive method, for describing the available technological solutions that could be used in vegetable growing, as well as desktop research for considering current situation and issues in observed topics, and static and dynamic methods commonly used in investment analysis. Used methods and derived results have, both, to inform and encourage the vegetable growers to accept presented or similar innovations as the one of instruments for reaching the overall sustainability of their farm business.

Presented technological alternatives, mainly linked to the use of renewables in plant (vegetable) production, have been developed, tested and implemented through few pilot projects in last several years by the scientific institutions from Serbia, national and regional leader in the field of energetics, robotics and IT technologies Institute "Mihajlo Pupin" (IMP) and Institute of Agricultural Economics (IAE) from Belgrade, regionally recognized scientific institution from the field of agro-economy and rural development. So, presented data and results mainly derive from techno-economic analyzes that were followed implemented projects.

Results and discussions

Research results derives from the primary goal of research. There will be presented certain technical characteristics and assessment of economic effects gained in investing into the development and implementation, or practical utilization of few modern technological solutions tested in vegetable growing and processing, that implies the use of RES. Presented technical solutions are: Solar mobile robotized electro-generator, "Smart farm" concept, "Agrokapilaris" irrigation system, and solar dryer.

Solar mobile robotized electro-generator

RES as a part of power supplying systems in agriculture could be used for performing many activities, such are irrigation, animal feeding, internal transportation, fish ponds aeration, greenhouses or stables heating, ventilation and lighting, drying, etc. (Ali, Dash and Pradhan, 2012; Chel and Kaushik, 2011). Among available RES, large attainability and relatively cheap access to the solar energy worldwide, found this energy source as one of the most used RES in many sectors of economy, including agriculture (Chikaire et al., 2010; Mekhilef, Saidur and Safari, 2011).

In line to global strivings to include science more deeply into the solving the actual issues related to climate change, GHGs emission, pollution, or sustainability of agriculture, imperative is development and implementation of new technologies. RES and clean technologies could consider the adequate solution as they support zero emission.

Previously, by the IMP Belgrade, developed mobile robotized solar electro-generator was tested in practice during the 2015, by the IMP and IAE Belgrade.

Testing encompasses power supplying of the sprinkler and drip irrigation systems used in vegetable growing, in order to assess the ecologic and economic benefits of such an innovation. Testing location were in villages near Belgrade and Pancevo, traditionally oriented to vegetable production in open field and protected area. The developed solar electro-generator represents highly efficient device strongly adjusted to ecological principles. Related to its capacity and characteristics, its mainly developed for the use at small to medium farms (Picture 1).

This is stand-alone device that doesn't require any energy infrastructure. Although it can be used in many economic activities its mainly constructed to serve in agricultural production. It provides energy supplying in noiseless and environmentally clean regime of work. It could be paired with soil and atmosphere sensors, and digital weather station towards the evaluating the available production conditions due to adequate and prompt producers responds. General benefits of its use are high mobility, small dimensions, independent, quite automatized and remote work, autonomy in work supported by permanent recharging of installed battery banks, users friendly use, cheap maintaining, long-lasting exploitation period for over 20 years (or up to 5,000 cycles of battery charging), possibility for sun tracking, many possibilities for different utilizations, etc.

Picture 1. Solar mobile robotized electro-generator
Source: IMP and IAE internal documentation.

Basic model assumes three-phase generator with maximal power of 5.5 KW and usual 3–8 hours working autonomy, while it could be upgraded to for 50 % more powerful model with stronger batteries. Besides, it could be hybridized into the electric aggregate that joins wind turbine and solar panels (Despotović et al., 2017; Jeločnik and Subić, 2022; Subić and Jeločnik, 2016, 2023).

Table 1. Resume of the investment in solar mobile robotized electro-generator used in vegetable irrigation (in EUR)

No.	Element	
1.	Estimated value of investment (EUR)	
1.1.	Total investment	7,700.00
1.2.	Investment in fixed assets	7,000.00
1.3.	Investment in PWC	700.00
2.	Financing sources	
2.1.	Sources – total	7,000.00
2.2.	Own resources	7,000.00
2.3.	Other sources (subsidies)	–
2.4.	Interest or discount rate (%)	5 %
3.	Object of investment	
3.1.	Investment	Investment into purchase of mobile robotized solar electro-generator
3.2.	Starting date of establishment	During 2015
3.3.	End date of establishment	During 2015
3.4.	Project economic life cycle (in line to usual credit line)	5 years
4.	Expected economic effects	
4.1.	*Static assessment*	
4.1.1.	Total Output-Total Input Ratio (%)	1.08
4.1.2.	Net Profit Margin (%)	6.37
4.1.3.	Accounting Rate of Return (%)	27.05
4.1.4.	Payback period	2 years and 7.88 months
4.2.	*Dynamic assessment*	
4.2.1.	Net present value	7,680.20
4.2.2.	Internal rate of return	31.99 %
4.2.3.	Payback period	2 years and 11.08 months
4.3.	*Break-even point analysis*	
4.3.1.	Break-even point (%)	6.80
4.3.2.	Rate of safety (%)	93.20

Described electro-generator could be economically, socially and environmentally desirable energy supplying option, especially at small and remote farms. Main elements derived from the investment analysis linked to purchase and use of mentioned power plant are presented in next table (Table 1). According to obtained indicators, the purchase and the use of the solar mobile robotized electro-generator sounds as reasonable business decision for the potential investor.

"Smart farm" concept implementation

In confronting the rise in awareness to negative natural trends and strivings to more comfortable performing of agricultural activities that will support farms sustainability, comes to establishment of smart farming concept and development of many smart farm systems, usually linked to certain sector of agriculture (Idoje, Dagiuklas and Iqba, 2021; Lytos et al., 2020; Wolfert et al., 2017). These concepts try to reach the full control of entire production processes, minimizing the occurrence of potential production risks (mostly natural).

Picture 2. "Smart farm" concept
Source: IMP and IAE internal documentation.

During the period 2018-2021. IMP and IAE were developed, implemented and tested in practice the "Smart farm" concept at the previously selected small family farm located in Belegis, municipality of Stara Pazova. Farm is focused to crop production, mainly to vegetable growing. Implemented "Smart farm" concept involves totally automatized and remote controlled fert-irrigation system (with mutually independent irrigation lines based on the use of electromagnet valves) that encompass integrated energy plant based on RES (hybrid system that transforms solar and wind energy into the electric energy), digital weather station, certain sensors, farm surveillance system, etc. Concept provides optimal and prompt irrigation and fertilizing adjusted to crops needs (Picture 2). Installed power plant includes cointegrated solar panels (with output power of 8 kW), wind generator (0.5 kW) and battery bank (48 Vdc/720Ah). Implemented system is scalable and could be adjusted to any size of farm (Despotović, 2022; Despotović, Rodić and Stevanović, 2022).

Presented concept for farm management is excellent solution for small farmers that strives to overall automatization and remote management under their production process, as well to leave shallow environmental footprint. Elements included in investment analysis in line to implementation of mentioned concept are presented in next table (Table 2).

Table 2. Resume of the investment in implementation of the "Smart farm" concept applied in vegetable growing (in EUR)

No.	Element	
1.	Estimated value of investment (EUR)	
1.1.	Total investment	75,240.00
1.2.	Investment in fixed assets	68,400.00
1.3.	Investment in PWC	6,840.00
2.	Financing sources	
2.1.	Sources – total	75,240.00
2.2.	Own resources	24,240.00
2.3.	Other sources (subsidies)	51,000.00
2.4.	Interest or discount rate (%)	5 %
3.	Object of investment	
3.1.	Investment	Investment in implementation of "Smart Farm" concept
3.2.	Starting date of establishment	During 2019
3.3.	End date of establishment	During 2020
3.4.	Project economic life cycle (in line to usual credit line)	5 years
4.	Expected economic effects	

Table 2. Continued

No.	Element	
4.1.	*Static assessment*	
4.1.1.	Total Output-Total Input Ratio (%)	1.59
4.1.2.	Net Profit Margin (%)	33.27
4.1.3.	Accounting Rate of Return (%)	24.20
4.1.4.	Payback period	4 years and 1.57 months
4.2.	*Dynamic assessment*	
4.2.1.	Net present value	64,724.60
4.2.2.	Internal rate of return	28.42 %
4.2.3.	Payback period	4 years and 4.27 months
4.3.	*Break-even point analysis*	
4.3.1.	Break-even point (%)	0.16
4.3.2.	Rate of safety (%)	99.84

In accordance to performed investment analysis there is strong belief that implementation and exploitation of described concept based on RES use could be a good alternative not just for small vegetable producers.

"Agrokapilaris" irrigation system

Modern crop production that strives to reach high yields of hi-quality fruits is unimaginable without the use of irrigation (De Pascale et al., 2011). No matter the type of vegetable growing (in the field, or in protected area), there are in use different (under)ground irrigation systems (Incrocci et al., 2020; Zinkernagel et al., 2020).

During the season 2020/21, implementing team of the IMP and IAE Belgrade have been established the "Agrokapilaris" underground irrigation system in the green house (size of 0.05 ha) located at the experimental farm of the high agriculture-chemistry school from Obrenovac municipality, Belgrade. Contrary to usually use subsurface irrigation systems, in this case is implemented innovative subsurface capillary irrigation system. In line to its technical constructions, system belongs to highly precise (strictly controlled optimal water consumption) and smart technologies (it has self-regulation mechanism active in moment of water transfer to the plant).

System characterizes innovation contained in construction that involve a network of underground channels of small dimensions, made from unbreakable plastic foil, that has the shape of Latin letter V carrying the water transmitters with built-in elements for letting water in the system. Requirements for optimal volume of water is in the form of capillary moisture available to plant at any moment (Picture 3). Water is moving radially, ascending and laterally, without water losses, along the overall irrigation system and during the entire vegetation period. Wet front around the root system has the shape of ellipse with the center in moisture transmitter (Kljajić and Kovačević, 2021). So, "Agrokapilaris" is different from other available irrigation systems in next characteristics: operates under extremely low pressures (up to 0.2 bar); has long utilization period as the clogging of water transmitters is too rare; and has self-regulation according to transferred volume of moisture to plants. Interesting is that irrigation system is run by the use of RES (hybrid power system based on solar and wind energy), what gives it eco-friendly label. Energy system that powers the irrigation is consisted of four photovoltaic panels with power of 275 W each and wind turbine with power of 500 W, while they are connected to adequate battery bank and inverter.

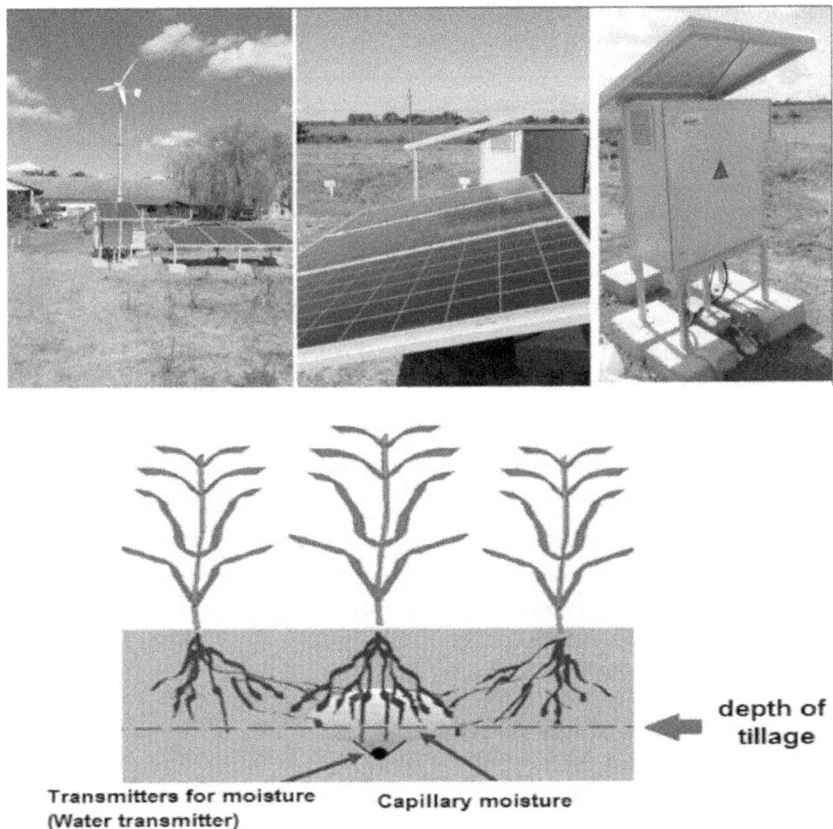

Picture 3. "Agrokapilaris" concept
Source: IMP and IAE internal documentation.

In order to avoid stopping of the irrigation in circumstances of unfavorable weather conditions, system could be supplied by electric power from the public electro grid (Despotović, Rodić and Stevanović, 2021; Despotović and Stevanović, 2021).

Table 3. Resume of the investment in irrigation system powered by the RES used in irrigation of vegetables grown in greenhouse (in EUR)

No.	Element	
1.	Estimated value of investment (EUR)	
1.1.	Total investment	10,167.00
1.2.	Investment in fixed assets	9,243.00
1.3.	Investment in PWC	924.00
2.	Financing sources	
2.1.	Sources – total	10,167.00
2.2.	Own resources	10,167.00
2.3.	Other sources (subsidies)	–
2.4.	Interest or discount rate (%)	7 %
3.	Object of investment	
3.1.	Investment	Investment into the subsurface irrigation system with power plant based on RES
3.2.	Starting date of establishment	During 2020
3.3.	End date of establishment	During 2021
3.4.	Project economic life cycle (in line to usual credit line)	5 years
4.	Expected economic effects	
4.1.	*Static assessment*	
4.1.1.	Total Output-Total Input Ratio (%)	1.73
4.1.2.	Net Profit Margin (%)	35.81
4.1.3.	Accounting Rate of Return (%)	22.29
4.1.4.	Payback period	3 years and 8.72 months
4.2.	*Dynamic assessment*	
4.2.1.	Net present value	6,345.00
4.2.2.	Internal rate of return	23.51 %
4.2.3.	Payback period	4 years and 1.53 months
4.3.	*Break-even point analysis*	
4.3.1.	Break-even point (%)	7.83
4.3.2.	Rate of safety (%)	92.17

Offered solution provides shallow environmental footprint and high profitability, so it could be considered as highly justified option for wider exploitation at small and medium, or even larger farms. Summary of investment in implemented irrigation system and power plant based on RES (solar and wind energy) is presented in following table (Table 3).

In line to gained results, it could be concluded that implementation and exploitation of described irrigation system and energy powerplant based on RES could be a good business decision for the producers that grown the vegetable in protected area.

Solar dryer

In a broader sense, as a part of vegetable production chain, processing of vegetables could be also considered (Siddiq and Uebersax, 2018). It enables creation of the value added even at the farm level (Yadav, Tiwari and Khare, 2023). Among the available processing methods, vegetable drying is largely used in practice, as it provides gaining of well-preserved food products that have longer shelf life with minimized risks for adverse food quality changes (Jayaraman and Gupta, 2020). Usually used equipment for that purposes are driven on electricity from public power grid or on certain type of fossil fuel, while today is not so rare the implementation of engine system of dryer run by the energy gained from the renewable energy sources (RES) (Orsat, Changrue and Raghavan, 2006). Practically, due to general RES availability, as well as the plant complexity and level of initially required investment, the drying of plant material, including vegetables, is perfectly match the use of solar energy (Janjai and Bala, 2012). Besides economic and environmental benefits for the processor and near surrounding, this method provides the eco-friendly aspect for the produced final product (Prakash and Kumar, 2013).

Picture 4. Implemented solar dryer for plant species
Source: IMP and IAE internal documentation.

One of processing solutions for plant species drying (fruits, vegetables, medicinal herbs, spices, fungi's, etc.) based on the use of solar energy has been established and implemented by the IMP and IAE Belgrade at the experimental agricultural holding of the secondary Chemistry-agricultural school from Obrenovac municipality – Belgrade during the season 2022/23 (Picture 4).

Table 4. Resume of the investment in small processing plant for vegetable drying (in EUR)

No.	Element	
1.	Estimated value of investment (EUR)	
1.1.	Total investment	20,876.00
1.2.	Investment in fixed assets	18,979.00
1.3.	Investment in PWC	1,897.00
2.	Financing sources	
2.1.	Sources – total	20,876.00
2.2.	Own resources	17,812.00
2.3.	Other sources (subsidies)	3,064.00
2.4.	Interest or discount rate (%)	5 %
3.	Object of investment	

Table 4. Continued

No.	Element	
3.1.	Investment	Investment into establishment of solar dryer
3.2.	Starting date of establishment	During 2022
3.3.	End date of establishment	During 2023
3.4.	Project economic life cycle (in line to usual credit line)	5 years
4.	Expected economic effects	
4.1.	*Static assessment*	
4.1.1.	Total Output-Total Input Ratio (%)	1.20
4.1.2.	Net Profit Margin (%)	30.81
4.1.3.	Accounting Rate of Return (%)	11.63
4.1.4.	Payback period	4 years and 3,58 months
4.2.	*Dynamic assessment*	
4.2.1.	Net present value	6,489,40
4.2.2.	Internal rate of return	13.65 %
4.2.3.	Payback period	4 years and 6,06 months
4.3.	*Break-even point analysis*	
4.3.1.	Break-even point (%)	3.06
4.3.2.	Rate of safety (%)	96.94

Dryer facility is fully automatized and remote controlled, while it has possibility for the energy storing, as it implies drying chamber with installed fans and chamber for storing (thermal buffer – space coated by chamotte) of unused heat energy in processing activities that could be later used for warming or cooling some other facilities at the holding (e.g. greenhouse, offices, veterinary clinic, stables, warehouse, etc.). Use of energy surplus derived from processing affects annual power costs cut for the holding equivalent to 7,200 KW. Contrary to that, in any time, short-term deficiency of solar (heat) energy could be changed with electricity from public power grid. General energy capacity of the installed mini solar dryer is about 3.5 KW, while the drying capacity is 100 kg of vegetables in one cycle of drying. Complete cycle of drying lasts for 48 hours. Established plant is initially projected for small farmers, but plant is scalable, so it could be projected for larger producers too. Installed plant could be used for over the 40 years, while its implementation requires several months.

Offered solution is subject of supporting measures at national level (maximally 50 % of the investment without VAT). In sense of ecology and economy it represents highly justified alternative that could be widely used at small scale farms. Summary of investment in installed processing plant is given by the next table (Table 4).

According to assumed elements of establishment and the use of solar dryer, and gained indicators for the assessment of the economic effectiveness of investing in the mentioned alternative, the implementation and exploitation of the solar dryer would be a reasonable business decision for the investor. Of course, it has be noted that smaller part of the investment is financially supported.

Conclusions

Vegetable growing has great importance for humans and sector of agriculture, as it supplies the human population with essential, high quality nutrients, contributes to crop rotation and making value added at the farm level, preserves the income and livelihood of farm members and rural population, etc. While performing the vegetable growing, farm should strive to keep all tree axes of sustainability. In other words, it has to secure continuity in volume and quality of produced food products, or to assure that farm's activities leave the shallow environmental foot print, as well as to adequately employs internal and external labor and contributes to development of local rural community.

Mentioned could be partly fulfilled by implementation and use of innovative technological and technical alternatives, adequate production procedures, and organizational and quality schemes, etc. In other words, farm could deeply lean on available knowledge, or could make stronger links to scientific and professional institutions towards the taking over existing, or designing and further development of particular elements of production base potentially used at farm estate. These strivings, and potentiation of making bridges between the knowledge and production, have to be also recognized by policy makers, providing further support for implementation of advanced tech-tech solutions that will boost the overall sustainability at the farms.

In line to primary goal of research, there are presented certain technical characteristics and economic effects gained from the investment in utilization of few hi-tech alternatives based on the RES use in vegetable production and processing. So, investment analysis of the implemented Solar mobile robotized electro-generator, "Smart farm" concept, "Agrokapilaris" irrigation system, or Solar dryer shows that all alternatives could be considered economically and environmentally justified to be used at farm level.

Acknowledgement

Paper is a part of research financed by the MSTDI RS, agreed in decision no. 451-03-47/2023-01/200009 from 3.2.2023.

References

Achikanu, C., Eze Steven, P., Ude, C. and Ugwuokolie, O. 2013. Determination of the vitamin and mineral composition of common leafy vegetables in south eastern Nigeria. *International Journal of Current Microbiology and Applied Sciences*, 2(11), pp. 347–353.

Ali, S., Dash, N. and Pradhan, A. 2012. Role of renewable energy on agriculture. *International Journal of Engineering Sciences & Emerging Technologies*, 4(1), pp. 51–57.

Ayyogari, K., Sidhya, P. and Pandit, M. 2014. Impact of climate change on vegetable cultivation-a review. *International Journal of Agriculture, Environment and Biotechnology*, 7(1), pp. 145–155.

Barnwal, B. and Sharma, M. 2005. Prospects of biodiesel production from vegetable oils in India. *Renewable and Sustainable Energy Reviews*, 9(4), pp. 363–378.

Chel, A. and Kaushik, G. 2011. Renewable energy for sustainable agriculture. *Agronomy for Sustainable Development*, 31, pp. 91–118.

Chikaire, J., Nnadi, F., Nwakwasi, R., Anyoha, N., Aja, O., Onoh, P. and Nwachukwu, C. 2010. Solar energy applications for agriculture. *Journal of Agricultural and Veterinary Sciences*, 2, pp. 58–62.

De Pascale, S., Dalla Costa, L., Vallone, S., Barbieri, G. and Maggio, A. 2011. Increasing water use efficiency in vegetable crop production: From plant to irrigation systems efficiency. *HortTechnology*, 21(3), pp. 301–308.

Despotović, Ž. 2022. Hybrid "Off-grid" power supply systems and their applications in agriculture: Practical realizations. *MKOIEE*, 10(1), pp. 17–33.

Despotović, Ž. and Stevanović, I. 2021. Hybrid power supply of the Agrokapilaris® system for irrigation of vegatable crops on the plot grabovac-obrenovac. In: *Power plants 2021*, Beograd: Društvo Termičara Srbije, Srbija, pp. 1–14.

Despotović, Ž., Majstorović, M., Jovanović, M. and Stevanović, I. 2017. The pressure control in irrigation "off-grid" photovoltaic system based on mobile solar generator. *MKOIEE*, 5(1), pp. 245–251.

Despotović, Ž., Rodić, A. and Stevanović, I. 2021. Inovativna rešenja sistema navodnjavanja uz primenu obnovljivih izvora energije: Tehnički aspekti. In: N. Kljajic, ed. *Tehno i agroekonomska analiza prednosti i nedostataka šire primene inovativnog načina podpovršinskog kapilarnog navodnjavanja u poljoprivrednom sektoru*. Beograd, Srbija: IEP, pp. 25–62.

Despotović, Ž., Rodić, A. and Stevanović, I. 2022. Power supply system and smart management of agriculture land using renewable energy sources. *Energija, ekonomija, ekologija*, 24(1), pp. 28–39.

Dias, J. and Ortiz, R. 2021. New strategies and approaches for improving vegetable cultivars. In: P. Nath, ed. *The basics of human civilization*. Boca Raton: CRC Press, USA, pp. 349–381.

Dinham, B. 2003. Growing vegetables in developing countries for local urban populations and export markets: Problems confronting small-scale producers. *Pest Management Science*, 59(5), pp. 575–582.

Dong, J., Gruda, N., Li, X., Cai, Z., Zhang, L. and Duan, Z. 2022. Global vegetable supply towards sustainable food production and a healthy diet. *Journal of Cleaner Production*, 369, 133212.

Feder, G. and Umali, D. 1993. The adoption of agricultural innovations: A review. *Technological forecasting and social change*, 43(3–4), pp. 215–239.

Finch, L., Cummings, J. and Tomiyama, A. 2019. Cookie or clementine? Psychophysiological stress reactivity and recovery after eating healthy and unhealthy comfort foods. *Psychoneuroendocrinology*, 107, pp. 26–36.

Guan, R., Van Le, Q., Yang, H., Zhang, D., Gu, H., Yang, Y. and Peng, W. 2021. A review of dietary phytochemicals and their relation to oxidative stress and human diseases. *Chemosphere*, 271, p. 129499.

Hazra, P., Chattopadhyay, A. and Karmakar, K. 2011. *Modern technology in vegetable production*. New Delhi: New India Publishing Agency, India.

Idoje, G., Dagiuklas, T. and Iqbal, M. 2021. Survey for smart farming technologies: Challenges and issues. *Computers & Electrical Engineering*, 92, p. 107104.

Incrocci, L., Thompson, R., Fernandez Fernandez, M., De Pascale, S., Pardossi, A., Stanghellini, C. and Gallardo, M. 2020. Irrigation management of European greenhouse vegetable crops. *Agricultural Water Management*, 242, p. 106393.

Janjai, S. and Bala, B. 2012. Solar drying technology. *Food Engineering Reviews*, 4, pp. 16–54.

Jayaraman, K. and Gupta, D. 2020. Drying of fruits and vegetables. In: A.S., Mujumdar ed. *Handbook of industrial drying*. Boca Raton, USA: CRC Press, pp. 643–690.

Jeločnik, M. and Subić, J. 2022. Irrigation costs management at the family farms. In: X International scientific-practical conference – *Innovative aspects of the development service and tourism*. Stavropol: Faculty of Social and Cultural Service and Tourism, Stavropol state agrarian university, Russian Federation, pp. 116–128.

Jena, A., Deuri, R., Sharma, P. and Singh, S. 2018. Underutilized vegetable crops and their importance. *Journal of Pharmacognosy and Phytochemistry*, 7(5), pp. 402–407.

Joshi, P., Joshi, L. and Birthal, P. 2006. Diversification and its impact on smallholders: Evidence from a study on vegetable production. *Agricultural Economics Research Review*, 19, pp. 219–236.

Kalmpourtzidou, A., Eilander, A. and Talsma, E. 2020. Global vegetable intake and supply compared to recommendations: A systematic review. *Nutrients*, 12(6), p. 1558.

Kljajić, N. and Kovačević, V. 2021. Inovativno podpovršinsko kapilarno navodnjavanje: Prednosti i perspektive razvoja. In: N. Kljajic, ed. *Tehno i agroekonomska analiza prednosti i nedostataka šire primene inovativnog načina podpovršinskog kapilarnog navodnjavanja u poljoprivrednom sektoru*. Beograd, Srbija: IEP, pp. 93–116.

König, B., Janker, J., Reinhardt, T., Villarroel, M. and Junge, R. 2018. Analysis of aquaponics as an emerging technological innovation system. *Journal of Cleaner Production*, 180, pp. 232–243.

Liu, R. 2013. Health-promoting components of fruits and vegetables in the diet. *Advances in Nutrition*, 4(3), pp. 384S–392S.

Locascio, S. 2005. Management of irrigation for vegetables: Past, present, and future. *HortTechnology*, 15(3), pp. 482–485.

Lumpkin, T., Weinberger, K. and Moore, S. 2005. *Increasing Income Through Fruit and Vegetable Production Opportunities and Challenges*. CGIAR Meetings – Agenda Documents, Montpellier: CGIAR, France, pp. 1–10.

Lytos, A., Lagkas, T., Sarigiannidis, P., Zervakis, M. and Livanos, G. 2020. Towards smart farming: Systems, frameworks and exploitation of multiple sources. *Computer Networks*, 172, p. 107147.

Mariyono, J. 2018. Profitability and determinants of smallholder commercial vegetable production. *International Journal of Vegetable Science*, 24(3), pp. 274–288.

Mason D'Croz, D., Bogard, J., Sulser, T., Cenacchi, N., Dunston, S., Herrero, M. and Wiebe, K. 2019. Gaps between fruit and vegetable production, demand, and recommended consumption at global and national levels: An integrated modelling study. *The Lancet Planetary Health*, 3(7), pp. e318–e329.

Mekhilef, S., Saidur, R. and Safari, A. 2011. A review on solar energy use in industries. *Renewable and Sustainable Energy Reviews*, 15(4), pp. 1777–1790.

Morin-Crini, N., Lichtfouse, E., Torri, G. and Crini, G. 2019. Applications of chitosan in food, pharmaceuticals, medicine, cosmetics, agriculture, textiles, pulp and paper, biotechnology, and environmental chemistry. *Environmental Chemistry Letters*, 17(4), pp. 1667–1692.

Oguntibeju, O., Truter, E. and Esterhuyse, A. 2013. The role of fruit and vegetable consumption in human health and disease prevention. *Diabetes Mellitus-Insights and Perspectives*, 3(2), pp. 172–180.

Onwude, D., Chen, G., Eke Emezie, N., Kabutey, A., Khaled, A. and Sturm, B. 2020. Recent advances in reducing food losses in the supply chain of fresh agricultural produce. *Processes*, 8(11), p. 1431.

Orsat, V., Changrue, V. and Raghavan, V.G.S. 2006. Microwave drying of fruits and vegetables. *Stewart Postharvest Review*, 2(6), pp. 1–7.

Ozkan, B., Kurklu, A. and Akcaoz, H. 2004. An input–output energy analysis in greenhouse vegetable production: A case study for Antalya region of Turkey. *Biomass and Bioenergy*, 26(1), pp. 89–95.

Prakash, O. and Kumar, A. 2013. Historical review and recent trends in solar drying systems. *International Journal of Green Energy*, 10(7), pp. 690–738.

Rao, A. and Rao, L. 2007. Carotenoids and human health. *Pharmacological research*, 55(3), pp. 207–216.

Razin, A., Taktarova, S. and Semenov, V. 2018. Innovative and investment development of vegetable growing. In: *MATEC Web of Conferences, ICRE 2018, proceedings*, vol. 212/07010, Les Ulis: EDP Sciences, France, pp. 1–6.

Richards, T. and Rickard, B. 2020. COVID-19 impact on fruit and vegetable markets. *Canadian Journal of Agricultural Economics*, 68(2), pp. 189–194.

Rickman, J., Bruhn, C. and Barrett, D. 2007. Nutritional comparison of fresh, frozen, and canned fruits and vegetables II. Vitamin A and carotenoids, vitamin E, minerals and fiber. *Journal of the Science of Food and Agriculture*, 87(7), pp. 1185–1196.

Robačer, M., Canali, S., Kristensen, H., Bavec, F., Mlakar, S., Jakop, M. and Bavec, M. 2016. Cover crops in organic field vegetable production. *Scientia Horticulturae*, 208, pp. 104–110.

Rodriguez-Casado, A. 2016. The health potential of fruits and vegetables phytochemicals: Notable examples. *Critical reviews in food science and nutrition*, 56(7), pp. 1097–1107.

Rolle, R. 2006. Improving postharvest management and marketing in the Asia-Pacific region: Issues and challenges. *Postharvest management of fruit and vegetables in the Asia-Pacific region*, 1(1), pp. 23–31.

Sabir, N. and Singh, B. 2013. Protected cultivation of vegetables in global arena: A review. *Indian Journal of Agricultural Sciences*, 83(2), pp. 123–135.

Shahbandeh, M. 2023. *Global production of vegetables in 2021*. Portal Statista, NY, USA. [online] Available at: <www.statista.com/statistics/264066/global-vegegable-production-by-region/> [Accessed 9 October 2023].

Sharma, N., Acharya, S., Kumar, K., Singh, N. and Chaurasia, O. 2018. Hydroponics as an advanced technique for vegetable production: An overview. *Journal of Soil and Water Conservation*, 17(4), pp. 364–371.

Shock, C., Pereira, A., Hanson, B. and Cahn, M. 2007. Vegetable irrigation. *Irrigation of Agricultural Crops*, 30, pp. 535–606.

Siddiq, M. and Uebersax, M., eds. 2018. *Handbook of vegetables and vegetable processing*. New York: John Wiley & Sons, USA.

Stoian, M. 2021. Renewable energy and adaptation to climate change. *Western Balkan Journal of Agricultural Economics and Rural Development*, 3(2), pp. 111–121.

Subić, J. and Jeločnik, M. 2016. Economic effects of new technologies application in vegetable production. In: D. Tomic, ed. *152nd EAAE Seminar – Emerging technologies and the development of agriculture*. Belgrade: SAAE, Serbia, pp. 15–35.

Subić, J. and Jeločnik, M. 2023. Economic aspects of the innovative alternatives use in agriculture. In: L. Chivu, I. De Los Ríos Carmenado, and J. Andrei, eds. *Crisis after the Crisis: Economic development in the new normal, ESPERA 2021*. Springer proceedings in business and economics. Cham, Germany: Springer, pp. 91–105.

Surucu-Balci, E. and Tuna, O. 2021. Investigating logistics-related food loss drivers: A study on fresh fruit and vegetable supply chain. *Journal of Cleaner Production*, 318, p. 128561.

Villalobos, J., Soto Silva, W., González Araya, M. and González Ramirez, R. 2019. Research directions in technology development to support real-time decisions of fresh produce logistics: A review and research agenda. *Computers and Electronics in Agriculture*, 167, p. 105092.

Vojkovska, H., Myšková, P., Gelbíčová, T., Skočková, A., Koláčková, I. and Karpíšková, R. 2017. Occurrence and characterization of food-borne pathogens isolated from fruit, vegetables and sprouts retailed in the Czech Republic. *Food Microbiology*, 63, pp. 147–152.

Wolfert, S., Ge, L., Verdouw, C. and Bogaardt, M. 2017. Big data in smart farming: A review. *Agricultural Systems*, 153, pp. 69–80.

Wu, L., Hu, K., Lyulyov, O., Pimonenko, T. and Hamid, I. 2022. The impact of government subsidies on technological innovation in agribusiness: The case for China. *Sustainability*, 14(21), p. 14003.

Yadav, M., Tiwari, S. and Khare, P. 2023. Value added products in vegetable crops. In: R. Pandey, R. Jaiswal, and A. Dhanaraj, eds. *Scientific approach for self-reliant India,* Agra: Shree Vinayak Publication, India, pp. 61–67.

Zapucioiu, L., Sterie, M. and Dumitru, E. 2023. Economic analysis of potato and tomato trade in Romania: The Gini coefficient. *Western Balkan Journal of Agricultural Economics and Rural Development*, 5(1), pp. 15–28.

Zinkernagel, J., Maestre Valero, J., Seresti, S. and Intrigliolo, D. 2020. New technologies and practical approaches to improve irrigation management of open field vegetable crops. *Agricultural Water Management*, 242, p. 106404.

Andreea Lidia Jurjescu, Alina Șimon, Florin Sala

Modeling and multicriteria analysis of soybean production variation – Case study in Romania

Abstract: *Proteic crops are interesting from a food, economic and agronomic perspective, through the supply of proteins, the high value on the market, and the benefits for the environment through the fixation of molecular nitrogen. Soybean is a crop plant with multiple functionalities, a pillar of sustainability that integrates into the concept of circular economy, the "ESE triangle" (Economy – Society – Environment). In accordance with the considered purpose, the study analyzed the situation of soybean crop in Romania in the period 2015 – 2020. The cultivated areas (S, thousand ha), the yield (Y, t ha^{-1}), and the total production (TP, thousand t) were considered in the study. The variation of the areas cultivated with soybean, and the yield, during the study period was analyzed, and the variation of the total production (TP) was evaluated in relation to the cultivated area and the yield, TP=f(S,Y). Multicriteria analysis (PCA) was used to detect the variable (S, Y) that had the strongest influence on total soybean production. Spline models most faithfully described, under statistical safety conditions, the variation of the cultivated areas $\left(\overline{\varepsilon} = 0.00073\right)$ and the yield $\left(\overline{\varepsilon} = 0.00161\right)$ in the soybean crop, in the study period 2015-2020. The multicriteria analysis (PCA) facilitated the obtaining of distribution and grouping diagrams of the years considered in the study, in relation to the considered elements (S, Y and TP) in the soybean culture. PC1 explained 83.12 % of variance and PC2 explained 16.866 % of variance. The cluster analysis facilitated the grouping based on the similarity of the years, in relation to the S, Y and TP values for the soybean crop (Coph.corr=0.881). Models in the form of equation and graphic models (3D, isoquants) resulted from the regression analysis, which described the TP variation in relation to the S and Y variables.*

Keywords: Circular economy, legume, modeling, proteic crops, soybean, spline model, yield.

Introduction

Soybean, *Glycine max* (L.) Merr. is a crop plant of major importance in the world, along with wheat, corn, rice, cotton, a crop with high versatility (Pedrozo, de Oliveira and Alberton, 2018; Karges et al., 2022).

Soy is an important crop due to its high protein and fat content (35–52 % protein; 14–24 % oil), varied content of nutritious and bioactive principles, with importance for human nutrition and animal feed, for the biofuel industry, as well as for new directions of nutraceutical interest (Awuni et al., 2020).

Soy-based products are increasingly present in the human diet, especially in the case of plant-based diets, as a result of the nutritional principles and bioactive compounds it contains (proteins, essential amino acids, antioxidants, phenols, flavonoids, isoflavones, salts minerals – Ca, Mg, K, etc.) (Messina et al., 2022; Robbani et al., 2022).

Soy is a plant with multiple functionalities, which can be considered the "key crop" in the "Economic-Social-Ecological" triad (the ESE triangle), in the context of sustainable agricultural systems, or of the circular economy.

Therefore, soybean crop needs to be approached on the basis of a complex of benefits that it generates, and not only on the basis of the classic "cost-benefit" analysis, which largely focuses only on the direct financial benefits, and omits the complex benefit, the "key benefit" in the agroecosystem (ecological benefit) of soybean crop. Studies in this direction confirmed the way in which the efficiency of the soybean culture changed under the conditions of a complex analysis, which considered both direct financial benefits and ecological ones (nitrogen fixation) and which made the difference in the efficiency of this culture. In cross-sectional data studies, from 271 farmers, Asodina et al. (2020) found that farmers' motivation for soybean cultivation increased especially associated with ancillary benefits (fixed organic nitrogen, benefits for the following crops, conservative crops, green environment), considered as non-market benefits.

In different agricultural areas around the world, soybean has established itself as a crop of major importance through the production of protein and oil it provides, but also through the favorable agronomic influences it has on other agricultural crops and on the soil (Agarwal et al., 2013).

The importance of soybean cultivation has been studied through the prism of food production (proteins, oil, nutritional principles), but also through the prism of the role at the ecosystem level, through the complex rhizospheric functions it has with species of microorganisms – mycorrhizae, rhizobia etc. (Pagano and Miransari, 2016).

In relation to the symbiotic fixation of nitrogen, numerous studies were carried out, and the functionality of the nodulation process, the N fixation process, the amount of N fixed, physiological indices of soybean plants, elements of productivity and production quality were evaluated (Pedrozo et al., 2018).

Nutrient management represents an important element of soybean crop technology for efficient production, as a result of the importance of nutrients in plant nutrition in relation to the complete and complex functionality of plants (photosynthesis, symbiotic N fixation) and the better response of plants to environmental stress factors (Bagale, 2021). However, the efficient production of soybean crops represents a constraint in relation to various factors of biotic, abiotic nature, applied technology and crop management (Bagale, 2021).

Pagano and Miransari (2016) appreciated the importance of knowing the relationship and response of soybean plants to different stress factors for an adequate crop management.

According to the biological requirements of soybean plants, the climatic conditions, in terms of water and temperature, present the main stress factor that affects the production, the yield of the soybean crop and the functionality of this ecosystem (Siamabele and Moral, 2021). The influence of climatic conditions on the functionality and productivity of the soybean crop (as well as other crops) strongly affects food security, especially in areas around the globe where people's incomes are very low. In such areas (e.g. countries in Africa), soybeans have a double functionality, as a source of food (especially protein), but also as a fertilizer for the soil through the symbiotic fixation of nitrogen and plant residues supplied to the soil, with all the logistical problems that have certain categories of farmer ideas (Siamabele and Moral, 2021).

In order to increase the productivity of soybean crops, the production systems assume adaptations of the technologies and the provision at the right time of the inputs that the soybean crop needs in the context of the "genotype×environment×technology" interaction for the purpose of a high photosynthetic yield and efficient metabolic processes (Singh et al., 2023).

As a result of the content of active principles in the soybean grains (proteins, oil, minerals, etc.) and its importance for human nutrition, animal feed, the biofuel industry and other economic sectors, soybeans are cultivated on a large scale in the world (Wójcik-Gront et al., 2022).

According to statistical data, the main seven soybean producing countries are Brazil (152,000,000 Metric Tons), United States (118,266,000 Metric Tons), Argentina (49,500,000 Metric Tons), China (18,400,000 Metric Tons), India (12,000,000 Metric Tons), Paraguay (10,000,000 Metric Tons) and Canada (6,543,000 Metric Tons) (Cook, 2023; World Agricultural Production.com, 2023).

Soybean is an important crop on a global level, it belongs to the first five crops (along with corn, wheat, rice, cotton), but in Europe soybean shows a minor role (Karges et al., 2022). The authors analyzed the agronomic potential and the limitations of soybean crop for feed and food in conditions of higher latitudes, in the North-East of Germany. The authors analyzed three factors regarding soybean culture (varieties, the effect of irrigation, and the agro-economic potential) and their interaction in a study between 2015 and 2017. It was highlighted the response of the varieties and their differentiation in terms of yield, the important role of irrigation in conditions of precipitation deficit (irrigation generated a 41 % increase in production), and the fact that in conditions of sufficient precipitation regime irrigation was not necessary. Analyzing the gross margin (irrigated/non-irrigated) the authors of the study demonstrated the agro-economic potential of soybean crop within the culture systems and protein production in the conditions of Central Europe (Karges et al., 2022).

In the countries of Central and Eastern Europe, soybeans have variable economic importance, relatively small compared to other areas around the world (Wójcik-Gront et al., 2022). In this context, Wójcik-Gront et al. (2022) based on a study carried out in the conditions of Poland, communicated the most important agronomic and environmental variables for soybean production. Based on data over a period of 10 years, the authors of the study analyzed different variables for the management of soybean crops (e.g. genotypes, fertilizers, herbicides, sowing, preceding crops, environmental conditions etc.). Based on complex analyzes (multiple regression models; regression trees) it turned out that the soybean yield showed a high variability, in direct relation to the water available for the plants and the physical properties of the soil. The authors concluded that environmental variables had a stronger effect compared to management variables. The other variables considered in the study and analysis also showed a reduced influence (e.g. nutrients, except phosphorus with higher importance; preceding crop, plant protection) (Wójcik-Gront et al., 2022).

In the context of the importance of soybean culture, the aspects identified by different studies regarding influencing factors and technological and natural variables, and the market and ecological benefits of soybean culture, the present study analyzed the situation of soybean culture in Romania for the period 2015–2020, and used multicriteria analysis and modeling to describe the variation in soybean production.

Methodology

Soybean crop was analyzed based on specific elements with reference to cultivated area and production, in the period 2015–2020 in Romania.

In relation to the purpose of the study, statistical data were taken regarding the soybean culture in the period 2015–2020, Table 1 (MADR, 2023).

Table 1. Evolution of the areas and production of soybean culture in Romania, 2015–2020 period

Analyzed parameters	Study period					
	2015	2016	2017	2018	2019	2020
Area (S, thousand ha)	128.156	127.266	165.143	169.422	158.149	164.686
Average production (Y, t ha^{-1})	2.045	2.070	2.383	2.748	2.630	1.857
Total production (TP, KT)	262.061	263.380	393.495	465.609	415.942	305.816

Source: MADR, 2023.

The evolution of the area cultivated with soybeans (S, thousand ha), the yield (Y, t ha^{-1}) and the total production (TP, thousand tons, kiloton – Kt) during the study period, 2015–2020, was analyzed.

To capture the variation of the considered elements (S, Y, TP) appropriate mathematical and statistical tools were used (Hammer, Harper and Ryan, 2001; Wolfram Research Inc., 2020).

The evolution of the cultivated area (S) and the yield (Y) was analyzed in relation to the time factor, during the study period. Multicriteria analysis (PCA, CA) was used to highlight the association of the elements considered for the analysis of soybean culture with the years of study and to quantify the degree of similarity at the level of years during the study period. Regression analysis was applied to evaluate the variation of total soybean production (TP) in relation to cultivated area (S) and yield (Y) during the study period.

Equation models, graphic models, and diagrams were generated to describe and represent the obtained results. For the statistical safety of the results, established statistical safety parameters were used (p, R^2, Coph.corr, RMSEP).

Results and discussions

Based on the statistical data considered regarding the soybean culture in Romania (MADR, 2023), Table 1, and the analysis of these data, the variation of the areas cultivated with soybean (S, thousand ha) and the average production (Y, t ha^{-1}) for the period 2015–2020 were analyzed, Table 1.

Based on the mathematical and statistical analysis, Spline models most faithfully described the variation of the surface S (thousand ha) and the average production Y (t ha^{-1}) in relation to the time factor, and the errors associated with the Spline model were deduced according to equation (1). In the case of the area cultivated with soybeans (S, thousand ha), the statistical data related to the spline model are presented in Table 2, and the graphic distribution is presented in Figure 1. In the case of the average soybean production (Y, t ha⁻¹), the statistical data related to the spline model are presented in Table 3, and the graphic distribution is presented in Figure 2.

$$\bar{\varepsilon} = \left(\sum_{i=1}^{n} \varepsilon_i\right)\bigg/n = \left(\sum_{i=1}^{n} \left|\frac{ys_i - y_i}{y_i}\right|\right)\bigg/n \qquad (1)$$

Table 2. Statistical values related to the spline model, to describe the variation of the areas cultivated with soybeans in relation to the time factor in Romania, period 2015–2020

Trials	data	Area (thousand ha) in relation to time (years)			
Year	x_i	y_i	ys_i	e_i	$I_{i/1}$
2015	1	128.16	127.01	−0.00897	1.000
2016	2	127.27	130.56	0.02585	1.028
2017	3	165.14	162.41	−0.01653	1.279
2018	4	169.42	169.07	−0.00207	1.331
2019	5	158.15	159.64	0.00942	1.257
2020	6	164.69	164.14	−0.00334	1.292
Average error				$\bar{\varepsilon} = 0.00073$	

Table 3. Statistical values related to the spline model, to describe the variation of average soybean production in relation to the time factor in Romania, period 2015–2020

Trials	data	Average production (t ha⁻¹) in relation to time (years)			
Year	x_i	y_i	ys_i	e_i	$I_{i/1}$
2015	1	2.045	2.0251	−0.00973	1.000
2016	2	2.070	2.1052	0.01700	1.040
2017	3	2.383	2.4060	0.00965	1.188
2018	4	2.748	2.7188	−0.01063	1.343
2019	5	2.630	2.5780	−0.01977	1.273
2020	6	1.857	1.8999	0.02310	0.938
Average error				$\bar{\varepsilon} = 0.00161$	

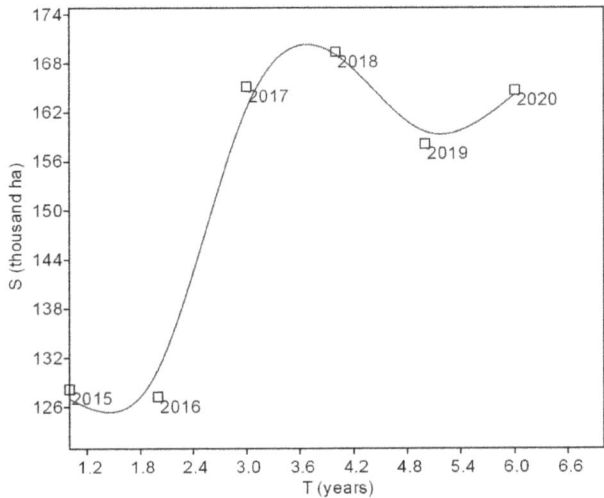

Figure 1. Graphical distribution of the area cultivated with soybeans according to the time factor in Romania, period 2015–2020

According to the PCA, the distribution diagram of the years during the study period was obtained, in relation to parameters evaluated in the soybean culture (S, Y and TP). Independent position was observed in the case of 2015, 2016 and 2020 years. In the case of 2015 and 2016 years, they are associated with reduced areas cultivated with soybeans and the lowest average productions per hectare, for the study period.

In the case of 2020 year, although the cultivated area was high (S=164,686 thousand ha), the similar area at the level of 2017 and 2018 years, the average production (Y) had the lowest value during the study period (Y=1.857 t ha^{-1}).

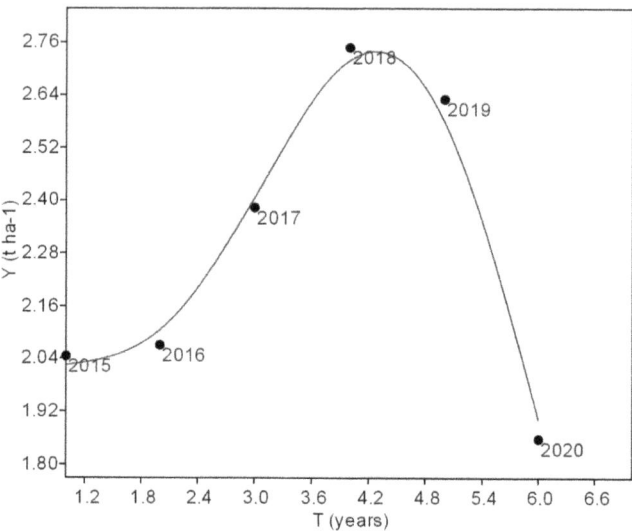

Figure 2. Graphic distribution of the average soybean production according to the time factor in Romania, period 2015–2020

This low value for the average production can be associated with the poor climatic conditions in the culture areas (especially the water factor), the lack of irrigation, even if appropriate technologies were made. The year 2017 occupied an intermediate position between the area (S) and the total production (TP). It was estimated that the total production (TP) in the case of 2017 was largely due to the total cultivated area and the average production per ha (Table 1). The years 2018 and 2019 were positioned intermediate, between the average production (Y) and the total production (TP). In the case of the years 2018 and 2019, it was estimated that the total production was mainly due to the higher average production per ha, compared to the other years of the study period, and secondarily to the areas cultivated with soybeans.

The graphic representation of the resulting diagram is given in Figure 3. PC1 explained 83.12 % of variance, and PC2 explained 16.866 % of variance.

Cluster analysis led to the dendrogram in Figure 4, under statistical safety conditions (Coph. corr.=0.881).

The formation of two distinct clusters was found, within which the time variables (years, study period) were grouped based on similarity in relation to the considered parameters (S, Y, TP).

Based on the calculated SDI values, Table 4, it was found that the highest level of similarity was recorded in the case of 2015 and 2016 (SDI=1.591). The years 2017 and 2019 (SDI=23.513) and the years 2018 and 2019 (SDI=50.930) followed.

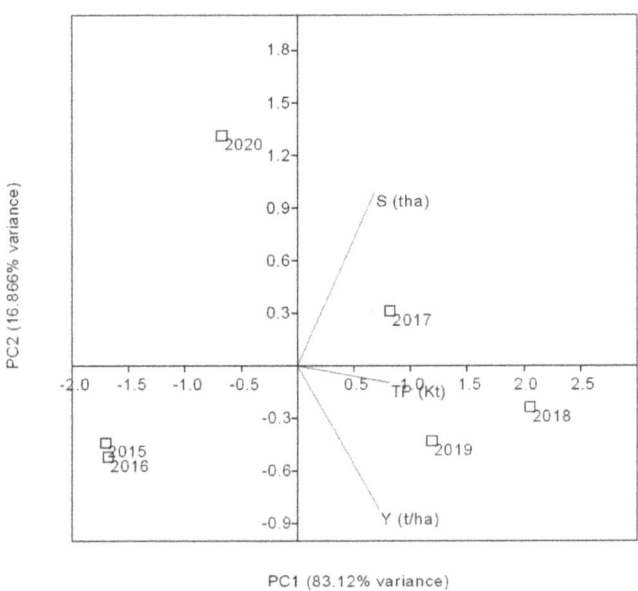

Figure 3. PCA diagram regarding the distribution of years in relation to the elements considered for soybean culture in Romania, period 2015–2020

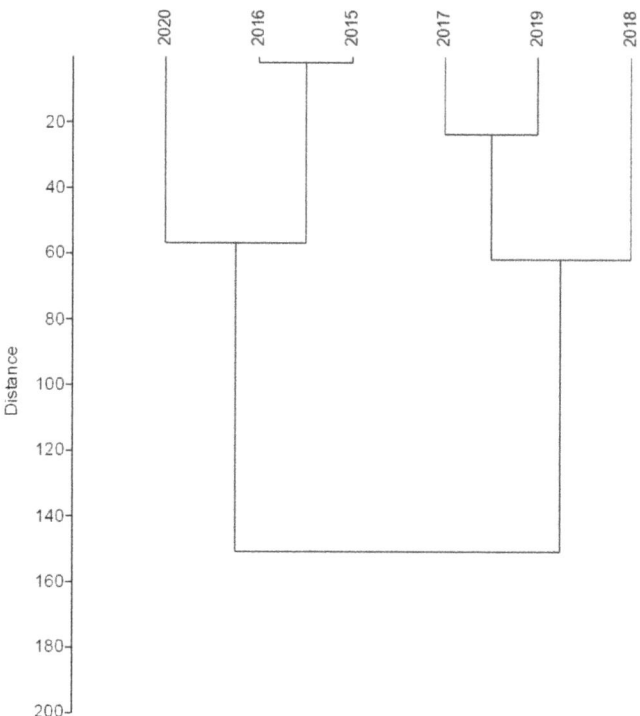

Figure 4. The cluster grouping of the years during the study period in relation to parameters considered in soybean culture (S, Y, TP) in Romania, period 2015–2020

Table 4. SDI values to describe the situation of soybean cultivation in relation to considered parameters, Romania 2015–2020

	2015	2016	2017	2018	2019	2020
2015		1.591	136.540	207.690	156.780	57.000
2016	1.591		135.520	206.580	155.660	56.578
2017	136.540	135.520		72.242	23.513	87.682
2018	207.690	206.580	72.242		50.930	159.870
2019	156.780	155.660	23.513	50.930		110.320
2020	57.000	56.578	87.682	159.870	110.320	

Ranking scaling analysis facilitated the ranking of the years of the study period, in relation to the values of the considered parameters regarding the soybean crop (S, Y, TP). The result is the diagram in Figure 5, in which the study years are ranked (Event, 2017 to 2022), and the interevent distance is presented.

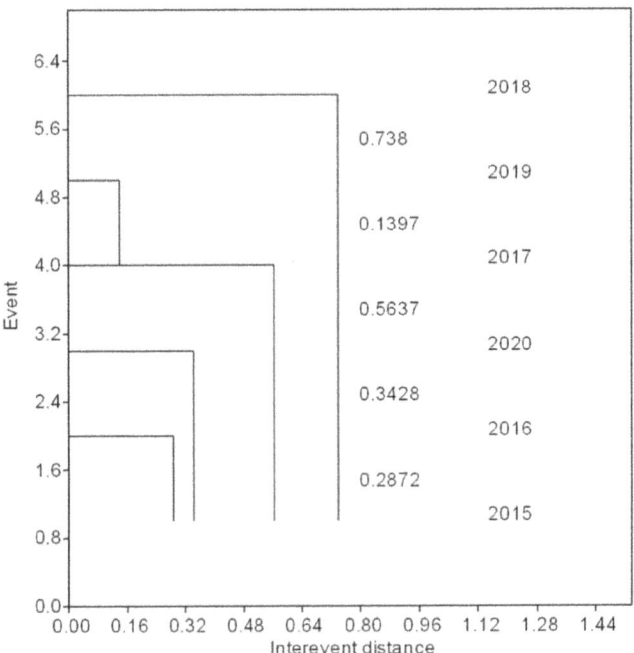

Figure 5. Scaling dendrogram for the years of the study period in relation to parameters considered in soybeans, the situation in Romania, period 2015–2020

Regression analysis was used to evaluate how the surface (S, thousand ha) and yield (Y, t ha^{-1}) variables influenced the total soybean production (TP, thousand t) during the study period.

Equation (2) was found, which described the variation of TP in relation to S and Y, under conditions of R^2=0.999, $p<0.001$. From the analysis of equation (2) and the graphic representation, it emerged that the total production was influenced in a different way by the two variables, both by increasing the surfaces, but in conditions of low yield, and in conditions of high yield, at the same surfaces or at lower surfaces.

The optimization of the two variables was not possible under the given conditions, their relation to the achievement of the total soybean production, under the given conditions, being of the "scissors" type, Figure 6 a, b.

$$TP = -0.000486 \cdot x^2 + 0.324021 \cdot y^2 + 0.088401x - 5.731674y \\ + 1 \cdot 026109xy + 0.15851 \quad (2)$$

where: TP – total production (Kt); x – cultivated surface area (S, thousand ha); y – yield (Y, t ha⁻¹)

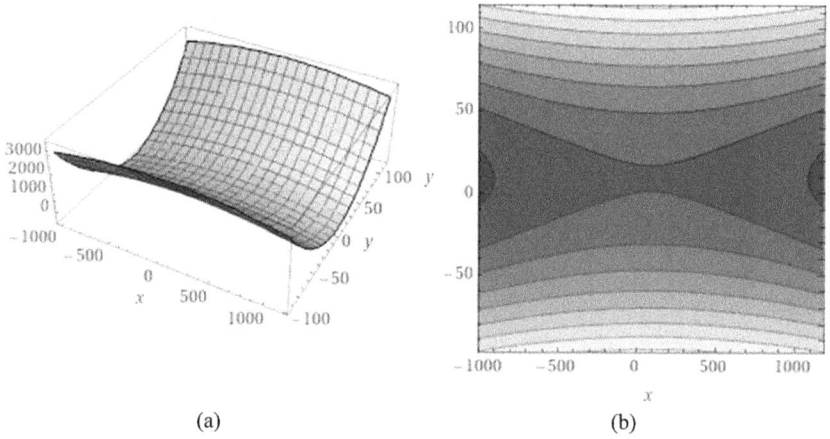

(a)   (b)

Figure 6. Variation of total soybean production (TP) in relation to cultivated area (S, x-axes) and yield (Y, y-axes)

From the analysis of the coefficients of equation (2), it was found that the yield (variable y in the equation) is much more important in the formation of total soybean production (TP), compared to the cultivated area (variable x in the equation). The graphic representation, Figure 6, in the form of 3D and isoquants, shows how the two variables, surface (S, thousand ha) and yield (Y, t ha⁻¹), contributed to the formation of total production (TP, thousand t) of soybeans during the study period.

The analysis of the production systems in soybean culture, the identification of the limiting factors (e.g. the water factor), and their correction in agricultural technologies could significantly contribute to the increase of soybean production, both the yield and the total production.

Conclusion

The variation of the elements considered for the analysis of soybean culture in Romania (area – S, thousand ha; yield – Y, t ha^{-1}) in the period 2015–2020 was most faithfully described by Spline models, under statistical safety conditions.

According to PCA, PC1 explained 83.12 % of variance, and PC2 explained 16.866 % of variance. The generated PCA diagram presented the distribution and association of the considered elements (S, Y, TP) in relation to the years during the study period. According to the cluster analysis, a dendrogram was obtained in which the years of the analysis period were grouped based on similarity in relation to the values of the elements considered in the study (S, Y, TP). A ranking diagram of the years (scaling dendrogram) was also generated. The SDI values quantified the level of similarity between years, in relation to the values of the elements associated with soybean culture, in the present study.

The regression analysis facilitated obtaining models in the form of equations and graphic models, which described the variation of the total production of soybean (TP) in relation to the area and the yield (S, Y). According to the generated results, the yield had a more pronounced influence in the variation of the total production, compared to the cultivated area, and the relationship between S and Y in the TP description was scissor type.

The analysis of the production systems in soybean culture, the identification of the limiting factors (e.g. the water factor), and their correction in agricultural technologies could significantly contribute to the increase of soybean production, both the yield, and the total production.

Acknowledgements

The authors thank SCDA Lovrin for facilitating this study. This research is supported by the Ministry of Agriculture and Rural Development, through the ADER 1.4.1 project.

References

Agarwal, D.K., Billore, S.D., Sharma, A.N., Dupare, B.U. and Srivastava, S.K. 2013. Soybean: Introduction, improvement, and utilization in India – Problems and prospects. *Agricultural Research*, 2, pp. 293–300. https://doi.org/10.1007/s40 003-013-0088-0.

Asodina, F.A., Adams, F., Nimoh, F., Weyori, E.A., Wongnaa, C.A. and Bakang, J.E.-A. 2020. Are non-market benefits of soybean production significant? An extended economic analysis of smallholder soybean farming in Upper West region of northern Ghana. *Agriculture & Food Security*, 9, p. 13. https://doi.org/10.1186/s40066-020-00265-7

Awuni, G.A., Reynolds, D.B., Goldsmith, O.G., Tamimie, C.A. and Denwar, N.N. 2020. Agronomic and economic assessment of input bundle of soybean in moderately acidic Savanna soils of Ghana. *Agrosystems, Geosciences & Environment*, 3(1), p. e20085. https://doi.org/10.1002/agg2.20085.

Bagale, S. 2021. Nutrient management for soybean crops. *International Journal of Agronomy*, 2021, p. 3304634. https://doi.org/10.1155/2021/3304634.

Cook, R. 2023. *World Soybean Production (Ranking by Country)*. [online] Available at: <https://beef2live.com/story-world-soybean-production-ranking-country-0-164836> [Accessed 12 September 2023].

Hammer, Ø., Harper, D.A.T. and Ryan, P.D. 2001. PAST: Paleontological statistics software package for education and data analysis. *Palaeontologia Electronica*, 4(1), pp. 1–9.

Karges, K., Bellingrath-Kimura, S.D., Watson, C.A., Stoddard, F.L., Halwani, M. and Reckling, M. 2022. Agro-economic prospects for expanding soybean production beyond its current northerly limit in Europe. *European Journal of Agronomy*, 133, p. 126415. https://doi.org/10.1016/j.eja.2021.126415.

MADR, Ministry of Agriculture and Rural Development. 2023. *Soia [Soybean]*. [online] Available at: < https://www.madr.ro/culturi-de-camp/plante-tehnice/soia.html> [Accessed 12 September 2023], [in Romanian].

Messina, M., Duncan, A., Messina, V., Lynch, H., Kiel, J. and Erdman, J.W.Jr. 2022. The health effects of soy: A reference guide for health professionals. *Frontiers in Nutrition*, 9, 970364. https://doi.org/10.3389/fnut.2022.970364.

Pagano, M.C. and Miransari, M. 2016. The importance of soybean production worldwide. In: M. Miransari, ed. *Abiotic and biotic stresses in soybean production, soybean production: Volume 1*. Academic Press, pp. 1–26. https://doi.org/10.1016/B978-0-12-801536-0.00001-3.

Pedrozo, A., de Oliveira, N.J.G. and Alberton, O. 2018. Biological nitrogen fixation and agronomic features of soybean (*Glycine max* (L.) Merr.) crop under different doses of inoculant. *Acta Agronomica*, 67(2), pp. 297–302. https://www.researchgate.net/publication/322732970_Biological_nitrogen_fixation_and_agronomic_features_of_soybean_Glycine_max_L_Merr_crop_under_different_doses_of_inoculant

Robbani, R.B., Hossen, M.M., Mitra, K., Haque, M.Z., Zubair, M.A., Khan, S. and Uddin, M.N. 2022. Nutritional, phytochemical, and in vitro antioxidant activity analysis of different states of soy products. *International Journal of Food Science*, 2022, p. 9817999. https://doi.org/10.1155/2022/9817999.

Siamabele, B. and Moral, M.T. (Reviewing editor). 2021. The significance of soybean production in the face of changing climates in Africa. *Cogent Food & Agriculture*, 7, p. 1. https://doi.org/10.1080/23311932.2021.1933745.

Singh, R.K., Upadhyay, P.K., Dhar, S., Rajanna, G.A., Singh, V.K., Kumar, R., Singh, R.K., Babu, S., Rathore, S.S., Shekhawat, K., Dass, A., Kumar, A., Gupta, G., Shukla, G., Rajpoot, S., Prakash, V., Kumar, B., Sharma, V.K. and Barthakur, S. 2023. Soybean crop intensification for sustainable aboveground-underground plant–soil interactions. *Frontiers in Sustainable Food Systems*, 7, p. 1194867. https://doi.org/10.3389/fsufs.2023.1194867.

Wójcik-Gront, E., Gozdowski, D., Derejko, A. and Pudełko, R. 2022. Analysis of the impact of environmental and agronomic variables on agronomic parameters in soybean cultivation based on long-term data. *Plants*, 11, p. 2922. https://doi.org/10.3390/plants11212922

Wolfram Research, Inc. 2020. *Mathematica*, Version 12.1, Champaign, Illinois: Wolfram Research, Inc.

World Agricultural Production.com. 2023. *World Agricultural Production*, [online] Available at: https://apps.fas.usda.gov/psdonline/circulars/production.pdf [Accessed 12 September 2023].

Ildikó Kolozsvári, Árpád Székely, Noémi Valkovszki,
Ágnes Kun, Mihály Jancsó, György Dajcs

Efficiency of slurry application in winter wheat with special reference to earthworm population trends

Abstract: *Today, sustainable agricultural production requires the use of alternative land management, and water-saving irrigation systems, and the use of organic and diluted fertilizers has once again come to the fore due to the increasing fertilizer prices. Besides the deteriorating soil biological parameters of our production areas, the use of organic and liquid fertilizers can provide a potential solution, where their careful agricultural use can increase soil fertility and have a positive effect on the average yield and soil life. In the course of our investigation, we utilized the slurry of a pig farm in the cultivation of winter wheat. The winter wheat culture grown in an untreated control area and one treated with liquid fertilizer were compared. During the growing season, liquid manure was applied twice to the soil surface of the sample area with a quantity of 53–60 m^3 per hectare. During the experiment, we monitored the changes in the phenological and physiological state of the plants, recorded the weight of the harvested grain, and after the harvest, we determined the number of earthworms per square meter of the sample areas in accordance with the ISO 23611-1:2006 standard. The obtained results support our hypothesis that the use of liquid manure positively affects the development of phenological parameters, the amount of the crop, and the biological condition of the soil. A significantly higher average yield was measured in the area treated with slurry fertilizer (4.1 t/ha) compared to the control treatment (2.5 t/ha). Similar trends were also observed in the density of earthworms, where the number of earthworms in the treated area (74 worms/ m^2) was significantly higher (p<0.01) compared to the control area (32 worms/m^2).*

Keywords: winter wheat, pig slurry fertilizer, earthworm, phenology, plant physiology, grain yield.

Introduction

Agriculture plays a key role in the world's food production and sustainability. Among the essential nutrients for plants, nitrogen (N) plays a role in food production. A wide range of nitrogen sources used in crop production include organic and inorganic fertilizers, and nitrogen that is mineralized in the soil. However, the excessive use of nitrogen poses significant environmental challenges (Sánchez and González, 2005).

Fertilizers are the main sources of nitrogen in agriculture today. No matter how efficiently they serve the nutrient needs of plants, they can have significant environmental consequences, which have already been demonstrated in many areas of the European Union. For this reason, the EU formulated a directive on the protection of waters against nitrate pollution (European Union, 1991). As a result of the excessive use of fertilizers, nitrogen can easily enter the groundwater through the capillaries, thus contaminating the water resources. A further significant proportion of the energy used by agriculture is used to produce N fertilizers (nitrous oxide N2O), with N emissions to the atmosphere from fertilizer use causing a greenhouse effect and thus contributing to climate change (Zamanian, Zarebanadkouki and Kuzyakov, 2018).

At the global level, due to population growth, the efficient use of nitrogen appears as an increasingly urgent problem (Keeler et al., 2016). The population explosion and climate change encourage all participants in all areas to find solutions for the sustainable agricultural use of N fertilizer (Clark, Hill and Tilman, 2018; Langholtz et al., 2021). This may involve technological approaches or simply reducing the amount of N fertilizer applied (Balafoutis et al., 2017; Pedersen et al., 2004). Furthermore, it should be emphasized that the cultivation of plants fertilized with synthetic N-fertilizer contributes significantly to meeting the food needs of the world population (Keeler et al., 2016; Zhang et al., 2015). The introduction of appropriate agricultural practices, the reasonable use of artificial fertilizers, and the development of the management of organic fertilizers can provide a solution. These not only serve to protect the environment, but also promote the economic sustainability of agriculture.

It is indisputable that agriculture plays a key role in the world's food production and sustainability. However, rising fertilizer prices pose a challenge to farmers who need to find efficient and economical ways to increase soil fertility. One alternative is the application of slurry to cropland, which not only reduces costs but also has a positive impact on soil-dwelling communities.

The development of new agricultural practices and technologies will enable the smart and more sustainable use of organic fertilizers for crop production. Therefore, our aim to raise awareness of the importance of more efficient and sustainable use of slurry for future food production and environmental protection.

Literature review

Nowadays, the amount of organic fertilizers used is far behind that of artificial fertilizers, despite the fact that the manure produced during animal husbandry has high agricultural and environmental potential. Organic fertilizers are often not used efficiently, despite the fact that they contain nitrogen (N), phosphorus (P) and potassium (K), which are essential for plants, as well as varying amounts of calcium (Ca), magnesium (Mg), sulfur (S) and other also trace elements. Pig slurry has an extremely rich macro- and meso-element content, so its application of it may be a good agriculture practice (Akinrinde Olubakin, Omotoso and Ahmed, 2006). A number of studies have investigated the use of pig slurry in field crop production under different climatic and soil conditions. Some of them showed that the weight of the harvested grain yield did not decrease when pig manures were used, and that their use is suitable for partial or complete replacement of artificial fertilizers (Bocchi and Tano, 1994; Motavalli et al., 1993; Nevens and Reheul, 2005; Zebarth et al., 1996).

Earthworms living in the soil can be used as bioindicators to the assessment of soil quality (Sheperd et al., 2008). Due to their limited living space and mobility, they react sensitively to changes in soil properties and the presence of pollutants (Paoletti, 1999). The application of organic fertilizers generally affects the biological activity of the soil (Sharma et al., 2017; Yagüe, Bosch-Serra and Jaume, 2012), especially the earthworm population (Murchie et al., 2015; Pérès et al., 2011). Earthworms play a key role in the sustainability of the agricultural system (Singh, 2018). This is mainly due to their role in the decomposition of organic matter, the water, nutrients and gas flow in the soil, their contribution to soil structure and their ability to improve soil porosity, including size, distribution, pore structure and morphology (Blouin et al, 2013). The use of bioindicators related to earthworm communities (e.g. abundance, biomass, species richness, diversity) well reflects the agrotechnical changes in crop production, providing information on soil quality (Paoletti, 1999). In particular, the use of farming practices lasting several years, such as the incorporation of organic waste into the soil, can result in the accumulation of certain compounds, which pose a risk to soil-dwelling organisms.

Methodology

The study area is located in Hungary, at the border of Kardoskút village (Békés county, GPS: 46.511222, 20.670185). The climate is characterized by a continental influence resulting from the proximity of the Eastern European steppes, with little and uneven rainfall distribution. The annual precipitation is between 450–500 mm. Kardoskút is one of the warmest landscapes in the Carpathian Basin. During the growing season, the average temperature in July was 23 °C, and the precipitation was only 242 mm in the vegetation period (from November to

June). The meteorological data was recorded by the Meteobot Pro meteorological station.

In the 90-hectare field, the early winter wheat variety "Bologna" was sown on 11.11.2021, with a seed quantity of 200 kg/ha. Even with the low thousand-seed weight, the variety is tillering. The resistance to ear fusarium was average, but its tolerance to powdery mildew was outstanding. The variety is characterized by stable productivity, excellent endurance and winter resistance, as well as good stem strength, so it is less prone to bending. It has very good drought tolerance, resulting in a uniform, plump and firm grain yield. It has an excellent gluten content, with moisture content exceeding 34 %.

Before the measurements, two sample areas of 100 m^2 per treatment were selected, which contained 10 plots of 1 m^2. No slurry fertilizer was applied in the control area. In the slurry treated area, before sowing and tillage, 60 m^3 per hectare and 53 m^3/ha of slurry was during the tillering period using the dribble bar technology. The slurry was analyzed by the Environmental Analysis Laboratory of the Hungarian University of Agriculture and Life Sciences (MATE), in accordance with Hungarian standard methods for all parameters.

After the tillering, we measured the growth rate of the plants on a weekly basis, and we determined the green mass of the plants and the length of their above-ground parts. The MetripondPlus SC scale was used to determine the plant weight, and a tape measure was used to record the plant height values. In addition to the phenological parameters, we also performed the total chlorophyll content of the plants (CPHLT) and the normalized vegetation index (NDVI) by a CID Bio-Science CI-710s SpectraVue Leaf Spectrometer instrument. The phenological studies were carried out with 10 replicates, and the spectrometric studies with 30 replicates

The sample areas were harvested using a hand sickle. The grains were separated from the ears with the WINTERSTEIGER LD 350 laboratory threshing machine, then the weight of the grain yield was measured with a MetripondPlus MWP-1500 scale, which was converted to t/ha value.

The assessment of soil quality and soil biological status is a key step in the management of sustainable agriculture and ecosystems. To measure earthworm abundance, an in situ manual sampling method according to ISO 23611-1:2006 was used, with the following sampling procedures:

- For each treatment, 25x25x30 cm soil samples were taken along with the selection of 5 sample areas. The samples were carefully placed into a plastic sheet. The purpose of this step was to preserve the integrity of the samples. As

a further analysis of the soil samples, the earthworms were collected by hand sorting.
- The collected earthworms were fixed in 70 % ethanol. This procedure helps preserve the condition of the worms and preserve them for later laboratory analysis.
- After soaking in 4 % formalin for one week, the earthworms were repeatedly stored in 70 % ethanol under laboratory conditions. This storage method ensures the long- term stability of the samples.
- Finally, under laboratory conditions, we determined the number of earthworms in units/m² and their biomass in g/m². We used the Sartorius analytical balance for mass measurement, and we took photographs of the individuals with a Nikon P-400Rv/P- 400R digital microscope (Nikon Corporation, Tokyo, Japan).

The IBM SPSS Statistics 27.0 software was used for the statistical evaluation of the results. We determined the effect of treatments on the phenological and physiological parameters of winter wheat, the development of the amount of grain yield, and soil activity with the one-way analysis of variance (ANOVA). Differences were found to be significant if they reached the Tukey test.

Results and discussions

Composition of slurry

Depending on the holding technology, the slurry has a solid and a dilute fraction. The handling of the solid phase is the same as the handling and placement of conventional manure. During the sedimentation and separation of the dilute phase, the placement of the settled fraction is also the same as the disposal of the solid phase of the litter technology. Depending on the technology used at the animal farm, the sex ratio and age composition of the animals, and the proportion of washing, leachate and rinsing water mixed with the excrement, there may be differences in the chemical composition. The chemical parameters of the liquid manure of the processing plant we examined are listed in Table 1, where it contains 1784 mg/dm³ of nitrogen, 360 mg/dm³ of phosphorus and 756 mg/dm³ of potassium in terms of the main macroelements. In the case of mesoelements, it had 900 mg/dm³ calcium and 552 mg/dm³ magnesium. These parameters are the same as the values described in the literature.

Table 1. The chemical parameters of our pig slurry composition

Examined parameters	Unit of measure	Chemical analysis of slurry
Water temperature (laboratory)	°C	22
pH (laboratory)		7.3
Ammonium ion	mg/dm^3	875
Ammonium-N	mg/dm^3	679
Total nitrogen	mg/dm^3	1,784
Total phosphorus	mg/dm^3	360
Total dry matter content	mg/dm^3	2,183
Total solutes	mg/dm^3	2,963
Sodium	mg/dm^3	436
Potassium	mg/dm^3	756
Calcium	mg/dm^3	990
Magnesium	mg/dm^3	552
Arsenic	mg/dm^3	0.01

Source: Authors' own research.

There is a difference between slurry fertilizers and stable fertilizers both in terms of their nutrient content and their effect. Slurry fertilizers have a high ammonia-nitrogen content, which is readily absorbed by the plants, but a low C:N ratio, which determines the conditions for using slurry fertilizers. However, it is important to note that under unfavorable conditions, such as warm weather or late incorporation of slurry, significant losses of ammonia can occur, which can take up to 100 % of the amount applied. These losses not only cause economic damage but can also result in environmental pollution. For this reason, it is important to choose the right timing and technology for slurry application to minimize ammonia losses and to use these fertilizers in a sustainable way in arable production.

Phenological parameters development of green mass

A significant difference ($p<0.001$) was observed between the control and slurry-treated samples when determining the biomass of winter wheat (Figure 1). For both treatments, the highest values were measured between sampling dates 4–7, where the average biomass per acre ranged from 10–17 g for the control and 15–28 g for the slurry-treated samples. In the measured growing season, significantly the highest ($p<0.001$) green mass for both treatments was recorded during the measurements on 5 May and 9 May compared to the first measurement results.

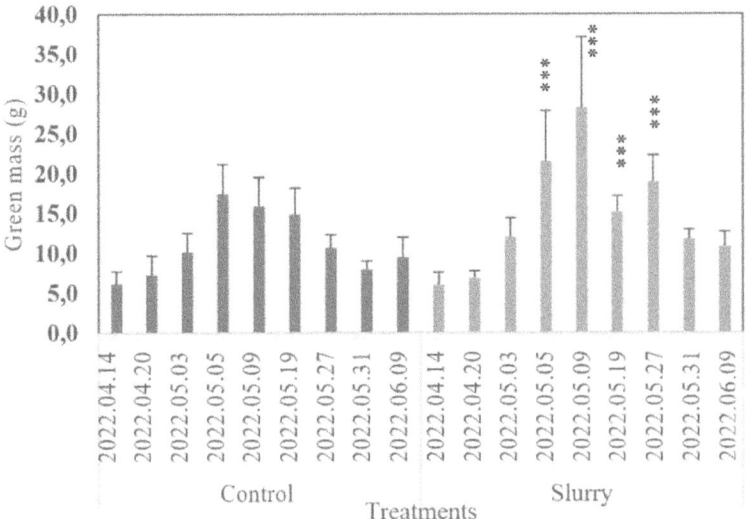

Figure 1. Development of green mass values of winter wheat between 14.04.-09.06.2022
*** $p<0.001$
Source: Authors' own research.

The results are in line with the research of Abubaker et al. (2020) and Sholly et al. (2010), who found that slurry treatment increased the green weight of wheat.

Evaluation of plant length value

The values of winter wheat above plant length showed developed similarly for both treatments (Figure 2). For untreated samples, the lowest values were recorded at the first measurement (29 cm) and the highest at the last measurement (61 cm), which was significantly higher ($p<0.001$) compared to the values at the other measurement times. For the slurry-treated plants, the measured values ranged from 23 to 75 cm. In this case, the last three measurement dates were significantly higher than the other dates. However, it was observed that the difference between the two treatments at the last test date was 14 cm, which was also reflected in the development of the biomass weight.

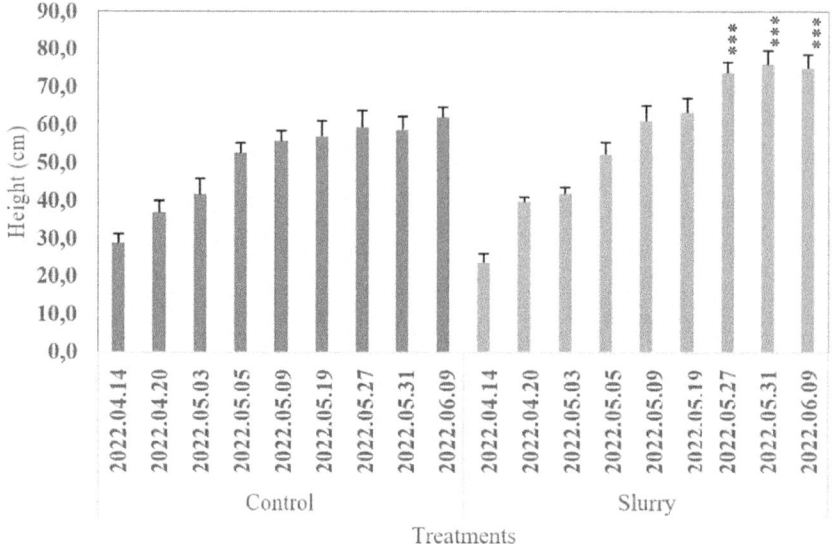

Figure 2. Development of the plant height of winter wheat 14.04.-09.06.2022, at the significance level $p<0.001$ (***)
Source: Authors' own research.

Our results also support the study by Basso et al. (2016), where slurry use positively influenced plant height.

Evaluation of the weight of the grain yield

In view of the exceptional drought conditions in 2022, the main arable crops had significantly lower yields than in the previous years. Barley yields were down by around 10 %, while the largest wheat harvest was 21 % lower. The effect of drought is most evident in the average harvested maize yield, which is down by 57 %. In the case of sunflowers, although the area under sunflower increased, yields fell by 29 %, and the area under rapeseed fell by 31 %. The loss of yields and the uncertainty of supply have led to a significant increase in purchase prices (KSH, 2022).

In our studies, there was a significant difference in grain yield weight between treatments (Figure 3). Application of slurry significantly increased ($p<0.001$) the grain yield weight. A value of 2.5 t/ha was detected for untreated control samples and 4.1 t/ha for the treated area.

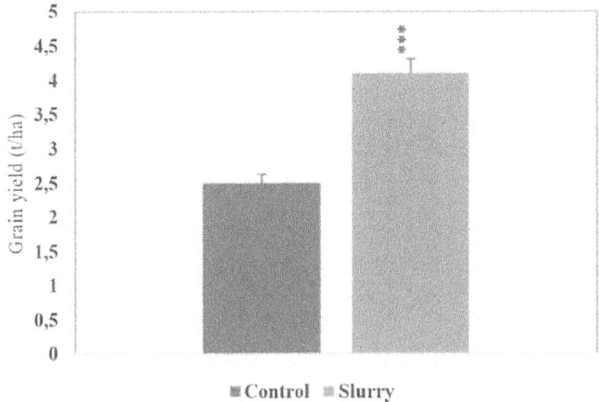

Figure 3. At harvest, the weight of the grain yield of the control and the area treated with pig slurry in t/ha, at the significance level $p<0.001$ (***)
Source: Authors' own research.

Plant physiological measurements

Monitoring the variation in chlorophyll content gives an overall estimation of plantation health. Since we can determine not only the chlorophyll content, but also the nitrogen content of the leaves. There is a linear relationship between nitrogen content and chlorophyll content in plant leaves. In the study, significantly higher ($p<0.05$) values were measured for both CPHLT (Figure 4a) and NDVI (Figure 4b) in the slurry-treated samples.

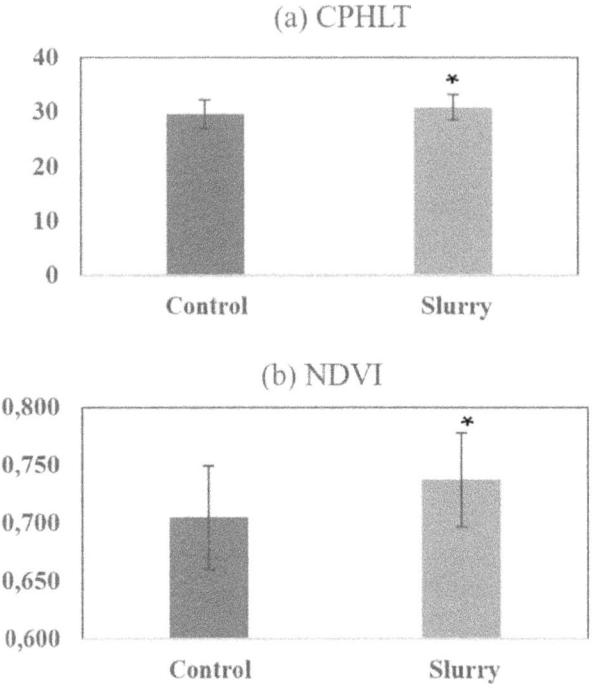

Figure 4 (a, b). Evaluation of the total chlorophyll content of plants (CPHLT) and the value of the normalized vegetation index (NDVI), at the significance level $p<0.001$ (*)
Source: Authors' own research.

The NDVI is a measure that can be calculated from satellite or aerial imagery and provides important information on vegetation health and vegetation change. The higher NDVI value, the darker green the area, i.e. the greater the green mass, indicating a healthy, well supplied with water and nutrients, strong and growing vegetation. The NDVI provides a way to monitor stress effects on plants and to better monitor their health. The results obtained can provide relevant and useful information for agricultural production and land use decisions.

Soil biology – Measuring earthworm abundance

Figure 5 shows the values of earthworm abundance in the experimental areas. There was a significant difference in the number of earthworms between treatments, with significantly more ($p<0.01$) individuals in the slurry treated samples

(74 worms/m²). In the untreated control area, there were only 32 earthworms per 1 m² of area. The same trend was observed in the biomass weight. Biomass values were 6 g/m² in the control area and 13 g/m² in the slurry treated area.

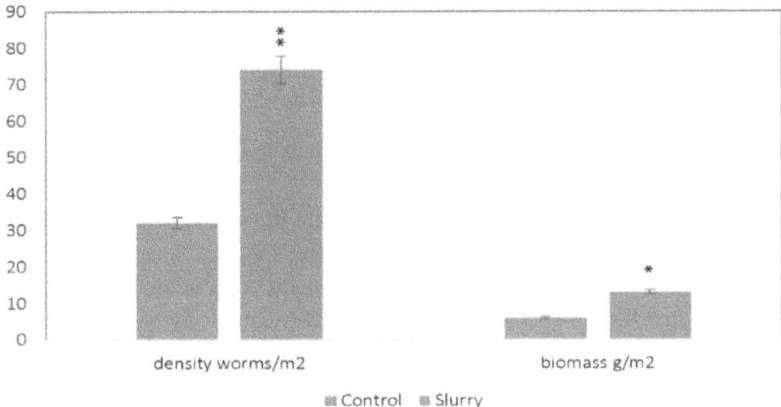

Figure 5. Evaluation of the number and mass of earthworms per 1m² area, at the significance level $p<0.001$ (**)

Source: Authors' own research.

Our studies are also representative of the fact that alternative nutrient supply, such as slurry application, could be key to the future of agriculture. Such solutions can have a positive impact for farmers. Through sustainability, we can reduce the challenges of rising fertilizer prices while maintaining the quality and health of our soils.

Conclusions

This research aims to raise awareness of the importance of more efficient and sustainable slurry use for future food production and environmental protection. The application of innovative agricultural practices and technologies will enable smarter and more sustainable use of organic fertilizers in agriculture.

During our studies, we observed a positive effect of slurry treatment on both phenological parameters and plant physiological values of the plants, which was also reflected in the grain yield. This was also consistent with the values of soil worm abundance.

Earthworms play an important role in agriculture, as their feeding activity improves soil structure and promotes the uniform distribution of nutrients in the soil. Crop management practices such as crop rotation, reduced tillage and the use of organic fertilizers, green manures and slurries can increase the economics and sustainability of production. International studies also confirm that in areas where organic or slurry fertilizers are used in crop production, soil activity increases, with a concomitant increase in earthworm incidence (Unwin and Lewis, 1986), as illustrated in the three areas below:

- Nutrients: Slurry is rich in nutrients such as nitrogen, phosphorus and potassium.
  During the introduction of liquid fertilizers into the soil, earthworms promote a more efficient distribution of nutrients in the soil, which results in a more easily accessible macro-, meso- and microelement content for plants.
- Soil structure: The burrowing activity of earthworms loosens the soil structure and increases soil respiration. In addition, the slurry applied to the soil increases the organic matter content of the soil, which contributes to improving the soil's structure and water retention capacity.
- Microbial activity: The feces of earthworms promote the activity of microorganisms in the soil, which contributes to increasing the biological viability of the soil. Spores in an inactive state survive the digestive processes of the alimentary canal of earthworms and are returned to the soil together with the excrement. Good soil biological activity supports plant vitality and nutrient uptake.

In addition, it is important to note that the benefits of slurry use and earthworm density are highly dependent on local soil and environmental conditions and the method of fertilization. It is also important not to use xenobiotics when fertilizing, which could be harmful to earthworms or soil biodiversity. Overall, it can be stated that, in addition to the applied agrotechnical elements, the soil improvement activity of earthworms is an essential element of successful plant cultivation.

References

Abubaker, J., Ibrahim, N., Alkanami, M., Alaswd, A. and El-Zeadani, H. 2020. Response of winter wheat to the application rate of raw and digested sheep manure alone and supplemented with urea in Libyan desert soil. *Scientific African*, 8, p. e00332. https://doi.org/10.1016/j.sciaf.2020.e00332.

Akinrinde Olubakin, S.O., Omotoso, S.O. and Ahmed, A.A. 2006. Influence of zinc fertilizer, poultry manure and application levels on the performance of sweet corn. *Agricultural Journal*, 1, pp. 96–103.

Balafoutis, A., Beck, B., Fountas, S., Vangeyte, J., Wal, T.V. der Soto, I., Gómez-Barbero, M., Barnes, A. and Eory, V. 2017. Precision agriculture technologies positively contributing to ghg emissions mitigation, farm productivity and economics. *Sustainability*, 9, p. 1339. https://doi.org/10.3390/su9081339.

Basso, C.J., Pinto, M.A.B., Santi, A.L., da Silva, R.F. and da Silva, D.R.O. 2016. Pig slurry as a nutrient source in wheat/corn succession. *Revista Ceres*, 63, pp. 412–418. https://doi.org/10.1590/0034-737X201663030019.

Blouin, M., Hodson, M.E., Delgado, E.A., Baker, G., Brussaard, L., Butt,K.R., Dai, J., Dendooven, L., Peres, G., Tondoh, J.E., Cluzeau, D. and Brun, J.-J. 2013. A review of earthworm impact on soil function and ecosystem services: Earthworm impact on ecosystem services. *Eur J Soil Sci*, 64, pp. 161–182. https://doi.org/10.1111/ejss.12025.

Bocchi, S. and Tano, F. 1994. Effects of cattle manure and components of pig slurry on maize growth and production. *European Journal of Agronomy*, 3, pp. 235–241. https://doi.org/10.1016/S1161-0301(14)80088-9.

Clark, M., Hill, J. and Tilman, D. 2018. The Diet, Health, and Environment Trilemma. *Annual Review of Environment and Resources*, 43, pp. 109–134. https://doi.org/10.1146/annurev-environ-102017-025957.

European Union. 1991. Council Directive 91/676/EEC of 12 December 1991 concerning the protection of waters against pollution caused by nitrates from agricultural sources, OJ L.

Keeler, B.L., Gourevitch, J.D., Polasky, S., Isbell F., Tessum, C.W., Hill, J.D. and Marshall, J.D. 2016. The social costs of nitrogen. *Science Advances*. 2, p. e1600219. https://doi.org/10.1126/sciadv.1600219.

KSH. 2022. *Főbb növénykultúrák terméseredményei.* [online] Available at: <https://www.ksh.hu/s/kiadvanyok/fobb-novenykulturak-termeseredmenyei-2022/index.html> [Accessed 28 August 2023].

Langholtz, M., Davison, B.H., Jager, H.I., Eaton, L., Baskara, L.M., Davis, M. and Brandt, C.C. 2021. Increased nitrogen use efficiency in crop production can provide economic and environmental benefits. *Science of The Total Environment*, 758, p. 143602. https://doi.org/10.1016/j.scitotenv.2020.143602.

Motavalli, P.P., Kelling, K.A., Syverud, T.D. and Wolkowski, R.P. 1993. Interaction of Manure and Nitrogen or Starter Fertilizer in Northern Corn Production. *Journal of Production Agriculture*, 6, pp. 191–194. https://doi.org/10.2134/jpa1993.0191.

Murchie, A.K., Blackshaw, R.P., Gordon, A.W. and Christie, P. 2015. Responses of earthworm species to long-term applications of slurry. *Applied Soil Ecology*, 96, pp. 60–67. https://doi.org/10.1016/j.apsoil.2015.07.005.

Nevens, F., Reheul, D. 2005. Agronomical and environmental evaluation of a long-term experiment with cattle slurry and supplemental inorganic N applications in silage maize. *European Journal of Agronomy*, 22, pp. 349–361. https://doi.org/10.1016/j.eja.2004.05.003.

Paoletti, M.G. 1999. The role of earthworms for assessment of sustainability and as bioindicators. *Agriculture, Ecosystems & Environment*, 74, pp. 37–155. https://doi.org/10.1016/S0167-8809(99)00034-1.

Pedersen, S.M., Fountas, S., Blackmore, B.S., Gylling, M. and Pedersen, J.L. 2004. Adoption and perspectives of precision farming in Denmark. *Acta Agriculturae Scandinavica, Section B - Soil & Plant Science*, 54, pp. 2–8. https://doi.org/10.1080/09064710310019757.

Pérès, G., Vandenbulcke, F., Guernion, M., Hedde, M., Beguiristain, T., Douay, F., Houot, S., Piron, D., Richard, A., Bispo, A., Grand, C., Galsomies, L. and Cluzeau, D. 2011. Earthworm indicators as tools for soil monitoring, characterization and risk assessment. An example from the national Bioindicator programme (France). *Pedobiologia*, 54, pp. 77–87. https://doi.org/10.1016/j.pedobi.2011.09.015

Sánchez, M. and González, J.L. 2005. The fertilizer value of pig slurry. I. Values depending on the type of operation. *Bioresource Technology*, 96, pp. 1117–1123. https://doi.org/10.1016/j.biortech.2004.10.002.

Sharma, B., Sarkar, A., Singh, P. and Singh, R.P. 2017. Agricultural utilization of biosolids: A review on potential effects on soil and plant grown. *Waste Management*, 64, pp. 117–132. https://doi.org/10.1016/j.wasman.2017.03.002.

Sheperd, G., Stagnari, F., Pisante, M. and Benites, J. 2008. *Visual soil assessment. Field guides. annual crops.* Rome: FAO.

Sholly, D.M., Richert, B.T., Sutton, A.L. and Joern, B.C. 2010. Effects of nitrogen and phosphorus application from swine manure on winter wheat growth and nutrient utilization. *Communications in Soil Science and Plant Analysis*, 41, pp. 1797–1815. https://doi.org/10.1080/00103624.2010.492437.

Singh, J. 2018. 3-Role of earthworm in sustainable agriculture. In: C.M. Galanakis, ed. *Sustainable food systems from agriculture to industry.* Academic Press, pp. 83–122. https://doi.org/10.1016/B978-0-12-811935-8.00003-2.

Unwin, R.J. and Lewis S. 1986. The effect upon earthworm populations of very large applications of pig slurry to grassland. *Agricultural Wastes*, 16, pp. 67–73. https://doi.org/10.1016/0141-4607(86)90037-5.

Yagüe, M., Bosch-Serra, À. And Jaume, B. 2012. Measurement and estimation of the fertiliser value of pig slurry by physicochemical models: Usefulness and constraints. *Biosystems Engineering*, 111, pp. 206–216. https://doi.org/10.1016/j.biosystemseng.2011.11.013.

Zamanian, K., Zarebanadkouki, M. and Kuzyakov, Y. 2018. Nitrogen fertilization raises CO2 efflux from inorganic carbon: A global assessment. *Global Change Biology*, 24, pp. 2810–2817. https://doi.org/10.1111/gcb.14148.

Zebarth, B.J., Paul, J.W., Schmidt, O. and McDougall, R. 1996. Influence of the time and rate of liquid-manure application on yield and nitrogen utilization of silage corn in south coastal British Columbia. *Canadian Journal of Soil Science*, 76, pp. 153–164. https://doi.org/10.4141/cjss96-022.

Zhang, X., Davidson, E.A., Mauzerall, D.L., Searchinger, T.D., Dumas, P. and Shen, Y. 2015. Managing nitrogen for sustainable development. *Nature*, 528, pp. 51–59. https://doi.org/10.1038/nature15743.

Bashiru Dahiru Magaji, Yusuf Usman Oladimeji,
Ado Yakubu, Henry Egwuma, Benjamin Ahmed,
Abubakar Abdullahi Hassan, Miroslav Raicov

Risk attitudes of micro, small and medium agribusiness enterprises in Nigeria

Abstract: *The study determines the risk attitudes of micro, small and medium-scale agribusiness entrepreneurs (MSMAEs) in North-west, Nigeria. A sample of 334 MSMAEs was selected using multi-stage sampling procedure. Data were collected with the use of Open Data Kit (ODK). Both descriptive statistics and principal component analysis (PCA) were employed to describe and analyze the risk attitudes of MSMAEs. The result of the analysis revealed that there are three attitudes towards risk among the respondents. Specifically, the respondents were found to be risk lovers, neutral and averse. However, it was revealed that the majority of the respondents were risk averse, irrespective of their business status. In particular, 54% of the respondents in micro agribusiness, 65% in small agribusiness, and 49% in medium agribusiness were risk averse. There was a statistically significant difference in the proportions of risk lovers among the males across the three groups of micro agribusinesses ($\chi_1^2(2)=8.56$, $p<0.05$; $\chi_1^2(2)=25.73$, $p<0.01$). This indicates that the majority of risk lovers among males were in micro agribusinesses while the minority was in small agribusinesses. However, there was no statistically significant difference in the proportions of risk-neutral and risk averse among males across the three groups of micro agribusinesses. Agribusiness enterprises' risk attitudes is very crucial in this study, since they will have a greater influence on business owners'/managers' decisions to execute risk management strategies.*

Keywords: risk lover, risk-neutral, risk-averse, entrepreneurs, Nigeria

Introduction

Micro, small and medium enterprises (MSMEs) have received considerable attention from policymakers across the world in recent years due to the critical role they play in creating job opportunities, ensuring food security, reducing poverty and sustaining the overall growth of economies (Andrei et al., 2021; Ciolac et al., 2022; Raicov and Feher, 2017; Suguna et al., 2022). In Nigeria, for example, MSMEs constitute a significant part of the economy accounting for 96 % to the total number of businesses and about 50 % to the GDP (PWC, 2020). MSMEs have been investigated as the major catalyst of innovation, industrialization as well as inclusive and sustainable economic growth (Li and Rama, 2015; Gherghina et al., 2020; Endris and Kassegn, 2022). In particular, MSMEs directly related to the agribusiness sector have been identified as a key vehicle for economic growth and the bedrock for export-led industrialization (Ali, 2016; Ng'ang'a and Gichira, 2017; Matkovski et al., 2022). More so, with economic progress the provision of food is achieved through longer value chains and thus agribusiness MSMEs become critical in linking farmers and consumers.

At the same time, agribusiness MSMEs face multiple challenges such as vulnerability to economic shocks, unstable policy environment, lack of funding, complex tax process and excessive regulatory burden (Prioteasa et al., 2020). This shows that MSMEs are dogged by various risks which constitute a major threat to their survival especially with little resources at their disposal. The risk sources to agribusiness enterprises have been categorized into social, market, political, financial, production and foreign exchange risk (Abrudan et al., 2022; Asravo and Sarpong, 2022; Nto, Mbanasor and Nwaru, 2011; Sekumade and Ogunro, 2013). These risks influence decisions made by entrepreneurs concerning appropriate strategies to manage the risks they encounter.

Sulewski et al. (2020) used the income variance approach to analyze farmers' production decision behavior under risk and categorized them as: (i) Risk-loving/preferring/taking: the farmer is open to taking the risk of performing above expectations while being cognizant of the chances of performing below expectations; (ii) Risk-neutral: the farmer is indifferent between end results that are predictable and those that are not with the same estimated returns; and (iii) Risk-averse: The farmer is regarded as risk-averse if he favors conditions where a given income is certain to one where similar or same estimated income is presented albeit with uncertainty.

According to economic theory (Danso et al., 2016), risk-averse entrepreneurs may be prepared to settle for a reduced profit in exchange for a lower exposure to risk, whereas entrepreneurs who are more willing to take risks may be rewarded with higher estimated earnings. In agribusiness settings, there are several ways the risk attitudes of entrepreneurs may affect the growth and success of firms (Lien et al., 2022). For instance, corporate strategy, commodity category offered, and price may all exhibit signs of a risk-taking mindset that results in profitability. To illustrate, a bread vendor may choose to market a bread that is commonly accepted and enjoyed (Lammers, Willebrands and Hartog, 2010). The ease of getting fresh stock, coupled with persistent demand and more secure pricing of such commodities, make it a low-risk business approach. It could be riskier to trade bread, which is seldom consumed but yield a profitable return when sold, raising the likelihood that the entrepreneur will be left with excess unsold inventory that goes bad. An alternate approach would be to vary the stock among several types of loaves, and effectively distribute the risks (Lammers,

Willebrands and Hartog, 2010). In this light, Boermans and Willebrands (2017) opined that the emergence of risk attitude and choice criteria are therefore linked to the entrepreneur's risk perception and risk propensity and these culminate in risk management.

Several studies have analyzed risk attitudes in agriculture (Musyoki et al., 2022; Ozsayin, 2022; Shah and Alharthi, 2022). Oladimeji et al. (2019a, 2019b) analyzed risks preferences and the factors that influence risk preferences, using the three risk categories as dependent variables. The results of Oladimeji et al. (2019a) showed that out of a total of 123 respondents in the concrete pond system, 52 % prefer risk, 34 % are risk neutral and 16 % are risk averse. This result indicates a better acceptance of risk compared to 154 respondents in the earthen pond system, where 21.4 % of respondents are risk-preferring, 59.7 % are risk-neutral, and 18.8 % are risk-averse. Shadbolt and Olubode-Awosola (2016) investigated farmers in Hamilton, New Zealand, classified according to their risk attitude. Based on how they performed across six years of volatility, the study determined that farmers who seek risk are best regarded as gamblers. The most profitable farmers' group was risk-neutral, had a stronger market focus, and was adept at handling considerable debt. They fulfilled the larger definition of entrepreneur since they had a constructive attitude towards change and the capacity to properly adjust to changing surroundings. With high cash returns and retained earnings, the risk-averse group surpassed the risk-seeking group. Farmers cannot be presumed to be effective change agents based just on their risk-taking attitude and belief in their risk-handling capacity; rather, they are those whose outcomes demonstrate that they are successfully taking a risk, have great entrepreneurial skills, and operate cost-effective farm enterprises.

However, although the risk perception and attitude of farmers have been demonstrated by these studies, there is little information on the risk attitude of MSMEs agribusiness entrepreneurs in Nigeria, which presents a challenge for policymakers in developing appropriate risk management strategies. Therefore, this study is an attempt to fill the gap by analyzing the risk attitude of agribusiness MSMEs.

Literature review

Agribusiness refers to the subdivision of agriculture that encompasses the production, manufacturing and delivery of farm inputs, equipment and supplies in one way and the processing, storage and distribution of agricultural goods in another dimension. This implies that the whole agricultural production, processing, delivery and consumption ranges from farm inputs all-encompassing wood producers, furniture manufacturers, food processors, food packers, food transporters and food marketing companies to restaurants and shopping malls. It entails input industries for agricultural production, post-farm gate industries; including the commodity processing, food manufacturing, and distribution industries and third-party firms that facilitate agribusiness operations. In other words, agribusiness includes the entire the operation involved in the manufacture and distribution of off-farm commodities and other items made from them (Igbokwume, Essien and Agunnah, 2015). In Gandhi's (2014) conception, agribusiness is defined as a science that coordinates the supply of agricultural production inputs and subsequently the production, processing and distribution of food and fibre.

Risk is defined as a probable negative shock to wealth, savings, or success of investments in the agricultural industry that may emanate from some present process or future event. Risk has been defined as the result of physically distinct hazards interacting with exposed systems taking into consideration the properties of the systems, such as their sensitivity or social vulnerability. Risk can also be described as the combination of an event, its likelihood and its consequences. The perceptive approach considers risk as a set of all destructive consequences that are believed to be possible by a person who has evidence about the frequency, severity, and variability of the effects (Ndamani and Watanabe, 2017).

The risk could also refer to the chance that some unfavorable events will occur and in this respect risk describes a situation where there is not just one possible outcome of returns to investment but an array of potential returns. Risk could therefore be viewed as uncertainty of financial loss (Kagwathi et al., 2014). This research work focuses on risks to agribusiness investors rather than risks to smallholder farmers, although, the two are not independent in some cases.

Risk management can be defined as an organized approach to identify possible or probable financial harm and take steps to minimize the financial impact to acceptable levels (Vaughan, 2016). Risk Management is a process through which risks can be measured, exploited, governed, financed and monitored from all sources by business organizations operating in any sector of the economy to increase the value of shareholders or owners. The enterprise risk management views risks as opportunities exemplified in the overall business strategy of an enterprise which must be identified, measured, responded to, prevented and monitored (Kehinde et al., 2017).

Risk sources to agribusiness enterprises, generally, can be grouped into social, market, institutional, financial, production and foreign exchange risk. Social risk is suggestive that the risks or hazards have their origin in man. The risk could be due to fire outbreak, burglary or theft, kidnapping of investors or workers for ransom, embezzlement, strike, civil commotion and changes in social structure e.g., divorce and dissolution of partnership which can lead to an unexpected decline in the efficient operation of enterprise and loss of useful man-day (Akinola, 2015). Market risk arises due to fluctuation in input and output prices which may occur when the farmer has committed to produce. It can also be a result lower offer prices or entry of big external players. It includes risks that result from unpredictable exchange rates. Market and production risks are linked. In developing countries, it is observed that local rural markets are isolated from national and international markets due to the high cost of transportation and marketing of products. Local prices are volatile driven by yield fluctuations, production uncertainty and local demand. Therefore, farmers face production and market risks that are correlated with the level of regional market integration. Market variability makes planning difficult by introducing uncertainties that, in turn, lead to an inefficient allocation of resources. (Akinola, 2015).

Scarborough (2011) and (Kazungu et al., 2014) define and classify micro, small and medium enterprises (MSMEs) into formal, informal and survival enterprises. These classifications are based on several criteria, such as the degree of informality, turnover, capital investment and number of employees. Formal enterprises are enterprises that comply with regulations, have relatively (and often absolutely) high levels of human capital and are integrated into the structure of the formal economy (as opposed to informal enterprises). In most countries, criteria for defining Micro Small and Medium Scale Enterprises (MSMEs) include indicators such as annual turnover and employment level. In Nigeria, the definition of MSMEs is according to the Small and Medium Enterprises Investment Scheme (SMIEIS) which states that an MSME is any enterprise with a maximum asset base of ₦200 million excluding land and working capital and with a several staff employed not less than 10 or not more than 300 (Agwu and Emeti, 2014).

Conceptual framework

The conceptual framework shows the interrelationships of the study variables informed by the enterprise risk management framework guided by the theory of enterprise risk management (Nocco and Stulz, 2006; Ng'ang'a, Muthusi and Nassiuma, 2015). As defined by Nocco and Stulz (2006), enterprise risk management refers to an approach under which all risks are viewed together within a coordinated and strategic framework, assessed and measured to mitigate or exploit the opportunities behind the risk put in place. The authors further argue that enterprise risk management creates value, because it strengthens the firm's ability to carry out its strategic plan, by minimizing costs and maximizing profitability of the organization. Hence, the conceptual model in Figure 1 assumes that every organization micro, small and medium operates in an environment with risks and uncertainties of different dimensions. Therefore, the management strategies adopted play a key role in determining the enterprise's performance. The level of application of risk management strategies between micro and small-scale enterprises and that of medium-scale enterprises vary significantly.

Micro, small and medium-level risks that characterise agribusiness can be categorized based on their origin as; economic risks; social risks; Political risks and environmental risks. Variations however arise in the way these categories of agribusiness enterprises manage the risks facing them based on their size (dependent variables). Hence, it depends on agribusiness entrepreneurs' risk attitudes or the way they view and characterises a risk based on uncertainty, and loss levels, the measures put in place can be categorized as: risk retention, risk reduction, risk transfer or risk avoidance. Thus, decisions on risk management options are based on enterprises' risk attitudes and ratings (Ng'ang'a, Muthusi and Nassiuma, 2015).

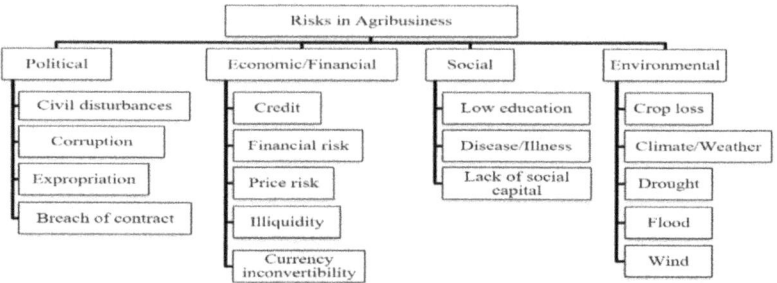

Figure 1. Risk to Agribusiness
Source: Authors' own research based on Theuvsen, 2013 and Oladimeji et al., 2019b.

Methodology

The study was conducted in the North-West geopolitical zone of Nigeria. The zone comprises Jigawa, Kaduna, Kano, Katsina, Kebbi, Sokoto and Zamfara States (Figure 2). It has a projected population of 55,820,957 people in 2022 at a growth rate of 3.2 % per annum based on the National Bureau of Statistics (NBS, 2023). The zone is characterized by a tropical climate with temperature varying at different times. High temperature is normally recorded between April and

September with the daily minimum and maximum temperatures of 14º and 39º Celsius respectively. Agriculture is considered the major economic activity of the zone with over 80 % of the population found in the rural areas and predominantly engaged in farming and animal husbandry. Also, trade and commerce are undertaken on micro, small and medium scale, especially in agricultural and other consumer goods (NBS, 2023).

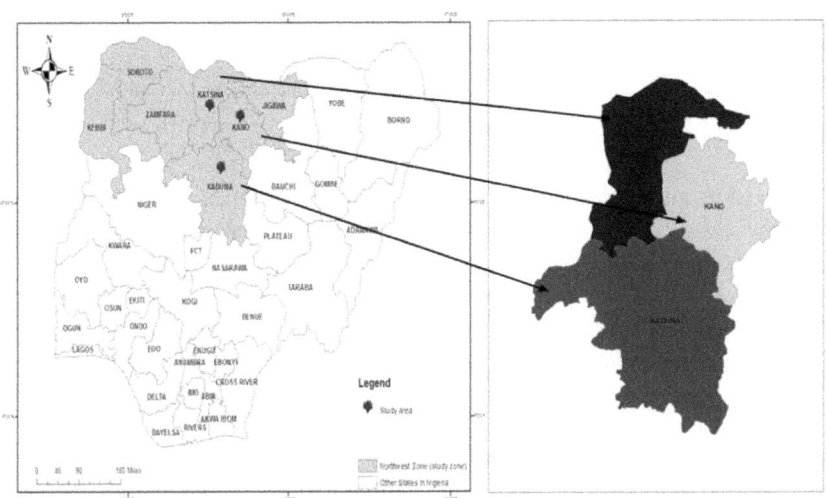

Figure 2. Map of Nigeria showing the study area
Source: Adapted and modified from NBS, 2023.

A three-stage sampling procedure was adopted for the purpose of this study. The first stage involved a selection of three out of seven States of Northwest Nigeria using a random sampling technique. These are Kano, Kaduna and Katsina States. The second stage involved purposive selection of all agribusiness firms from the three States. Thereafter, a stratified sampling technique was used to group the sampling unit based on agribusiness membership status: sole proprietor, cooperative and partnership. All the registered agribusiness enterprises were selected from each stratum (Table 1). Finally, the snowball sampling technique was also employed to identify a sampling unit of 181 who were not registered with any of the ministries and agencies in the States. This was achieved using lists of the micro, small and medium-scale enterprise owners obtained from the institutions to serve as referrer to them. The registered MSMEs sample size is 334. The total sample size was therefore 515.

Table 1. Sampling Frame and Size Selection Plan of the Study

Form of Organization	Kaduna	Kano	Katsina	Pooled
Sole proprietorship	107	123	81	311
Cooperative	29	36	46	111
Partnership	27	25	41	93
Total	163	184	168	515

Source: SMEDAN, 2017 and SMCI, 2018.

Data were collected using primary and secondary sources. This was achieved with the aid of well-structured questionnaire. Principal component analysis (PCA) was used to achieve the objective of the study. In achieving this objective, the first task was to identify the most important risks faced by the respondents. The next task involved the estimation of the risk based on political, economic, social and environmental risk scores. The Political Risk Score (PRS), Economic Risk Score (ERS), Social Risk Score (SRS) and Environmental Risk Score (ENRS) were estimated using PCA technique as follows:

$$PRS = \beta_1 Q_1 + \beta_2 Q_2 + \beta_3 Q_3 \tag{1}$$

$$ERS = \beta_4 Q_4 + \beta_5 Q_5 + \beta_6 Q_6 + \beta_7 Q_7 + \beta_8 Q_8 + \beta_9 Q_9 + \beta_{10} Q_{10} \tag{2}$$

$$SRS = \beta_{11} Q_{11} + \beta_{12} Q_{12} + \beta_{13} Q_{13} + \beta_{14} Q_{14} + \beta_{15} Q_{15} \tag{3}$$

$$ENRS = \beta_{16} Q_{16} + \beta_{17} Q_{17} \tag{4}$$

Where: Q_1 through Q_{17} were measured in a 5-point Likert scale such as: SA = Strongly Agreed (1), AD = Agreed (2), U = Undecided (3), D = Disagreed (4) and SD = Strongly Disagreed (5).

$\beta_1 - \beta_{17}$ = Weight/factor loading of the $Q_1 - Q_{17}$ factors.

The determination of the risk attitudes of micro, small and medium agribusiness enterprises was carried out in two steps as follows:

Step 1: Estimation of risk level

This was achieved by estimating the following equation using PCA:

$$RL = PRS + ERS + SRS + ERS \qquad (5)$$

Where: RL = Risk level; PRS = Political Risk Score; ERS = Economic Risk Score; SRS = Social Risk Score and ENRS = Environmental Risk Score.

Step 2: Estimation of risk attitude

The following diagram shows the estimation procedure that enabled the categorization of the respondents into risk averter, risk neutral and risk lover. The estimated RL in step 1 will be the defined scale on which the categorization was made. This is as follows:

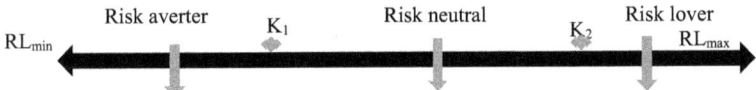

Where: RL_{min} = Minimum Risk Level; RL_{max} = Maximum Risk Level; K_1 = Lower threshold value; K_2 = Upper threshold value.

Risk averter = 1 if $RL_{min} \leq RL < K_1$, 0 = otherwise
Risk neutral = 1 if $K_1 \leq RL < K_2$, 0 = otherwise
Risk lover = 1 if $K_2 \leq RL \leq RLmax$, 0 = otherwise.

Results and discussion

Risk Attitudes of Agribusiness entrepreneurs

This section described the risk attitudes exhibited by micro, small, and medium agribusinesses in Northwest Nigeria. Specifically, results in Table 2 presented the frequency distribution of the respondents and was first disaggregated in terms of their sex which were further dichotomized to uncover potential attitude differentials of the two demographic characteristics across the categories of agribusinesses. Also, the distribution of respondents across the three groups of agribusinesses is based on the three levels of their risk attitudes: risk lover, neutral, and averse. Furthermore, in an attempt to gane an insight into the observed significant differences in risk attitudes that existed among the three groups of agribusinesses, the non-parametric Kruskal Wallis chi-square test with and without ties were equally conducted.

Distribution of agribusiness entrepreneurs' risk attitudes by sex

The results of the distribution of risk attitudes exhibited by micro, small, and medium agribusinesses in Northwest Nigeria are presented in Table 2. In terms of risk lover male respondents, 46% were in micro agribusiness, 17% were in small agribusiness, and 37% were in medium agribusiness. There was a statistically significant difference in the proportions of risk lovers among the males across the three groups of micro agribusinesses ($\chi_1^2(2)=8.56$, $p<0.05$; $\chi_1^2(2)=25.73$, $p<0.01$) (Table 2). This indicates that the majority of risk lovers among males were in micro agribusinesses while the minority was in small agribusinesses. For the proportion of risk-neutral among the males, 33% were in micro agribusiness, 55% were in small agribusiness, and 12% were in medium agribusiness. There was no statistically significant difference in the proportions of risk-neutral among males across the three groups of micro agribusiness ($\chi_1^2(2)=1.51$, $p>0.1$; $\chi_2^2(2)=2.27$, $p>0.1$).

Table 2. Frequency distribution of the respondents' risk attitudes based on sex across agribusiness status

	Business status								
	Micro (n_1= 190)		Small (n_2= 190)		Medium (n_3=69)		Pooled (n=515)		$\chi_1^2(2)$	$\chi_1^2(2)$
Variable	Freq	%	Freq	%	Freq	%	Freq	%		
Male										
Risk lover	19	46	7	17	15	37	41	100	8.56**	25.73***
Risk neutral	35	33	59	55	13	12	107	100	1.51	2.27
Risk averse	60	34	94	54	21	12	175	100	2.96	3.97
Total	114	35	160	50	49	15	323	100		
Female										
Risk lover	15	71	2	10	4	19	21	100	4.58	15.41***
Risk neutral	18	42	22	51	3	7	43	100	0.42	0.80
Risk averse	41	33	70	56	13	10	124	100	4.49	6.67**
Total	74	39	94	50	20	11	188	100		

***<0.01; **<0.05; *<0.1; $\chi_1^2(2)$ and $\chi_1^2(2)$ = Kruskal Wallis Chi-square test without and with ties and 2 degree of freedom, respectively.

Similarly, for the proportions of risk-averse among males, 34% were in micro agribusiness, 54% were in small agribusiness, and 12% were in medium agribusiness. There was no statistically significant difference in the proportions of risk-neutral among males across the three groups of micro agribusiness ($\chi_1^2(2)=2.96$, $p>0.1$; $\chi_2^2(2)=3.97$, $p>0.1$). Similarly, Boakye (2017) observed that male entrepreneurs were more risk-seeking compared to females.

For the proportions of risk lovers among the females, 71% were in micro agribusiness, 10% were in small agribusiness, and 19% were in medium agribusiness. There was a statistically significant difference in the proportions of the risk lovers among the females across micro, medium and medium agribusinesses ($\chi_1^2(2)=4.58$, $p<0.05$; $\chi_2^2(2)=15.41$, $p<0.01$). The implication is that most of the females who were risk lovers were in micro agribusinesses while the minority were in small agribusinesses. For the proportions of risk-neutral, 42% of the females were in micro agribusiness, 51% in small agribusiness, and 7% in medium agribusiness. There was no statistically significant difference in the proportions of those who were risk-neutral among the females across the three groups of agribusiness ($\chi_1^2(2)=0.42$, $p>0.1$; $\chi_2^2(2)=0.80$, $p>0.1$).

Out of the 124 females that were risk averse, 33% were in micro agribusiness, 56% in small agribusiness, and 10% in medium agribusiness. Hence, there was a statistically significant difference in the proportions of females that were risk averse across micro, small, and medium agribusiness ($\chi_1^2(2)=4.49$, $p>0.1$; $\chi_2^2(2)=6.67$, $p<0.05$). This implies that the majority of the females that were risk averse were in the small agribusiness while the minority were in the medium agribusiness. Boakye (2017), Lammers, Willebrands and Hartog (2010), and Boermans and Willebrands (2017) show consistency in this regard.

Distribution of agribusiness entrepreneurs' risk attitudes by State

The distribution of agribusiness in terms of State is presented in Table 3. Out of the 33 respondents that were risk lovers in Kano State, 36% were in micro agribusiness, 21% in small agribusiness, and 42% in medium agribusiness. There was no statistically significant difference in the proportions of respondents in Kano State that were risk lovers ($\chi_1^2(2)=0.44$, $p>0.1$; $\chi_2^2(2)=0.13$, $p>0.1$).

Out of the 56 respondents that were risk neutral in Kano State, 34% were in micro agribusiness, 45% in small agribusiness, and 21% in medium agribusiness. There was a statistically significant difference in the proportions of the respondents in Kano State that were risk-neutral across the three groups of agribusiness ($\chi_1^2(2)=5.84$, $p<0.1$; $\chi_2^2(2)=9.74$, $p<0.01$). This implies that the majority of the respondents in Kano State who were risk-neutral were in small agribusiness.

Hence, out of the 114 respondents that were risk averse in Kano State, 50% were in micro agribusiness, 23% in small agribusiness, and 27% in medium agribusiness. There was a statistically significant difference in the proportions of the respondents that were risk averse across the three groups of agribusiness ($\chi_1^2(2)=$ 4.22, mp>0.1; $\chi_2^2(2)=$ 5.72, p<0.1). This indicates that the majority of the respondents that were risk averse in Kano State were in micro agribusiness.

The result also shows that out of the 24 respondents that were risk lovers in Kaduna State, 92% were in micro agribusiness and 8% in medium agribusiness. There was a statistically significant difference in the proportions of the respondents that were risk lovers across the three groups of agribusiness ($\chi_1^2(2)=5.36$, $p<0.1$; $\chi_2^2(2)=13.3$, $p<0.01$). This shows that the majority of the respondents who were risk lovers in Kaduna State were in micro agribusiness. Out of the 34 respondents in Kaduna State that were risk-neutral, 89% were in micro agribusiness, 5% were in small and medium agribusiness. There was a statistically significant difference in the proportions of the respondents that were risk neutral in Kaduna State across the three groups of agribusiness ($\chi_1^2(2)=8.41$, $p<0.05$; $\chi_2^2(2)=14.82$, $p<0.01$). This reveals that the majority of the respondents that were risk-neutral in Kaduna State were in micro agribusiness.

Table 3. Frequency distribution of the respondents' risk attitudes based on state across agribusiness status

	Business status						Pooled (n=515)		Chi-square test	
	Micro (n_1=190)		Small (n_2=190)		Medium (n_3=69)				$\chi_1^2(2)$	$\chi_1^2(2)$
Variable	Freq	%	Freq	%	Freq	%	Freq	%		
Kano										
Risk lover	12	36	7	21	14	42	33	100	0.44	0.13
Risk neutral	19	34	25	45	12	21	56	100	5.84*	9.74***
Risk averse	57	50	26	23	31	27	114	100	4.22	5.72*
Total	88	43	58	29	57	28	203	100		
Kaduna										
Risk lover	22	92	0	0	2	8	24	100	5.36*	13.3***
Risk neutral	34	89	2	5	2	5	38	100	8.41**	14.82***
Risk averse	44	50	43	49	1	1	88	100	26.89***	36.96***
Total	100	67	45	30	5	3	150	100		
Katsina										
Risk lover	0	0	2	40	3	60	5	100	3.44*	37.42***
Risk neutral	0	0	54	96	2	4	56	100	0.10	0.15

Table 3. Continued

Risk averse	0	0	95	98	2	2	97	100	2.35	3.31*
Total	0	0	151	96	7	4	158	100		
Pooled										
Risk lover	34	55	9	15	19	30	62	100	12.50***	39.10***
Risk neutral	53	35	81	54	16	11	150	100	1.35	2.17
Risk averse	101	34	164	55	34	11	299	100	5.84*	8.02**
Total	188	37	254	50	69	14	511	100		

Source: Survey Data (2019)
***<0.01; **<0.05; *<0.1; $\chi_1^2(2)$ and $\chi_2^2(2)$ = Kruskal Wallis Chi-square test without and with ties and 2 degree of freedom, respectively.

Furthermore, of the 88 respondents in Kaduna State that were risk averse, 50% were in micro agribusiness, 49% were in small agribusiness, and 1% were in medium agribusiness. There was a statistically significant difference in the proportions of the respondents that were risk averse in Kaduna across the three groups of agribusiness ($\chi_1^2(2)=26.89$, $p<0.01$; $\chi_2^2(2)=36.96$, $p<0.01$), implying that majority of the respondents that were risk averse in Kaduna State were in micro agribusiness, followed by small agribusiness and finally medium agribusiness. Out of the 5 risk lovers in Katsina Sate, 40% were in small agribusiness and 50% were in medium agribusiness. There was a statistically significant difference in the proportions of the respondents who were risk lovers in Katsina across the three groups of agribusiness ($\chi_1^2(2)=3.44$, $p<0.1$; $\chi_2^2(2)=37.42$, $p<0.01$). This indicates that most of the risk lovers in Katsina State were in medium agribusiness.

Similarly, of the 55 risk-neutral respondents in Katsina State, 96% were in small agribusiness while the remaining 4% were in medium agribusiness. Despite the apparent proportional difference, it was not statistically significant ($\chi_1^2(2)=0.10$, $p>0.1$; $\chi_2^2(2)=0.15$, $p>0.1$). Out of the 97 risk-averse respondents in Katsina State, 98% were in small agribusiness and 2% were in medium agribusiness. There was a statistically significant difference in the proportions of the respondents that were risk averse in Katsina across the three groups of agribusiness ($\chi_1^2(2)=2.35$, $p>0.1$; $\chi_2^2(2)= 3.31$, $p<0.1$). This implied that virtually all risk-averse respondents in Katsina State were in small agribusiness.

Based on the pooled data, it was found that, of the 62 risk lovers (takers), 55% were in micro agribusiness, 15% in small agribusiness, and 30% in medium agribusiness (Table 3). There was a statistically significant difference in the proportions of the agribusiness enterprises that were risk lovers as a whole across the three groups of agribusinesses ($\chi_1^2(2)=12.5$, $p<0.01$; $\chi_2^2(2)=39.1$, $p<0.01$). This indicates that the majority of the risk lovers were in micro agribusiness, followed by medium agribusiness and small agribusiness. The predominance of risk lovers in micro agribusiness enterprises could be a result of their willingness to expand their businesses. Hence, the majority of these micro agribusiness entrepreneurs are newcomers with new business set up, and were envisaging more gains and thus willing to commit additional capital into their businesses.

Out of the 150 risk-neutral respondents, 35% were in micro agribusiness, 54% were in small agribusiness, and 11% were in medium agribusiness. However, there was no statistically significant difference in the proportions of risk-neutral across the groups of agribusinesses. Out of the 299 risk-averse, 34% were in micro agribusiness, 55% were in small agribusiness, and 11% were in medium agribusiness. There was a statistically significant difference in the proportions of risk-averse respondents as a whole across the three groups of agribusiness ($\chi_1^2(2)=5.84$, $p<0.1$; $\chi_2^2(2)=8.02$, $p<0.05$). This means that the majority of the risk-averse respondents were in small agribusiness while the minority were in medium agribusiness. This could be based on the fact that most of the small agribusiness respondents in the area are a start-up company where cash flow and funds are typically tight, so they are more likely to be averse to risk to protect the financial viability of the start-up organization. As in the case of medium enterprises, it could be a result of long existence, and high risks including becoming obsolete, stagnant, or too conservative with their business plan. The result can be aligned with the findings of Block, Sandner and Spiegel (2009), who discovered that there are strong differences of the risk attitude within the group of entrepreneurs.

Agribusiness enterprises' risk attitudes is very crucial in this study, since they will have a greater influence on business owners'/managers' decisions to execute risk management strategies. If an investor is particularly risk-averse, he or she will be hesitant to take on numerous risks and will strive to lessen, transfer, or even eliminate as many risks as possible. Conversely, a risk-seeking agribusiness entrepreneur will intentionally take a substantial portion of the business's risks and will mostly refrain from constantly executing tactics that may decrease, prevent, or transfer risks. Hence, the exposure of agribusiness enterprises to risks can be varied, depending on agribusiness entrepreneurs' risk attitudes (Oladimeji et al., 2019a, 2019b). However, Shadbolt and Olubode-Awosola (2016) discovered that the most successful businesses were risk-averse, had a solid business focus, and were adept at handling high amounts of debt. They satisfied the larger definition of an entrepreneur since they had a constructive attitude towards change and the capacity to successfully adjust to changing situations. With high cash outcomes and retained earnings, the risk-averse group beat the risk-seeking group. On the contrary, Zurriaga-Carda, Kageyama and Akai (2016) reported that risk averseness has a strong negative effect on entrepreneurial intentions.

Conclusions

In light of the outcome of this research, it can be inferred that micro, small and medium-scales agribusiness entrepreneurs were risk lovers, neutral and averse. However, it was revealed that the majority of the respondents were risk averse, irrespective of their business status. In particular, the majority of risk-averse respondents were in small agribusiness followed by micro with minority in medium agribusiness. Also, there was a statistically significant difference in risk attitude across micro, small, and medium agribusiness enterprises in terms of sex and State of the respondents.

References

Abrudan, D.B., Daianu, D.C., Maticiuc, M.D., Rafi, N. and Kalyar, M.N. 2022. Strategic leadership, environmental uncertainty, and supply chain risk: An empirical investigation of the agribusiness industry. *Agricultural Economics – Czech*, 68, pp. 171–179, https://doi.org/10.17221/55/2022-AGRICECON.

Agwu, M.O. and Emeti, C.I. 2014. Issues, challenges and prospects of small and medium scale enterprises (SMEs) in Port-Harcourt city, Nigeria. *European Journal of Sustainable Development*, 3(1), pp. 101–114, https://doi.org/10.14207/ejesd.2014.v3n1p101.

Akinola, B.D. 2015. Risk preferences and coping strategies among poultry farmers in Abeokuta metropolis, Nigeria. *Global Journal of Science Frontier Research: Agriculture and Veterinary*, 14(5), pp. 23–30.

Ali, J. 2016. Performance of small and medium-sized food and agribusiness enterprises: evidence from Indian firms. *International Food and Agribusiness Management Review*, 19(4), pp. 53–64, https://doi.org/10.22434/IFAMR2016.0024.

Andrei, J.V., Chivu, L., Gheorghe, I.G., Grubor, A., Sedlarski, T., Sima, V., Subic, J. and Vasic, M. 2021. Small and medium-sized enterprises, business demography and European socio-economic model: Does the paradigm really converge?. *Journal of Risk and Financial Management*, 14(2), p. 64, https://doi.org/10.3390/jrfm14020064.

Asravo, R.K. and Sarpong, D.B. 2022. Risk preferences and management strategies of farmers in Ghana: Does the type of crop grown matter?. *Journal of International Development*, https://doi.org/10.1002/jid.3719.

Boakye, A.A. 2017. *Risk attitudes, risk management and business success of micro and small informal agribusiness entrepreneurs in Ghana: The case of agri-food processors*. [online] Available at <http://ugspace.ug.edu.gh/handle/123456789/23542> [Accessed 10 July 2023].

Boermans, M.A. and Willebrands, D. 2017. Entrepreneurship, risk perception and firm performance. Working Papers, 17-04, Utrecht School of Economics. [online] Available at: <https://www.uu.nl/use/research> [Accessed 10 July 2023].

Block, J., Sandner, P. and Spiegel, F. 2009. *Do risk attitudes differ within the group of entrepreneurs?* MPRA Paper no. 17587. [online] Available at: https://mpra.ub.uni-muenchen.de/17587/1/MPRA_paper_17587.pdf> [Accessed 10 July 2023].

Ciolac, R., Iancu, T., Popescu, G., Adamov, T., Feher, A. and Stanciu, S. 2022. Smart tourist village – An Entrepreneurial necessity for maramures rural area. *Sustainabilty*, 14(14). https://doi.org/10.3390/su14148914.

Danso, A., Adomako, S., Damoah, J.O. and Uddin, M. 2016. Risk-taking propensity, managerial network ties and firm performance in an emerging economy. *The Journal of Entrepreneurship*, 25(2), pp. 155–183. https://doi.org/10.1177/0971355716650367.

Endris, E. and Kassegn, A. 2022. The role of micro, small and medium enterprises (MSMEs) to the sustainable development of sub-Saharan Africa and its challenges: A systematic review of evidence from Ethiopia. *Journal of Innovation and Entrepreneurship*, 11, p. 20. https://doi.org/10.1186/s13731-022-00221-8.

Gandhi, V.P. 2014. Growth and transformation of the agribusiness sector: Drivers, models and challenges. *India Journal of Agricultural Economics*, 69(1), pp. 1–31.

Gherghina, S.C., Botezatu, M.A., Hosszu, A. and Simionescu, L.N. 2020. Small and Medium-Sized Enterprises (SMEs): The engine of economic growth through investments and innovation. *Sustainability, 12*(1), p. 347. https://doi.org/10.3390/su12010347.

Igbokwume, M.C., Essien, B.A., Agunnah, M.U. 2015. The imperative of Nigeria agribusiness: Issues and challenges. *Journal of Business and Management*, 3(5), pp. 7–10. https://doi.org/10.11648/j.sjbm.s.2015030501.12.

Kagwathi, G.S., Kamau, J.N., Njau, M.M. and Kamau, S.M. 2014. Risks faced and mitigation strategies employed by small and medium enterprises in Nairobi, Kenya. *Journal of Business and Management*, 16(4), pp. 1–11. https://doi.org/10.9790/487X-16450111.

Kazungu, I., Ndiege, B.O., Mchopa, A. and Moshi, J. 2014. Improving livelihoods through micro and small agribusiness enterprises: analysis of contributions, prospects and challenges of nursery gardens in Arusha Tanzania. *European Journal of Business and Management*, 6(9), pp. 142–148.

Kehinde, A., Opeyemi, A., Benjamin, A., Adedayo, O. and Abel, O.A. 2017. Enterprise risk management and the survival of small scale businesses in Nigeria. *International Journal Accounting Research*, 5(2), pp. 1–8. https://doi.org/10.4172/2472-114X.1000165.

Lammers, J., Willebrands, D. and Hartog, J. 2010. *Risk attitude and profits among small enterprises in Nigeria*. Tinbergen Institute Discussion Paper no. 2010-053/3, Amsterdam and Rotterdam: Tinbergen Institute. [online] Available at: <https://www.econstor.eu/bitstream/10419/86795/1/10-053.pdf> [Accessed 10 July 2023].

Li, Y. and Rama, M. 2015. Firm dynamics, productivity growth, and job creation in developing countries: The role of micro- and small enterprises. *The World Bank Research Observer*, 30(1), pp. 3–38. https://doi.org/10.1093/wbro/lkv002.

Lien, G., Kumbhakar, S.C., Mishr, A.K. and Hardaker, J.B. 2022. Does risk management affect productivity of organic rice farmers in India? Evidence from a semiparametric production model. *European Journal of Operational Research*, 303(3), pp. 1392–1402. https://doi.org/10.1016/j.ejor.2022.03.051.

Matkovski, B., Zekic, S., Jurjevic, Z. and Dokic, D. 2022. The agribusiness sector as a regional export opportunity: Evidence for the Vojvodina region. *International Journal of Emerging Markets*, 17(10), pp. 2468–2489. https://doi.org/10.1108/IJOEM-05-2020-0560.

Musyoki, M.E., Busienei, J.R., Gathiaka, J.K. and Karuku, G.N. 2022. Linking farmers' risk attitudes, livelihood diversification and adoption of climate smart agriculture technologies in the Nyando basin, South-Western Kenya. *Heliyon*, 8(4), p. e09305. https://doi.org/10.1016/j.heliyon.2022.e09305.

NBS. 2023. National Bureau of Statistics. [online] Available at: <https://www.nigeriastat.gov.ng> [Accessed 10 July 2023].

Ndamani, F. and Watanabe, T. 2017. Determinants of farmers' climate risk perceptions in agriculture: a rural Ghana perspective. *Water*, 9(210), pp. 1–14. https://doi.org/10.3390/w9030210.

Ng'ang'a, D. and Gichira, R. 2017. Factors affecting growth of agribusiness micro and small enterprises in Embu County. *Strategic Journal of Business and Change Management*, 4(3), pp. 246–261.

Ng'ang'a, S.I., Muthusi, B.M. and Nassiuma, B. 2015. Comparative study of enterprise risks and management practices between micro and small industries (MSIs) and medium and large industries (MlIs) In Nakuru Municipality, Kenya. *European Journal of Business and Social Sciences*, 3(11), pp. 121–144.

Nocco, B.W. and Stulz, R.M. 2006. Enterprise risk management: Theory and practice. *Journal of Applied Corporate Finance*, 18(4), pp. 8–20. https://doi.org/10.1111/jacf.12490.

Nto, P.O.O., Mbanasor, J.A. and Nwaru, J.C. 2011. Analysis of risk among agribusiness enterprises investment in Abia State. *Nigeria Journal of Economics and International Finance*, 3(3), pp. 187–197.

Oladimeji, Y.U., Galadima, S.H., Abubakar, A., Sanni, A.A., Abdulrahman, S., Egwuma, H., Ojeleye, A.O. and Yakubu, A. 2019a. Risk analysis in fish farming systems in Oyo and Kwara States, Nigeria: A prospect towards improving fish production. *Animal Research International*, 16(1), pp. 3226–3237.

Oladimeji, Y.U., Hassan, A.A., Egwuma, H., Sani, A.A., Galadima, S.A. and Ajao, A.M. 2019b. Analysis of risks in honeybee production farms in Nigeria: A boost to food security. *Egyptian Academic Journal of Biological Sciences A. Entomology*, 12(1), pp. 163–176.

Ozsayin, D. 2022. Dairy farmers' risk attitudes and their perceptions towards environmental risks in Northwest Turkey. *Emirates Journal of Food and Agriculture*, 34(11), pp. 888–903. https://doi.org/10.9755/ejfa.2022.v34.i11.2967.

Prioteasa, A.L., Chicu, N., Bugheanu, A.M. and Dinulescu, R. 2020. Risk Management Practices in Small and Medium Enterprises: Evidence from Romania. *Management and Economics Review*, 5(1), pp. 1–15. https://doi.org/10.24818/mer/2020.06.01.

PWC. 2020. *PwC's MSME Survey 2020. Building to last. Nigeria report.* [online] Available at: <https://www.pwc.com/ng/en/assets/pdf/pwc-msme-survey-2020-final.pdf> [Accessed 10 July 2023].

Raicov, M. and Feher, A. 2017. Rural Romanian space between survival and business opportunities. *Lucrari Stiintifice Management Agricol*, 19(2), pp. 137–144.

Scarborough, N.M. 2011. *Effective small business management: An entrepreneurial approach.* 10th ed. Pearson College Div.

Sekumade, A.B and Ogunro, V.O. 2013. Risk assessment and management for agribusiness enterprises investment in Ondo State, Nigeria. *European Journal of Business and Management*, 5(1), pp. 199–209.

Shadbolt, N.M. and Olubode-Awosola, F. 2016. Resilience, risk and entrepreneurship. *International Food and Agribusiness Management Review*, 2(19), pp. 33–52.

Shah, J. and Alharthi, M. 2022. The association between farmers' psychological factors and their choice to adopt risk management strategies: The case of Pakistan. *Agriculture-Basel*, 12(3), p. 412. https://doi.org/10.3390/agriculture12030412.

SMEDAN-Small and Medium Scale Enterprises Development Agency of Nigeria. 2017. National survey of micro, small and medium enterprises (MSMEs) 2017. [online] Available at: <https://smedan.gov.ng/images/NATIONAL%20SURVEY%20OF%20MICRO%20SMALL%20&%20MEDIUM%20ENTERPRISES%20(MSMES),%20%202017%201.pdf> [Accessed 10 July 2023].

SMCI-States Ministry of Commerce & Industry. 2018. *1st, 2nd, 3rd and 4th Quarterly reports.*

Suguna, M., Shah, B., Sivakami, B.U. and Suresh, M. 2022. Factors affecting repurposing operations in Micro Small and Medium Enterprises during Covid-19 emergency. *Operations Management Research*, 15(3–4), pp. 1181–1197. https://doi.org/10.1007/s12063-022-00253-z.

Sulewski, P., Was, A., Kobus, P., Pogodzinska, K., Szymanska, M. and Sosulski, T. 2020. Farmers' attitudes towards risk – An empirical study from Poland. *Agronomy*, 10, p. 1555. https://doi.org/10.3390/agronomy10101555.

Theuvsen, L. 2013. Risk management in agriculture: Focus on risk assessment and risk-bearing capacity, In: R. Doluschitz ed. *Proceeding to the workshop: Transformation of entrepreneurship in agriculture – New challenges for farmers, associations, upstream and downstream partners from the value chain and for science.* Stuttgart: University of Hohenheim. [online] Available at: <https://doi.org/10.22004/ag.econ.190788> [Accessed 10 July 2023].

Vaughan, E.J. 2016. Agribusiness risk management. In: IRMI Emmett J. Vaughan Agribusiness Conference, 26–28 April 2016, Sacramento, California: International Risk Management Institute, Inc. pp. 24–45. Available at: <https://www.irmi.com> [Accessed 10 July 2023].

Zurriaga-Carda, A., Kageyama, K. and Akai, K. 2016. Effects of risk attitude, entrepreneurship education and self-efficacy on entrepreneurial intentions: A structure equation model approach to entrepreneurship. *International Review of Management and Business Research*, 4(5), pp. 1424–1433.

Nicoleta Mateoc-Sîrb, Teodora Mateoc-Sîrb, Ariana Velciov, Cosmin Lădariu, Zeno Gârban

Food safety and security from public desideratum to agrobiological and social aspects

Abstract: *Issues related to food safety and security represented/represent objectives of major interest at local and global level. These aspects concern, as a whole, the problems of agrobiology, the preparation/processing of foods in optimal conditions, followed by the organization of marketing. This article systematically approaches the so-called "defining dimensions" of food security from a conceptual and applied perspective. Expanding the field, the particularities of food and nutritional security are discussed. In a conjugated context regarding the previously mentioned aspects, the "safety-security interrelationship" in the food science is presented. For the understanding of applicative issues of national and international interest, references are made to the "traceability of food products" mentioned in Regulation EC 178/ 2002. Some available aspects regarding the food consumption at national level – resulted from the statistical data in our country. Statistics proved that currently the Romanian producers are unable to produce enough food products in order to ensure the necessary consumption of the population. In this situation, Romania is forced to resort to importing most of basic food products. Currently, the Romanian farmers can ensure only the consumption needs for cereals and cereal based food products. According to statistical data, our country imports significant quantities of all other basic food products.*

Keywords: food , safety, traceability of food products, food consumption.

Introduction

The current trends of globalization can be characterized with the help of descriptors such as: environmental protection policies, resource conservation, food policies (food safety, food and nutritional security), socio-economic policies, agricultural development, public health issues, etc. Regarding food and nutrition, the systematic approach to their study was possible due to the deepening of the notions related to the specific bio-constituents of the body and the constituents present in the composition of food. Along with the latter, chemical xenobiotics were detected in food and in the existential area (Figure 1). It is well known that nutrients enter the human body with the consumption of food.

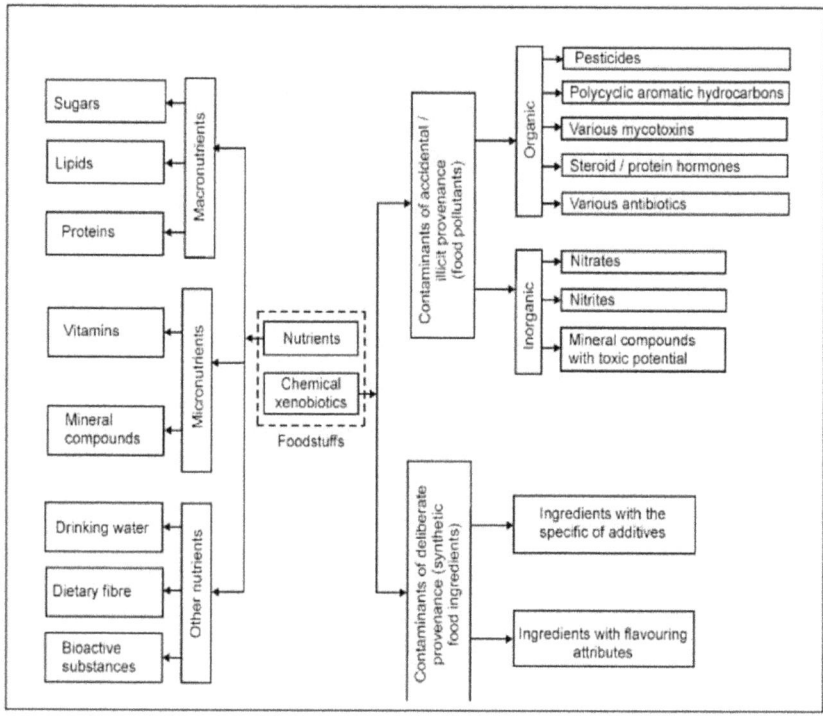

Figure 1. Nutrients end xenobiotiocs in foods
Source: Gârban, 2018.

Food is constantly exposed to possible contamination in the environment with foreign compounds (chemical, physical, biological xenobiotics) or through human intervention, such as, for example, the introduction of food ingredients (chemical xenobiotics) into the food composition that can, sometimes, affect the health of consumers. In this context, on the agri-food chain, regulations and control aim at the quality of food products to not be dangerous for the body (food safety).

Literature review

Food safety should be approached through the lens of food innocuousness based on the knowledge of food hygiene norms on which "good practices" in the field of the food industry depend.

At the international level, there are structures designed to coordinate aspects related to food safety. At the European level, there is, independently of the European Commission-EU, the European Food Safety Authority-EFSA and, in the USA, Food and Drug Administration – FDA coordinates the issue of food safety.

According to *Codex Alimentarius*, "food safety" is defined as *"the assurance that the food product will not be harmful to the consumer when it is prepared and/or consumed"* (FAO, 2023).

Food security is a concept based on four dimensions: availability, accessibility, utilization (circumstantial dimensions) and stability, which is an integrative dimension. For all four defining dimensions of food security, various indicators can be considered: in the case of availability, the assessment should be made according to the fertility of agricultural land, to population fluctuation, and to food production. Affordability is influenced by food cost, consumer income, and daily food consumption. The use has, as indicators, the status of the physical development of the consumers based on anthropometric measurements, possible morbidity states, as well as mortality rate. Stability, as an integrative dimension, is related to circumstantial dimensions (Figure 2).

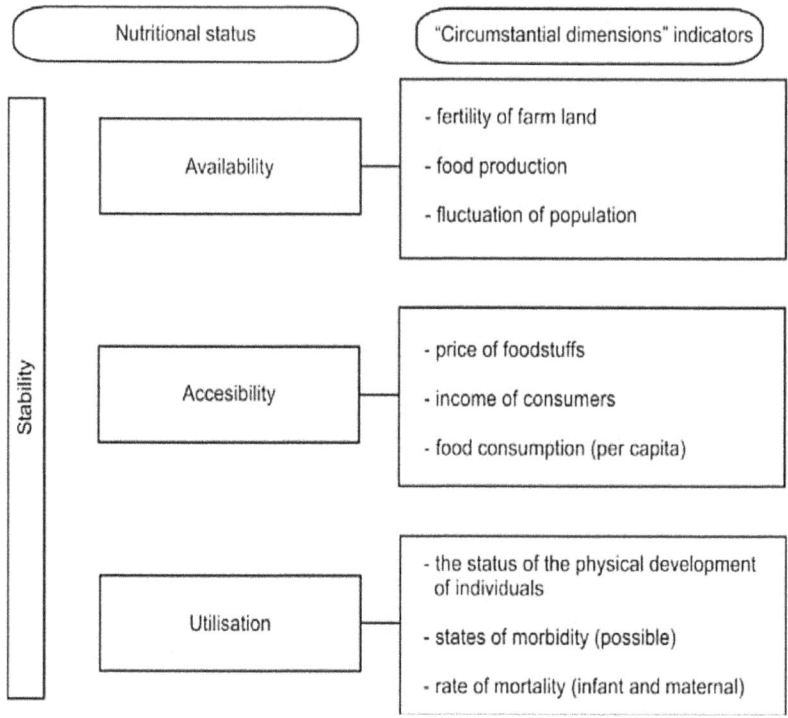

Figure 2. Dimensions of food and nutritional security
Source: Gârban, 2018.

Nutritional security is a concept that refers to three "determinants": adequate food consumption (the presence in the daily diet of foods with a high content of nutrients, appropriate from a qualitative and quantitative point of view), care and feeding practices and, finally, sanitation and health, a determinant that mainly refers to the provision of drinking water supply and the removal of waste water to maintain health status (Wüstefeld, 2013).

Since 2009, the two concepts have merged, the "standard" term of Food and Nutritional Security being agreed upon, as mentioned in 2012 within the World Food Security Committee (CSA).

Methodology

Food and nutritional security exist when all people, always, have physical, social, and economic access to food consumed in sufficient quantity and quality to meet their nutritional needs; it is supported by an adequate sanitation environment, health services and care for a healthy and active life (Gârban, 2018).

In this context, every country is bound to ensure the safety and food security of its citizens, because the problem of access to food for the population contributes to peace and social tranquillity, to stability, and prosperity (Mateoc et. al., 2022a, 2022b, 2022c).

The statistical data provided by the National Institute of Statistics (NIS) were used for the analysis of ensuring the necessary food consumption for the Romanian population. The following indicators were analyzed: *Average annual food consumption per inhabitant*, *Usable production* (Up), *Import* (I), *Export* (E), *Stock variation* (±Sv), *Available supply* (As), and *Degree of self-sufficiency* (Dss).

> **Usable production (Up)** – the quantities of primary products obtained during the reference period, including the quantities used by producers from their own production (self-consumption) and/or the quantities of transformed products. Losses incurred during the production process (e.g., losses during harvesting and/or in the manufacturing process) are not included in production.
> **Import (I)** – the quantities of agri-food products of foreign origin that entered the economic circuit of the country during the reference period, through international trade activity.
> **Export (E)** – the quantities of agri-food products from domestic production and, sometimes, of foreign origin (re-export) that left the economic circuit of the country, during the reference period, through international trade activity.
> **Stock variation (±Sv)** – the evolution of stocks in the reference period calculated as the difference between the final stock (the stock at the end of the reference calendar year) and the initial stock (the stock at the beginning of the reference calendar year).
> **Available supply (As)** – the quantities of products available to cover the internal needs of a country, established according to the following balance relationship:
> As = Up + I − E − (±Vs)
> **Degree of self-sufficiency (Dss)** – the extent to which domestic production covers domestic consumption requirements in the reference period, calculated according to the following relationship:
> **Dss = Pint/As * 100 (Pint= internal production)**

Results and discussions

Currently, there are concerns related to the fundamental and applicative aspects of the interrelationship between food safety and food and nutritional security, an interrelationship influenced both by the dimensions of food security and by the determinants of nutritional security. This far-reaching issue involves the most diverse aspects: food, agricultural, social, political, economic, medical, educational.

The approach to the theoretical and applicative aspects at the chemical nutrient – xenobiotic interface mainly refers to human nutrition and food safety (Figure 3).

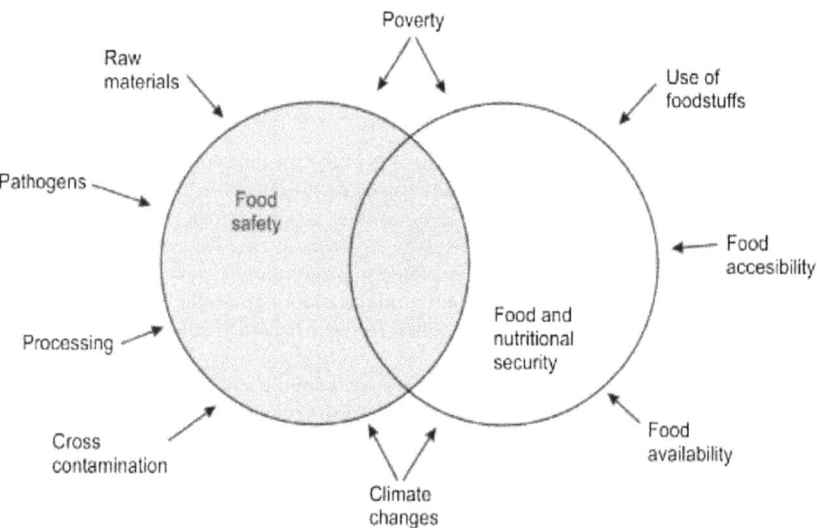

Figure 3. Interrelation between food safety and food and nutrition security
Source: Gârban, 2018.

Unlike nutrients, which are the basic constituents of food and which, once they reach the body via the enteral or parenteral route, participate in metabolic processes following the main native biochemical pathways, food chemical xenobiotics reach the body via the digestive route and are subjected to biotransformation processes on specific biochemical pathways (Ronis and Cunny, 2008; Wilson and Nicholson, 2003). Xenobiotics enter the body together with nutrients, both categories of compounds being present in food. The presence of xenobiotics in food can be deliberate, when the food contains certain ingredients with the role of food additives (antioxidants, dyes, sweeteners, emulsifiers, preservatives) or illicit, when food pollutants are present in the food (pesticides, hormones, antibiotics, mycotoxins, polycyclic aromatic hydrocarbons, nitrites, nitrates, compounds with toxicogenic potential). These compounds enter the body along the food chain, through translocation (Gârban, 2018). At the basis of food safety and of food and nutritional security are multiple aspects – social, biological, economic, medical, political – with implications on the raw materials needed for the food industry and biotechnologies (Peev-Otiman and Mateoc-Sîrb, 2023).

From the analysis of the statistical data made available by the National Institute of Statistics, the food consumption pattern of the Romanian population could be identified. For example, the structure of the average daily food consumption, net, per inhabitant, in the year 2021, was made up of products of plant origin in proportion of 59.8 % and of products of animal origin in proportion of 40.2 % (NIS, 2022a, 2022b, 2022c, 2022d). In the average daily net consumption per inhabitant, six groups of products had the largest shares, which represented over 90 % of the food consumption of the population: milk and dairy products 30.8 %, vegetables and vegetable products 17.5 %, cereals and cereal products 15.5 %, fruit and fruit products 10.8 %, potatoes 9.6 %, and meat and meat products 7.3 %.

Table 1. Average annual consumption per inhabitant of the main food groups in Romania

Main food group	Year	Average annual food consumption per inhabitant (kg)
1. Cereals and cereal products in grin equivalent	2017	208.2
	2018	205.4
	2019	204.2
	2020	199.9
	2021	200.6
2. Potatoes	2017	96.6
	2018	95.5
	2019	92.2
	2020	93.4
	2021	98.1
3. Vegetables and vegetable products, legumes, and melons	2017	187.8
	2018	202.2
	2019	196.6
	2020	196.0
	2021	203.4

(continued on next page)

Table 1. Continued

Main food group	Year	Average annual food consumption per inhabitant (kg)
4. Fruits and fruit products	2017	96.1
	2018	110.8
	2019	111.3
	2020	107.6
	2021	115.3
5. Milk and dairy products	2017	251.4
	2018	258.3
	2019	259.8
	2020	260.1
	2021	263.3
6. Meat and meat products	2017	71.5
	2018	76.7
	2019	77.7
	2020	77.4
	2021	78.0

Source: Authors' own research based on NIS 2023c, 2023d.

In this sense, it is important, from the point of view of food safety and security, to analyze the average annual consumption per inhabitant and the degree of self-sufficiency in the main food groups in Romania (Table 1 and Figures 4, 5, 6, 7, 8, and 9).

The data presented in Table 1 show that, during the five years under study, Romania was only able to ensure its consumption needs from its own production for cereals and cereal products; for the other food groups, it was forced to resort to imports to satisfy the consumption needs of the population (Department of Parliamentary Studies and EU Policy, 2022; Otiman et al., 2023; Țigan et al., 2021; Venig et al., 2023).

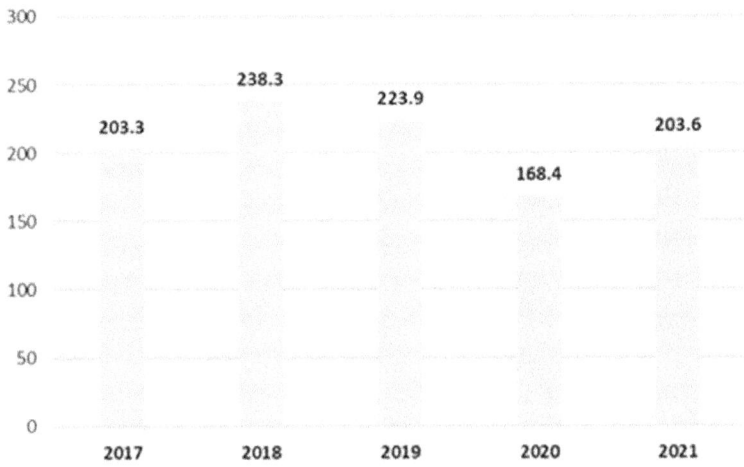

Figure 4. Degree of self-sufficiency in cereals and cereal products, %
Source: NIS 2023c, 2023d.

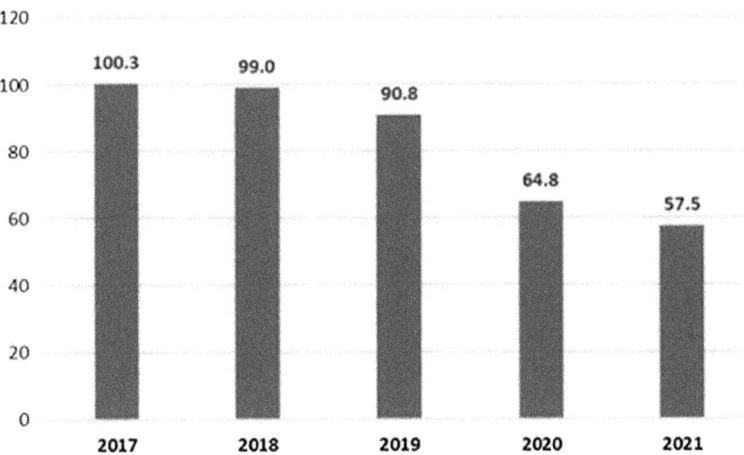

Figure 5. Degree of self-sufficiency in potatoes, %
Source: Source: NIS 2023c, 2023d.

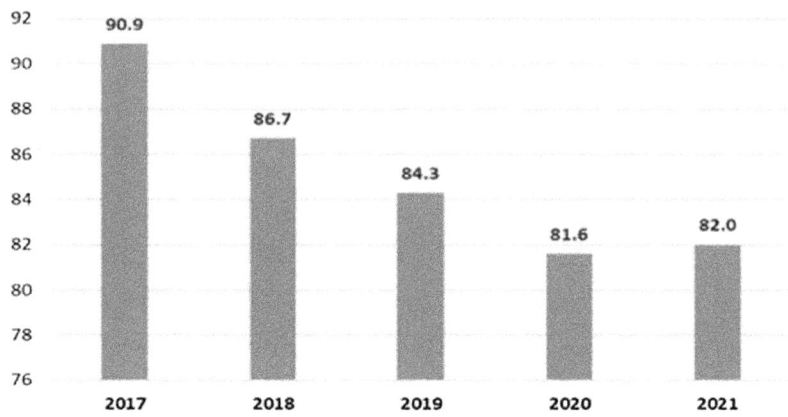

Figure 6. Degree of self-sufficiency in vegetables and vegetable products, %
Source: Source: NIS 2023c, 2023d.

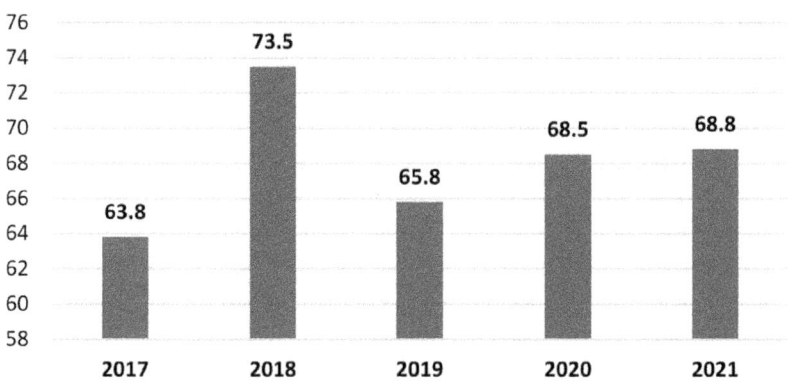

Figure 7. Degree of self-sufficiency in fruits and fruit products, %
Source: NIS 2023c, 2023d.

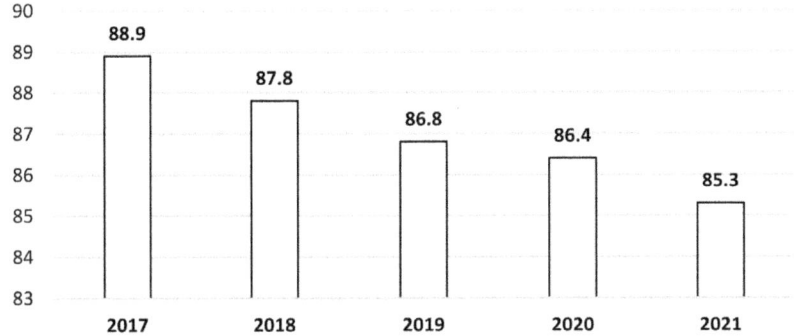

Figure 8. Degree of self-sufficiency in milk and dairy products, %
Source: NIS 2023c, 2023d.

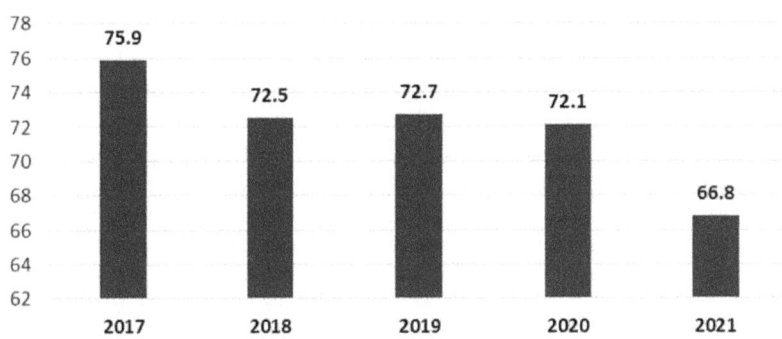

Figure 9. Degree of self-sufficiency in meat and meat products, %
Source: NIS 2023c, 2023d.

The largest imports, in descending order, were recorded in potatoes, meat and meat products, fruits and fruit products, vegetables and vegetable products, legumes, grains and melons, milk and dairy products, respectively (Figures 4, 5, 6, 7, 8, and 9).

In times of economic prosperity and peace, the impact of food system vulnerabilities may not be acutely felt (Romanian Government, 2023). But in times of political or socio-economic turmoil, national, regional and global food system strengthening efforts determine the impact on overall levels of food security (Bacău et al., 2021; Olteanu, 2023).

Conclusions

The inter-relational nature of the issue related to the concepts of "safety" and "security" requires an interdisciplinary approach involving the most varied aspects: political, social, economic, food, agricultural, educational, which have led to the emergence of new ways of evaluating effects induced by xenobiotics, one of which is the use of biomarkers.

Likewise, food and nutritional and food security concerns can be integrated into research that aims at identifying problems in the field of agriculture, forestry, veterinary medicine, and agricultural management.

Through food safety and security, every person's access to sufficient, nutritious, sustainable and and accessible food products, with reasonable/affordable prices, must be guaranteed.

References

Bacau, C., Mateoc-Sirb, N., Ciolac, R., Mateoc, T. and Tabara, V. 2021. Study on the production and use of biomass energy in the Timis County. *Romanian Biotechnological Letters*, 26, pp. 2434–2440.

Departamentul de studii parlamentare și politici UE. Direcția pentru Uniunea Europeană (Department of Parliamentary Studies and EU Policy. Directorate for the European Union). 2022. Planul de urgență pentru garantarea aprovizionării cu alimente și a securității alimentare în vremuri de criză [Contingency plan to guarantee food supply and food security in times of crisis]. [online] Available at: <https://www.cdep.ro/afaceri_europene/afeur/2022/fi_3335.pdf> [Accessed 10 October 2023].

FAO. 2023. *Codex alimentarius international food standards*. [online] Available at: <https://www.fao.org/fao-who-codexalimentarius/en/> [Accessed 6 October 2023].

Gârban, Z. 2018. *Quo vadis food xenobiochemistry*. 3rd ed. Bucharest: Romanian Academy Press.

Guvernul României (Romanian Government). 2023. *Program de guvernare 2023–2024. Viziune pentru națiune* [Government Program 2023–2024. Vision for the nation]. [online] Available at: <https://gov.ro/fisiere/pagini_fisiere/23-06-16-12-32-52Programul_de_Guvernare_2023-2024.pdf> [Accessed 10 October 2023].

Mateoc-Sîrb, N., Milin, A., Peț, E., Venig, A., Sârb, Gh.S., Mănescu, C., Gavrilescu, C. and Nan A. 2022a. Analysis of the consumption of agroo-food products in the current period. *Analele Universității din Oradea, Fascicula: Protecția mediului*, XXXIX(27), pp. 31–36.

Mateoc-Sîrb, N., Gavrilescu, C., Mateoc, T., Adamov, T. and Mănescu, C. 2022b. Agricultura ecologică – mit sau realitate. In: I.S. Bruma and C.D.Vasiliu, eds. *Provocări rurale contemporane. Studii de agro-economie și antropologie rurală.* Cluj-Napoca: Presa Universitară Clujeană, p. 230.

Mateoc-Sîrb, N, Bacău, C.-V., Duma-Copcea, A., Mateoc-Sîrb, T., Mănescu, C., Niță, S., Sicoe-Murg, O. and Suster, G. 2022c. Agricultural trends in Romania in the context of the current trends of the world economy. *Journal of Applied Life Sciences and Environment,* 55(3), pp. 335–350. https://doi.org/10.46909/alse-552068.

National Institute of Statistics (NIS). 2023a. Comunicat de presă. Domeniul Agricultură [Press release. Agriculture field]. [online] Available at: <https://insse.ro/cms/sites/default/files/com_presa/com_pdf/rga_2020r.pdf> [Accessed 8 October 2023].

National Institute of Statistics (NIS). 2023b. TEMPO online database. [online] Available at: <http://statistici.insse.ro:8077/tempo-online/> [Accessed 8 October 2023].

National Institute of Statistics (NIS). 2023c. Disponibilitățile de consum ale populației în anul 2021 [The consumption availability of the population in 2021]. [online] Available at: <www.insse.ro/cms/sites/default/files/field/publicatii/disponibilitatile_de_consum_ale_populatiei_anul_2021.pdf> [Accessed 8 October 2023].

National Institute of Statistics (NIS). 2023d. Disponibilitățile de consum ale populației în anul 2022 [The consumption availability of the population in 2022]. [online] Available at: <https://insse.ro/cms/ro/tags/disponibilitatile-de-consum-ale-populatiei_anul_2022.pdf> [Accessed 8 October 2023].

Olteanu, M. 2021. *România, pe locul 22 în lume în ceea ce privește securitatea alimentară* [Romania, in 22nd place in the world in terms of food security], *The Economist,* Umbrela Strategica, 7 March 2021. [online] Available at: <https://umbrela-strategica.ro/romania-pe-locul-22-in-lume-in-ceea-ce-priveste-securitatea-alimentara-the-economist> [Accessed 8 October 2023].

Otiman, P.I., Mateoc-Sîrb, N., Goșa, V., Băneș, A., Sălășan, C., Feher, A., Raicov, M. and Iucu Z. 2023. *Unde se află și încotro se îndreaptă agricultura României? 1991–2007–2023.* Bucharest: Ed. Academiei Române, p 190.

Peev-Otiman, P.D. and Mateoc-Sîrb, N. 2023, Analyzing the development possibilities of the mountain area of Banat, Caras-Severin County. *Sustainability,* 15(11), p. 8730, https://doi.org/10.3390/su15118730.

Ronis, J.J.M. and Cunny, H.C. 2008, Developmental effects on xenobiotic metabolism. In: R.C. Smart and E. Hodgson, eds. *Molecular and chemical toxicology.* 4th ed. Hoboken, USA: Wiley & Sons Inc, pp. 257–272.

Țigan, E., Brînzan, O., Obrad, C., Lungu, M., Mateoc-Sîrb, N., Milin, I.A. and Gavrilaș, S. 2021. The consumption of organic, traditional, and/or european eco-label products: Elements of local production and sustainability. *Sustainability*, 13(17), p. 9944.

Venig, A., Venig, A., Mateoc- Sîrb, N. and Peț, E. 2023. Analysis of the food insecurity generated by the war in Ukraine. *Lucrări Științifice Management Agricol*, 25(2), pp. 146–155.

Wilson, I.D. and Nicholson, J.K. 2003. Topics in xenobiochemistry: Do metabolic pathways exist for xenobiotics? The micrometabolism hypothesis. *Xenobiotics*, 33(9), pp. 887–901.

Wüstefeld, M. 2013. Food and nutrition security. In: United Nation System, *Meeting of the minds. Nutrition impact of food systems*, Geneva.

Natalia Mocanu, Vasile Secrieru

The role of financial audit in public institutions in the Republic of Moldova

Abstract: *The successful development of a state, implicitly of its institutions, depends on the availability of budget funds, their proper distribution and use. In this context, the issue of public financial audit must be given and given a decisive role. In close connection with the current developments of the economic situation, on the world level, as well as in our country, the financial auditor is the professional who contributes to a significant extent, to the evaluation of the climate and economic and to restore confidence in initiatives, honest and relevant measures to overcome current crises. Through professionalism, attitude, independence and transparency in carrying out financial audit missions, by preparing quality reports, the financial auditor satisfies the public interest celebrated with regard to the correctness of the affairs subject to its evaluation, fulfilling, at the same time, an important social role. The purpose of the present paper is to develop the institutional and organizational framework that determines and supports the financial audit activity at the level of public institutions in the Republic of Moldova in order to ensure an efficient management of public financial funds. Following the evaluation of the quality of the financial audit at the level of public institutions in the Republic of Moldova, the authors formulate the following conclusions: the financial audit as well as the public audit, here we are talking about the external and the internal, it is necessary to be perfected in order to satisfy, through all existing methods, the efficient use of public money. Parallel to this, there are many chances that the increase of the present and future indicators will need a more efficient audit. The three lines of defense model is a valuable framework that highlights the role of internal audit in ensuring effective risk management and the importance of fulfilling this position and functions in the corporate governance structure.*

Key words: financial audit, public institutions, efficiency, budgetary framework, public agency.

Introduction

The successful development of a state, implicitly of its institutions, depends on the availability of budget funds, their proper distribution and use. In this context, the issue of public financial audit must be given a decisive role.

At the same time, the permanent character characterizing the actuality and importance of the topics connected to the generic of public financial audit is determined by the following three factors: (1) changing the nature of public audit, which has expanded so as to include areas of control that considerably exceed financial assurance. Progressively, in order to strengthen the impact of financial audit on public financial management, it is linked to other forms of public audit. In this respect, financial audit either represents the primary, basic position, being accompanied by other forms of audit, or exercises a secondary position, in order to materialize the conclusion of the primary form of audit; (2) the public management reform introduced new, transparent requirements for public sector institutions, in this respect, public audit being also applied as a way of measuring performance; (3) increasing expectations from the state and responsible bodies in this field (Copaceanu, 2016). The expectations of the business community and, in general, the benefits of audit reports compared to financial auditors are mainly related to the achievement of a quality audit mission, of a nature to relevant assurances regarding the reality and correctness of the information contained in the financial statements. The objectives of the work consist in investigating the theoretical foundations associated with risks at the level of public institutions from the perspective of financial audit and evaluating the level of independence of the financial audit as a determining factor of its quality (Law 271/2017) (Official Monitor of the Republic of Moldova, 2018).

Literature review

The issue of the public audit, in general, and the financial one, in particular, was reflected in the works of foreign and authors in the country who have published in their works about financial audit in public institutions both within foreign research centers of excellence, and autochthonous. The methodological and theoretical-scientific support is the research of the following scholars: Copaceanu (2016), Lenz and Hahn (2015), Lois et al. (2021), Getie Mihret and Wondim Yismaw (2007), Medina and Schneider (2018), Lartey et al. (2019). The line of research continues through the writings of other scholars who study the particularities of internal control and governance important corporations for the economy of Republic of Moldova and in relation to the external international level environment (Millerand and Power, 2013).

Methodology

In order to achieve the objectives proposed in the article, the general scientific methods were applied: analysis, synthesis, history and logic, critical analysis of materials, but also methods of analysis and economic-financial diagnosis, statistical processing of empirical data and official data. Considering the complexity of the research topic, namely the intercorrelation of the public financial audit as a component element of public finances with the theory and practice of public administration, the principles of dialectical determinism were applied in the process of carrying out scientific investigations. The informational foundation of the research is provided by the statistical data obtained by the author from the Court of Accounts of the Republic of Moldova, the Ministry of Finance, the National Bureau of Statistics, the World Bank, the International Monetary Fund, the Organization for Economic Cooperation and Development, Transparency International-Moldova, World Justice Project, data obtained from the author's survey in the field of public audit (Ministry of Finance, 2023; NBS, 2022; Civil Code of the Republic of Moldova, Official Monitor, 2019).

Results and discussions

The provision of various public goods and services is a basic responsibility of the state exercised by it through its institutions. Although with the expansion of the market, many services are available in the private sector; many basic services continue and will continue to be provided only by public institutions, due to the nature of these services. These include regulatory services important for maintaining public order in society by ensuring that appropriate rules of public behavior are established by acts and rules, which must be respected by citizens. Public institutions provide goods and services important to protect human rights, as well as to enable everyone to enjoy certain freedoms and perform various social and economic functions in order to obtain income and ensure their well-being. The basket of private goods and services an individual receives depends on their ability to pay, and therefore many people who have inadequate income are unable to meet all their needs. The state, through its institutions, has a very important role to play in making essential public goods and services available, which ensure a certain minimum level of welfare to all those who need them. Financial and other resources under state management are always limited and therefore public services must be provided efficiently and effectively to ensure the desired level of well-being for all citizens and in the shortest possible time. Society has various forms of discrimination and deprivation, which should be reduced and therefore resources allocated so that there is justice and equity in the outcome of public service delivery. It is even more important for society to ensure that services are provided efficiently and effectively to achieve the desired result. It is therefore necessary to understand the different aspects of public service delivery, including those related to the quality of public financial management at the level of public institutions (Official Monitor of the Republic of Moldova, 2017, 2020).

Public institutions exist in ambiguous contexts. While, on the one hand, individual public institutions are often very stable, inflexible and, in some cases, even immortal; On the other hand, they exist in a very dynamic field and subject to continuous change and intense reform work (Lenz and Hahn, 2015).

Within the public finance management system, a special place is occupied by audit, which is one of the cornerstones of an efficient governance, which aims to improve the efficiency of the public institution. In this regard, we mention that the purpose of the audit is to support the achievement of the objectives of the public institution by providing recommendations for improving operations. The audit helps the institution to achieve its objectives by systematically evaluating governance, risk management and control processes.

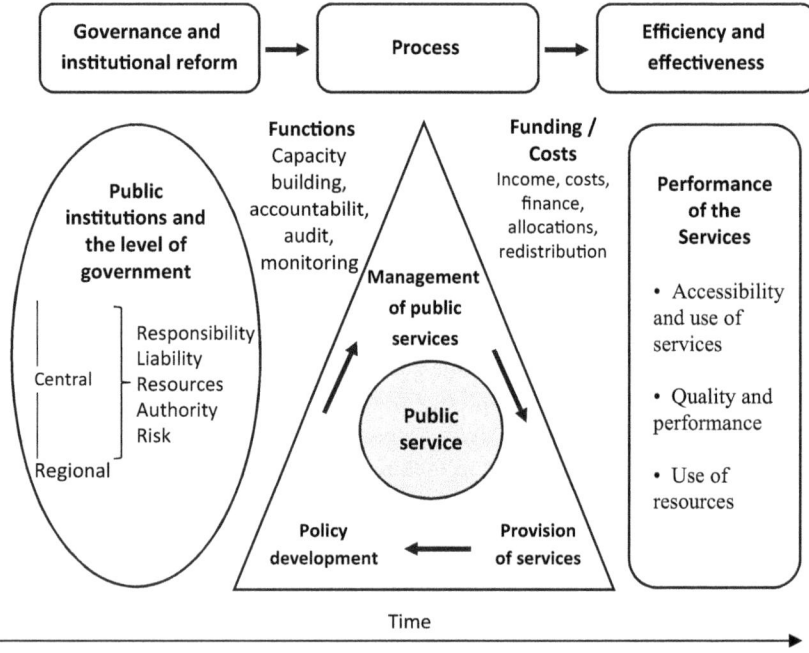

Figure. 1. The relationship between public institutions, public services, public finance management and audit

Source: Authors' own research.

In Figure 1 the author presents schematically the relationship between public institutions, public services, public finance management and public audit.

Thus, public institutions at central, regional and local level of government are taken into account, together with its functions and processes associated with public services (including management, delivery and policy development), resulting in an assessment of service performance in terms of efficiency and effectiveness.

At the international level, both external and internal financial audits have been developed to protect local and central public institutions. The internal financial audit according to multiple researches in public institutions and the external financial audit in the same institutions showed progress later in these institutions, unfortunately both audits were not concerned with the research of a specific new method of determining possible problems – risks in organizations publicly (Medina and Schneider, 2018).

Public finance management together with financial audit control having a management regime must take into account two shortcomings and risks:

(a) those characteristics that belong only to a country and its public sector in a certain period;
(b) specific to a public institution in this country.

If we are to refer to both economic and financial risks to which internal and financial control are subject, the author makes a detailed analysis, namely:

the wealth of a country can be assessed by public units for purposes other than the governing apparatus (Presidency, local public agencies, etc.) and the executive authority (Parliament, town hall, etc.);
public money can be used for other purposes, i.e. outside those of legislative administration (Parliament, local councils, etc.) and executive authority (Government, APL, etc.);
public units have the possibility, most often, to increase state credits beyond the limits determined by the legislative body (Parliament, local council, etc.) and the executive authority (Government, APL, etc.);
public institutions can often fall prey to corruption through the laundering of public money, the illicit enrichment of parliamentarians, persons in management or their relatives can also influence the application of anti-corruption laws;
there are cases when public institutions are not correctly exposed in the financial statements of the executive authority (Government, APL, etc.);

the legislation regarding liability, for example, the presentation of performance indicators, the reporting of the financial situation, the controls performed, all types of audit may be missing or with violations performed;

the right of individuals to submit complaints against the results of public institutions may be compromised;

there are cases when the executive management cannot have the experience to impose fines and sanctions when public units do not fulfill their responsibilities.

It should be highlighted that there may be risky situations, being admitted in some country, that is, it should be fully integrated in a permanent restructuring of the public institution: all well-trained employees, information center packages and public financial means may be missing or not be sufficient as necessary, to be required to carry out public financial controls, public audit and governance expressed in the legislation; besides the fact that the legislative power embodies certain types of guarantees established in time, their inconsistency can be like a failure.

The idea of specifying the risks that may arise or even face one or more countries including their public sectors (category b of problems and risks) taking into account the role and functions of the financial audit and the correct management of public funds, the author proposed the following algorithm:

(a) first of all, indices were chosen that include the risks and problems of public institutions and, at the same time, of the public sector and that can be trained or removed from internal and external financial audit operations.

The indicators have been identified so that they are credible. Credibility is ensured by the fact that these indicators are designed and evaluated at the level of international organizations with strong experience in the field concerned;

(b) the bibliography and other sources from which the used indexes were selected were exposed;

(c) the deviation limits of the values of the indicators used were selected;

(d) the countries were selected which, had the deviation of the indices and , will serve as examples for carrying out the analyzes.

We can say, regarding this, that all the sample countries were proposed from the list of those with developed economies; EU member countries that were socialist; economically weak countries but with democracies of the Commonwealth of Independent States;

(e) the sizes of the indices chosen for the countries used in the analysis, including the Republic of Moldova, were also selected;

(f) and for the indices of the pre-selected countries, the matrix of risks that are most often found in public institutions in these countries was elaborated and proposed, and the place of the Republic of Moldova in this composition was established.

According to Table 1 where the indicators selected by the authors are presented with their analysis, for the establishment and determination of legal risks are used:

> *Index of the rule of law.* This indicator is established, calculated and evaluated by the World Justice Project® (WJP) – an independent, multidisciplinary organization that works to visualize the rule of law around the globe. (WJP, 2023); *Regulatory Quality Index.* The index is designed and monitored by the World Bank Group (World Bank, 2023).

To perceive and evaluate the risk of failure of the government's activity, the following are used: The Government Effectiveness Index. This indicator is an index developed by the World Bank Group. For the perception and assessment of corruption risk are used: Corruption Control Index. The Corruption Control Index is constructed by the World Bank Group.

Table 1. Indicators of risk determination in public organizations

Indicators	Indicator definition	Limit of variation
Rule of law index	This indicator includes the information with which economic agents are measured in order to trust and not deviate from the community's orders, especially in order to execute and fulfill documents, contracts, property rights, as well as the idea of fighting crimes and violence.	Min. −2.5: (weak level) Max. + 2.5: (strong level)
The efficiency indicator of the Government's activity	The Government's efficiency indicator includes information about the quality of public services, and its safety from political influence, the efficiency of policy development and the confidence of management regarding the policies developed.	
Control of corruption index	The corruption check indicator includes ideas related to the evaluation of public power and is developed for illicit enrichment, including small and large methods of corruption, as well as state capture by elites and private interests.	
Regulatory quality index	The Regulatory Quality Index captures perceptions of the power of the country's leadership to develop and execute solid policies and regulations that enable and promote private sector development.	
Indicators of responsibility	The Opinions and the Accountability Indicator which includes ideas about the assessment of the population that has the possibility to choose the leadership of their country, as well as the option of free exposure, the possibility of association and free media	
Political stability index	The indicator of economic-political stability without violence or terrorism that evolves is the knowledge of the option that the leadership of the country can be destabilized or thrown out with certain non-constitutional methods, i.e. through demonstrations of strikes, uprisings, through the involvement of governmental or terrorist parties. This indicator is calculated as an average of several indicators, for example those from Forum and Political, Economist, World Economic Intelligence Unit, Risk Services.	
Corruption Perceptions Index	The Corruption Perceptions Index is an indicator of perceptions about corruption in the public institutions, or political-administrative corruption. The calculations of these indexes are determined using the materials of surveys and evaluations, of anti-corruption institutions and a lot of organizations that are recognized	100 = no corruption

Indicators	Indicator definition	Limit of variation
Shadow economy, percent of GDP	The shadow economy as a percentage of total annual GDP. The detailed methodology of the estimates can be obtained from the following IMF working paper	

Source: Authors' own research.

In the authors' opinions, all these types of risks, on the one hand, special principles that impose on the organizations responsible for the internal and external public financial audit, urgent determination and management, and, on the other hand, the development of the respective indicators and propose certain conclusions regarding the quality of internal and external financial audit.

Table 2. The matrix of risks faced by public institutions, 2021

Indicators	Finland	Norway	Denmark	Canada	Germany	Estonia	Lithuania	Slovenia	Latvia	Israel	Slovakia	Hungary	Poland	Romania	Georgia	Bulgaria	Armenia	Moldova	Kazakhstan	Azerbaijan	Ukraine	Russia	Belarus	Venezuela
Index of the Rule of Law																								
Government Effectiveness Index																								
Corruption check indicator																								
Corruption Perception Index																								
Regulation Quality Index																								
Index Opinions and responsibility																								
Political Stability Index																								
Shadow economy as a percentage of GDP																								

Legend: Critical risks, High risks, Moderate risks, Low risks, Acceptable risks

Source: Authors' own research based on the The Global Economy Statistics, 2023.

Next by the level of risks are some post-socialist countries, including countries of the former USSR: first of all, Estonia, Slovakia, Slovenia, Lithuania, Latvia, Poland. In this case, the risks faced by these Depending on the established results and the calculated sizes of the indices taken into account in this evaluation, the author determined the main risk categories: critical risk – represented in red color, high risk – represented in pink color, moderate risk – represented in beige color ; low risk – light green and acceptable risk – dark green.

According to the data presented in the tables above, we can say that the most promising public domains, directly public institutions, are those from countries with developed democracies, namely Denmark, Finland and Norway, followed by those from Germany and Canada. According to these resolutions, we can indicate the presence of the main situations related to the financial audit: – or the political-economic and social environment, being relatively safe, facing low-level, but acceptable risks, results in the spectrum of financial audit measures (the audit has a role reactive); – present of such a nature, respectively as a measure of internal and external audit operations (the audit has a proactive role); – or both situations occur at the same time countries are mostly moderate. This suggests the need to increase and streamline financial audits at the level of public institutions in these countries according to Miller and Power (2013).

The countries listed as Bulgaria, Hungary, Romania are the countries with public institutions that face particularly major risks with corruption problems and the lack of performance of the Government and the entire management. It is clear that such generalizations would be necessary in the evaluation of financial audit measures to remove these risks.

If we are to analyze Armenia and the Republic of Moldova, it is quite difficult. Both have the same particularly high risks in their sectors and public institutions. According to these findings, it can be concluded that the efficiency of internal and external public financial audit measures is quite low. At the same time, some countries such as Russia, Belarus, Ukraine, Kazakhstan, Azerbaijan have public domains and public institutions that face problems such as risks critical in certain areas (Statista, 2023).

Argumentation of the role of audit in the public sector of the Republic of Moldova through the unfavorable evolution of macroeconomic indicators. Considering the OECD conclusion based on the finding that 11 out of 35 of its member countries have debts above 100 % of GDP, and the consequences of the COVID-19 pandemic and the military conflict in Ukraine increase government spending, good public debt management is essential.

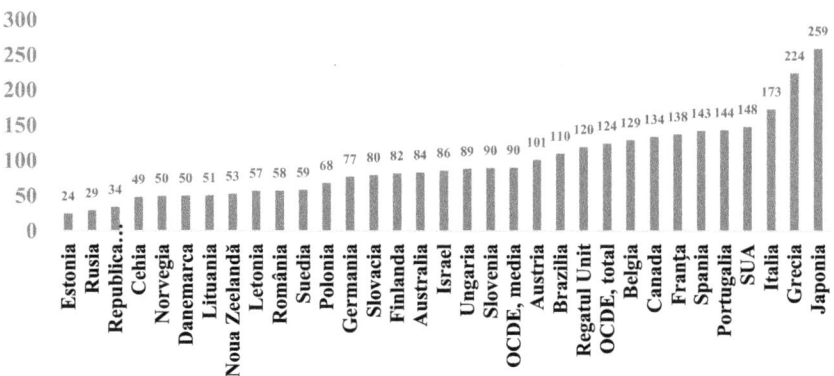

Figure 2. The public debt in % of GDP in some countries, 2021

The internal and external financial audit of public institutions intervenes in the accountability of the leadership of the countries, especially of the governments, but the planning shows us that it is inopportune in the management of public finances. The predictions are demonstrated on concrete examples and that finances are limited in the public domain, the population increasingly needs more resources, more qualitative and timely services, more qualitative infrastructure and common responses to crises (for example, COVID-19, war from Ukraine). If we are to do an analysis, then people through their contributions form the public revenues of the states. Often the leadership of countries to ensure the needs of the population often resort to debt methods – public debt, which from year to year increases globally; when requesting foreign financing through grants, extending the inflationary phenomenon.

Conclusions

Following the evaluation of the quality of financial audit at the level of public institutions in the Republic of Moldova, the authors formulate the following conclusions:

The mission of the Ministries of Finance in this sense consists in the elaboration, promotion and realization of the objectives of these regulations in accordance with the provisions of international standards in the field, ensuring of the reform and continuous development of accounting and auditing of financial situations in the corporate sector in accordance with EU Directives.

A financial management regime, including financial control and audit, must take into account the following two categories of problems and risks: risks inherent in the public institution model itself, but also those specific to a country and its public sector at any given time.

The audit entities perform the audit of the financial statements in accordance with the International Auditing Standards, issued by the Council for International Auditing and Assurance Standards, which become valid in The Republic of Moldova after their acceptance by the Government of the Republic of Moldova, with the application of the principles of professional ethics provided by the Code of Ethics to professional accountants and prepare the auditor's report to be presented to the audited entities. The unfavourable evolution of macroeconomic indicators (share of public debt to GDP, ratio of budget deficit to GDP, share of grants in total BPN revenues, dynamics of consumer price index) argues the insufficient role of public audit, including financial audit, in the public sector of the Republic of Moldova.

References

National Bureau of Statistics (NBS). 2022. *National and regional accounts of the Republic of Moldova*, p.67. Available at: <https://statistica.gov.md/files/files/publicatii_electronice/Conturi_nationale/Conturi_Nationale_editia_2022.pdf> [Accessed 17 September 2023].

Civil Code of the Republic of Moldova, no. 1107 of 06.06.2002. *Official monitor of the Republic of Moldova*, 01.03.2019, 66–75 art. 132. Available at: <https://www.legis.md/cautare/getResults?doc_id=112573&lang=ro> [Accessed 17 September 2023].

Copaceanu, C. 2016. Promoting and developing the internal audit profession through the Association of Internal Auditors of the Republic of Moldova. *Journal of Interdisciplinary Studies "C. Stere"*, 2(10), pp. 19–26.

Decision No. 33 of 12-23-2019 regarding the approval of the Report regarding compliance with audit quality control procedures. *Official Monitor of the Republic of Moldova*, 17–01–2020, pp. 7–13 art. 46.

Getie Mihret, D. and Wondim Yismaw, A. 2007. Internal audit effectiveness an Ethiopian public sector case study. *Managerial Auditing Journal*, 22(5), pp. 470–484. https://doi.org/10.1108/02686900710750757.

Lartey, P.Y., Kong, Y., Bah, F.B.M., Santosh, R.J. and Gumah, I.A. 2019. Determinants of internal control compliance in public organizations; Using preventive, detective, corrective and directive controls. *International Journal of Public Administration*, pp.1–13.

Law regarding the audit of financial situations no. 271 of 15.12.2017. *Official monitor of the Republic of Moldova*, 2018, 7–17, 66. Available at: <https://www.legis.md/cautare/getResults?doc_id=110387&lang=ro> [Accessed 17 September 2023].

Law of the Republic of Moldova regarding the activity of banks no. 202 of 06.10.2017. *Official monitor of the Republic of Moldova*, 15.12.2017, no. 434–439.

Lenz, R. and Hahn, U. 2015. A synthesis of empirical internal audit effectiveness literature pointing to new research opportunities. *Managerial Auditing Journal*, 30(1), pp. 5–33. https://doi.org/10.1108/MAJ-08-2014-1072.

Lois, P., Drogalas, G., Nerantzidis, M., Georgiou, I. and Gkampeta, E. 2021. Risk-based internal audit: Factors related to its implementation. *Corporate Governance: International Journal of Business in Society*, 21(4), pp. 645–662(18).

Medina, L. and Schneider, F. 2018. Shadow economies around the world: What did we learn over the last 20 years?. *International Monetary Fund*, Working Paper 2018/017. [online] Available at: <https://www.imf.org/en/Publications/WP/Issues/2018/01/25/Shadow-Economies-Around-the-World-What-Did-We-Learn-Over-the-Last-20-Years-45583>

Miller, P. and Power, M. 2013. Accounting, organizing and economizing: connecting accounting research and organization theory. *Academy of Management Annals*, 7(1), pp. 557–605.

Ministry of Finance, Republic of Moldova. 2023. Open data portal. *Data regarding the execution of budgets of local public authorities*. [online] Available at: <https://www.mf.gov.md/ro/content/catalogul-de-date-deschise-al-mf-pentru-anul-2023> [Accessed 17 September 2023].

Statista. 2023. Global economy – Statistics & facts. Available at: <https://www.statista.com/topics/1467/global-economy/#topicOverview> [Accessed 17 September 2023].

The Global Economy Statistics. 2023. Available at: <Economic indicators data for over 200 countries | TheGlobalEconomy.com> [Accessed 17 September 2023].

World Justice Project® (WJP). 2023. *Measuring the rule of law*. Available at: <https://worldjusticeproject.org/> [Accessed 17 September 2023].

World Bank. 2023. Worldwide governance indicators. Available at: <https://www.worldbank.org/en/publication/worldwide-governance-indicators> [Accessed 17 September 2023].

Marieta Nesheva, Leyda Todorova

Assessment of plum hybrids based on their viral symptoms in field conditions

Abstract: *The main objective of the plum breeding programs is obtaining cultivars resistant or tolerant to Plum pox virus (PPV). The current study aimed to evaluate the viral symptoms of plum hybrids and trace the inheritance of their reaction. The evaluation included 86 hybrids originating from the parental combinations "Ortenauer x Stanley", "Belle de Louvain x" Čačanska lepotica' and "Pulpudeva x H 1-22". The hybrids were planted in 2011 under a natural infection background. In 3 consecutive years, visual observations traced the presence of viral symptoms and their severity. The presence of Plum pox virus was proved by ELISA test. The highest number of symptomless hybrids was reported for the Ortenauer x Stanley hybrid family based on visual inspections. Symptoms of necrosis characterising the plants' hypersensitive reaction were observed only in the hybrid family Ortenauer x Stanley. The severity of the symptoms varied over the years.*

Keywords: breeding, evaluation, hybridological analyzes, Prunus domestica L.

Introduction

The European plum (*Prunus domestica* L.) is one of Bulgaria's main stone fruit species. It is widely grown in all regions of the country, and this species is the second after sweet cherry in areas occupied. According to Agrostatistics (MAF, 2022) in 2021, in Bulgaria, domestic plum ranks first in the quantity of fruit produced 65,123 tons. Along with the species' wide distribution, a major disease related to it is endemic in Bulgaria. The Sharka disease, caused by the Plum pox virus (PPV), is among the most harmful and a major limiting factors for plum production (Milusheva and Bozhkova, 2015).

Typical viral symptoms on leaves include chlorotic spots, rings or whole regions, and necrotic areas. In addition, deformations and premature fruit drop may be observed, reducing fruit quality and yield (García et al., 2014; Sochor et al., 2012). It was estimated that the losses cause by the disease had cost a total of ~10,000 million Euros over the previous 30 years worldwide (Barba et al., 2011; Barba, Ilardi and Pasquini, 2015; Cambra et al., 2006).

The most reliable approach to sustainable plum production is growing resistant or tolerant cultivars. As there is no effective chemical control against viral diseases, tolerance or resistance to PPV is one of the main breeding objectives nowadays. Due to the complex genetic structure of *Prunus domestica* L., limiting the spread of the viral particles is possible only through the hypersensitive reaction of plants (Neumüller et al., 2013). So far, this is the most promising way of resistance in plums. In the absence of resistant cultivars, tolerant cultivars, which display symptoms on leaves and symptomless fruit, have been used in some breeding programs.

The current study aimed to evaluate the viral symptoms of plum hybrids and trace the inheritance of their reaction. The evaluation included 86 hybrids originating from parental combinations "Ortenauer x Stanley", "Belle de Louvain x" Čačanska lepotica' and "Pulpudeva x H 1-22".

Literature review

The plum (*Prunus domestica* L.) is hexaploid and due to the high decay of the traits in the progeny, the breeding of this species is a difficult task. So far, a team of German scientists has made a breakthrough in this direction in 1999, when the plum cultivar "Jojo" was created. However, its resistance to Sharka is not absolute but is based on its hypersensitivity reaction to the virus (Kegler et al., 2000). Up to date, this phytopathogen limiting mechanism is still the most promising possible in plum.

In the past twenty-seven original plum cultivars and several "Kyustendilska sinya" cultivar clones were registered in Bulgaria. Most of them were obtained due to purposeful breeding activities and the rest of them – by open pollination of local plum cultivars or of the cultivars "Stanley" and "President". Nine Bulgarian cultivars resulted from crossing of "Kyustendilska sinya" x "Montfort". "Kyustendilska sinya" was used as a mother form in establishing twelve cultivars and in "Pop Hariton" and "Baleva", which are complex hybrids, it was also used as a mother form in a combination in F1 progeny (Zhivondov, Bozhkova and Milusheva, 2012). The choice of that cultivar to be used in the breeding is not random. It has several valuable economic and biological characteristics, such as high fertility, late flowering period, self-fertility, excellent taste, and non-adherent stone. However, the cultivar has been excluded from modern breeding programs because of its high susceptibility to PPV (Bozhkova, 2013).

The plum breeding program of the Fruit Growing Institute in Plovdiv was officially started in 1987. Its main objectives are the development of new cultivars highly tolerant or resistant to PPV, with good biological and economic characteristics (Zhivondov, 1994; Zhivondov, Bozhkova and Milusheva, 2012).

In 2004–2012, 61 parent combinations were developed at the Fruit Growing Institute in Plovdiv, obtaining 1568 hybrid seeds (Bozhkova, 2011). The most productive as parental cultivars under the conditions in our country proved to be "Stanley", "Althan's gage", "Ortenauer" and "Pacific" used as the mother parent, and the cultivars "Čačanska lepotica" and "Čačanska najbolja" – as the father parent (Bozhkova, 2011).

Methodology

The study was conducted in the period 2020–2022 in a breeding orchard at the Fruit Growing Institute – Plovdiv. A total number of 86 hybrids from the parental combinations "Ortenauer x Stanley", "Belle de Louvain x Čačanska lepotica" and "Pulpudeva x H 1-22" were evaluated based on the visually observed leaf symptoms typical for *Plum Pox Virus*. The presence of the virus in the orchard was proved by an ELISA test of hybrids with typical; symptoms.

The shape and type of the symptoms were visually determined and described. Based on the visual symptoms the degree of attack was determined for 15 leaves taken from each hybrid. The percentage of leaf area with symptoms of each leaf in relation to healthy leaf tissue was estimated visually. Area of observed external disease symptoms were scored for disease index using a 5-grade scale: 0 – no symptoms; 1 – symptoms observed on 1–25 % of the leaf surface; 2 – symptoms observed on 26–50 % of the leaf surface; 3 – symptoms observed on 51–75 % of the leaf surface; 4 – symptoms observed on 76–100 % of the leaf surface.

The severity data were processed by McKinney's formula, which generates a numeric disease severity index. $SI(\%)=(\Sigma vn)/(NV)\times 100$, where v represents the numeric value of the disease index scale, n is the number of leaves assigned to the disease index scale, N is the total number of the leaves and V is the numeric value of the highest disease index scale.

Descriptive statistics and hierarchical cluster analyzes using the between-groups linkage method of the IBM SPSS Statistics 26 statistical software were used for statistical data processing.

Results and discussions

Sharka disease is caused by the Plum pox virus (PPV) and is the phytopathogen of the most significant economic importance in plum orchards. It was first described in 1933 by Prof. D. Atanasov (Atanasov, 1934). The typical viral symptoms on leaves include chlorotic spots, rings or whole regions, and necrotic areas (Sochor et al., 2012). There are no immune plums, so a new strategy for obtaining resistance by breeding was required. The hypersensitive reaction is a defence reaction noted in some plum cultivars and hybrids by necrotic leaf spots (Hartman, 1998). Hartman (2002) assumed that the hypersensitivity trait originates from the cultivar "Ortenauer", and the cultivar is used as a donor of the hypersensitivity resistance in other breeding programs (Jacob, 2007; Lichtenegger et al., 2010; Neumüller, Treutter and Hartmann, 2010).

In the three hybrid families were observed different types of symptoms. Necrosis was observed only in the hybrid family "Ortenauer x Stanley" (Figure 1). Some of the hybrids showing this kind of symptoms died in the second and third year of the study. In this hybrid family, the highest number of symptomless hybrids was recorded, varying from 41.86 % in the first year of the study to 23.53 % in the second year. A lower percentage of hybrids with no symptoms was observed in the hybrid family "Belle de Louvain x" Čačanska lepotica' – from 20.69 % to 13.79 % in the different years (Figure 2). The higest number of hybrids in the three years of observations were the ones showing a combination of chloritic rings and regions. In the first two years, the hybrid family "Pulpudeva x H 1-22" had the highest number of hybrids with chlorotic rings. In the third year of the observations, the number of hybrids with combined chlorotic rings and regions was the highest (Figure 3). Hybrids with no viral symptoms in this hybrid family were observed only in the first year.

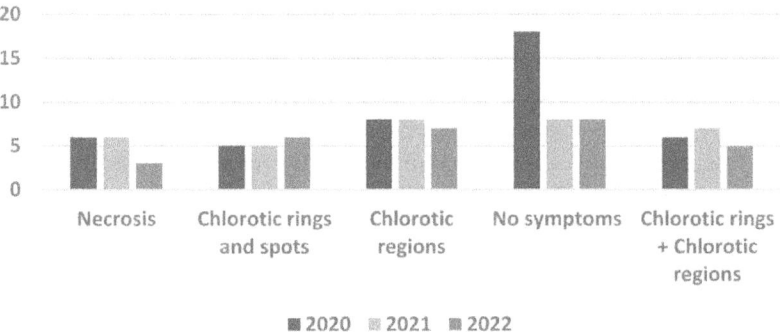

Figure 1. Number of hybrids with a different type of symptoms from "Ortenauer x Stanley" hybrid family

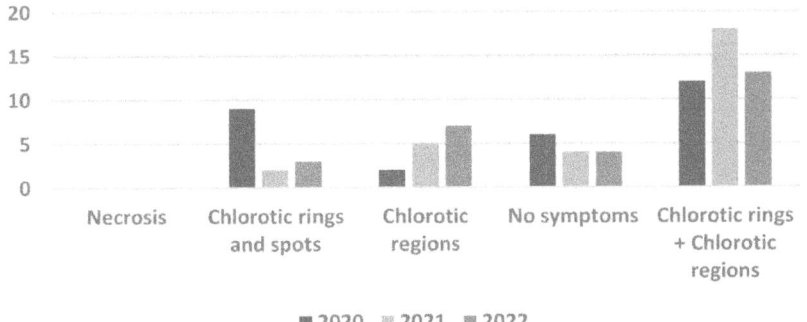

Figure 2. Number of hybrids with a different type of symptoms from "Belle de Louvain x" Čačanska lepotica' hybrid family

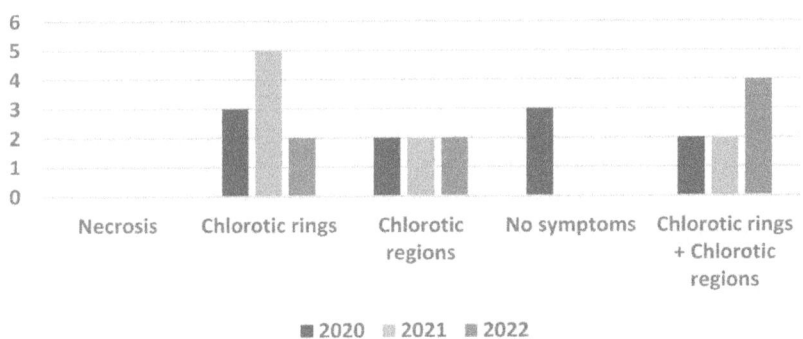

Figure 3. Number of hybrids with a different type of symptoms from "Pulpudeva x Hybrid 1–22" hybrid family

Disease severity is defined as the "area of a sampling unit (plant surface) affected by disease expressed as a percentage or proportion of the total area" (Nutter Jr., Teng and Shokes, 1991). Quantification of plant disease intensity is required for many different purposes, including understanding yield loss, evaluating the effects of treatments and comparing phenotypes for disease resistance (Cooke, 2006; Madden, Hughes and van den Bosch, 2007; Bock et al., 2010).

The calculated severity index for the symptoms observed on the leaves of the hybrids varied in the different years and hybrid families (Table 1). The estimated mean severity index for the hybrids in the "Ortenauer x Stanley" parental combination was the lowest in three consecutive years. The highest mean severity of the symptoms was observed for the hybrids obtained from the "Belle de Louvain" x "Čačanska lepotica" parental combination.

Table 1. Variation of the symptoms' severity index (%) in the hybrid families

"Ortenauer x Stanley"				
Year	Minimum	Maximum	Mean	Std. Deviation
2020	0.00	68.00	23.86	21.65
2021	0.00	77.33	22.87	27.87
2022	0,00	83.66	25.92	29.00
"Belle de Louvain x 'Čačanska lepotica"				
Year	Minimum	Maximum	Mean	Std. Deviation
2020	0.00	73.33	32.41	25.32
2021	0.00	76.00	34.76	26.75
2022	0.00	72.00	33.04	24.23
"Pulpudeva x Hybrid 1–22"				
Year	Minimum	Maximum	Mean	Std. Deviation
2020	0.00	72.00	24.13	29.71
2021	2.66	80.00	28.53	28.12
2022	2.67	76.00	34.76	28.25

Source: Authors' own research.

The three-year data for the severity index for each hybrid was classified using Hierarchical cluster analyzes. The hybrids from the parental combination "Ortenauer x Stanley" were divided into 5 clusters (Figure 4). The first cluster includes 13 hybrids with low severity index values estimated for each one of the three years – from 0 % to 6.67 %. The hybrids OS 12–30 and OS 12–31 had low severity indexes in two of the experimental years, increasing to 20 % and 22.22 % resp. in the third year. The hybrids in the other clusters had higher values of the calculated severity index of the symptoms observed. The hybrid family "Belle de Louvain x 'Čačanska lepotica" was also divided into 5 main clusters (Figure 5). In the first cluster are included 3 hybrids with a severity index 0 % in the three years and 3 hybrids with severity index varying from 0 % to 9.33 %. For the third hybrid family "Pulpudeva x Hybrid 1–22" the hybrids included in the cluster with the lowest values are only 3 with a severity index of 2.67 to 26.66 % (Figure 6). The most valuable for future breeding schemes aiming at resistance to the viral disease are the hybrid families "Ortenauer x Stanley" and "Belle de Louvain x Čačanska lepotica". In this hybrid, families were observed high number of plants with no symptoms or plants with low disease severity index in the three years.

Assessment of plum hybrids based on viral symptoms 231

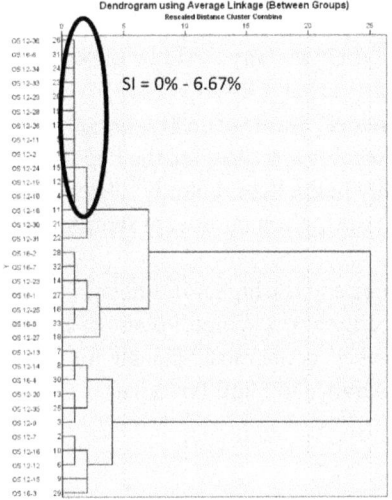

Figure 4. Hierarchical cluster analyzes of the three-year data for "Ortenauer x Stanley", using average linkage (Between Groups)

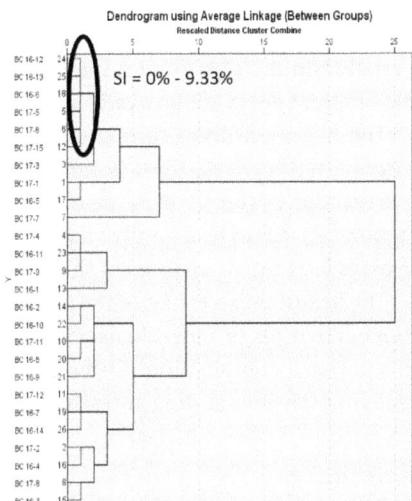

Figure 5. Hierarchical cluster analyzes of the three-year data for "Belle de Louvain x 'Čačanska lepotica", using average linkage (Between Groups)

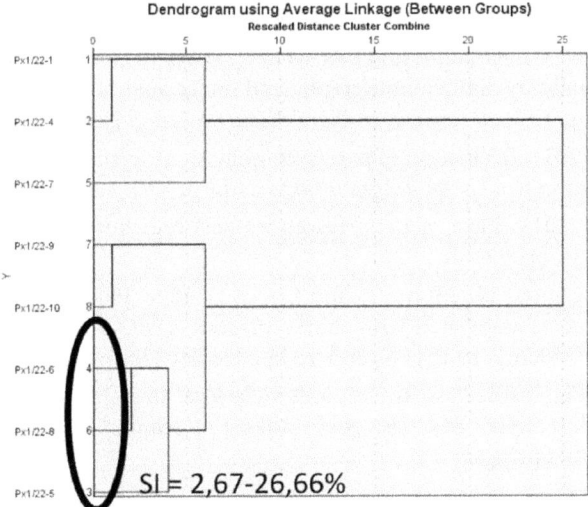

Figure 6. Hierarchical cluster analyzes of the three-year data for "Pulpudeva x Hybrid 1–22", using average linkage (Between Groups)

Conclusions

Variation in the typical viral symptoms and their severity could be observed in the different years of observation.

The hypersensitive plum cultivar "Ortenauer" transmit in the progeny this reaction to the virus. Viral symptoms with necrosis, typical for the hypersensitive reaction of the plants, were observed only in the hybrid family "Ortenauer x Stanley". In this hybrid family were observed the highest number of symptomless hybrids over the three years of the research.

The lowest mean values of the severity index and the highest number of symptomless hybrids or individuals with low severity of the symptoms shows us that this parental combination "Ortenauer x Stanley" is the most valuable for future breeding schemes aiming at resistance or tolerance to Plum Pox Virus.

References

Atanasov, D. 1934. *Diseases of cultivated plants.* Sofia: Pridvorna Pechatnica.

Barba, M., Hadidi, A., Candresse, T. and Cambra, M. 2011. Plum pox virus. In: A. Hadidi, M. Barba, T. Candresse, and W. Jelkmann, eds. *Virus and virus-like diseases of pome and stone fruits*, St. Paul, MN: APS Press, pp. 185–197.

Barba, M., Ilardi, V. and Pasquini, G. 2015. Control of pome and stone fruit virus diseases. *Advances in Virus Research*, 91, pp. 47–83. https://doi.org/10.1016/bs.aivir.2014.11.001.

Bock, C., Poole, G., Parker, P. and Gottwald, T. 2010. Plant disease severity estimated visually, by digital photography and image analysis, and by hyperspectral imaging. *Critical Reviews in Plant Science Science,* 29, pp. 59–107.

Bozhkova, V. 2011. Evaluation of parent combinations fertility in plum breeding (Prunus domestica L). *Acta Agriculturae Serbica*, XVI(31), pp. 43–49.

Bozhkova, V. 2013. Plum genetic resources and breeding. *AgroLife Scientific Journal*, 2(1), pp. 83–88.

Cambra, M., Capote, N., Myrta, A. and Llacer, G. 2006. Plum pox virus and the estimated costs associated with sharka disease. *EPPO Bulletin*, 36, pp. 202–204. https://doi.org/10.1111/j.1365-2338.2006.01027.x.

Cooke, B. 2006. Disease assessment and yield loss. In: B.M. Cooke, D. Gareth Jones, and B. Kaye, eds. *The epidemiology of plant diseases.* 2nd ed. The Netherlands: Springer.

García, J., Glasa, M., Cambra, M. and Candresse, T. 2014. Plum pox virus and sharka: A model potyvirus and a major disease. *Molecular Plant Pathology*, 15(3), 226–241.

Hartmann, W. 1998. New plum varieties form Hohenheim. *Acta Horticulturae*, 478, pp. 171–174.

Hartmann, W. 2002. The importance of hypersensitivity for breeding plums and prunes resistant to Plum pox virus (Sharka). *Acta Horticulturae*, 577, pp. 33–37. https://doi.org/10.17660/ActaHortic.2002.577.3.

Jacob, H. 2007. Ripening time, quality and resistance donors of genotypes of Prunus domestica and their inheritance pattern in practical plum breeding. *Acta Horticulturae*, 734, pp. 77–82. https://doi.org/10.17660/ActaHortic.2007.734.7.

Kegler, H., Schwarz, S., Fuchs, E. and Gruentzig, M. 2000. Sharka resistant plums and prunes by utilization of hypersensitivity. *Acta Horticulturae*, 538, pp. 391–395. https://doi.org/10.17660/ActaHortic.2000.538.69.

Lichtenegger, L., Neumüller, M., Treutter, D. and Hartmann, W. 2010. The inheritance of the hypersensitivity resistance of European plum (Prunus domestica L.) against the Plum Pox Virus. *Julius-Kühn-Archiv*, 427, p. 327.

Madden, L., Hughes, G. and van den Bosch, F. 2007. *The study of plant disease epidemics*. St. Paul, MN: APS Press.

Milusheva, S. and Bozhkova, V. 2015. Reaction of six Prunus rootstocks to Plum Pox Virusin Plovdiv, *Bulgaria. Acta Horticulturae*, 1063, pp. 111–116. https://doi.org/10.17660/ActaHortic.2015.1063.15.

Ministry of Agriculture and Food (MAF) of Republic of Bulgaria. 2022. *Agricultural statistics handbook*. [online] Available at: <https://www.mzh.government.bg/bg/statistika-i-analizi/izsledvane-rastenievadstvo/danni/> [Accessed 1 October 2023].

Neumüller, M., Treutter, D. and Hartmann, W. 2010. Breeding for sharka resistance and high fruit quality in European Plum (Prunus Domestica l.) at Weihenstephan: Breeding strategy and selection tools. *Acta Horticulturae*, 874, pp. 221–228. https://doi.org/10.17660/ActaHortic.2010.874.30.

Neumüller, M., Mühlberger, L., Siegler, H., Hartmann, W. and Treutter, D. 2013. New rootstocks with resistance to Plum Pox Virus for Prunus domestica and other stone fruit species: the 'Docera' and 'Dospina' rootstock series, *Acta Horticulturae*, 985, pp. 155–165. https://doi.org/10.17660/ActaHortic.2013.985.19.

Nutter, F. Jr., Teng, P. and Shokes, F. 1991. Disease assessment terms and concepts, *Plant Disease*, 75, pp. 1187–1188.

Sochor, J., Babula, P., Adam, V., Krska, B. and Kizek, R. 2012. Sharka: The past, the present and the future. *Viruses*, 4(11), pp. 2853–2901. https://doi.org/10.3390/v4112853.

Bianca-Florentina Nistoroiu, Ragif Huseynov,
Ștefan Laurențiu Prahoveanu

Sustainable development through gender equality

Abstract: *This article examines the role gender equality plays in promoting sustainable development and provides a thorough examination of the interactions between these two crucial elements of global growth. The long-term viability of the economy, society, and the environment has been deemed to be dependent on gender equality, an essential human right and one of the United Nations' Sustainable Development Goals (SDGs). The economic benefits of gender equality are discussed in this article, including how reducing gender-based inequities in labor, entrepreneurial activity, and accessibility to financial assets may help to fight poverty and spur economic growth. Along with their contributions to innovation and efficiency, women's engagement in leadership roles and decision-making processes is also emphasized.*

Keywords: gender equality, development, resilience, economy, sustainability.

Introduction

In a world that is changing swiftly, as evidenced by accelerated population growth and escalating environmental problems, achieving sustainable agricultural growth has become crucial for ensuring food security and the sustainability of the environment (Doss and Meinzen-Dick, 2015; Mumtaz, 1995). A key driver for attaining sustainability in agriculture is gender equality, which is a cornerstone of human rights. Individuals can uncover a wealth of advantages for rural communities as well as the environment by empowering women and guaranteeing their fair involvement in agricultural operations (Quisumbing et al., 2015).

In all cultures, women are systematically disadvantaged in relation to men, and gender equality is a fundamental part of sustainable development. The following are some of the ways in which gender equality can be an enabler of sustainable development. Promoting women's and girls' empowerment plays a pivotal role in fostering economic development. By ensuring equal rights to economic resources, including access to, control over, and ownership of property, technology, and the Internet, we can advance sustainable economic growth. The United Nations Development Programme (UNDP) emphasizes the significance of granting women these economic rights in its 2022 report (UNDP, 2022) and OECD (2022).

Sustainable development can be advanced by providing all students with the knowledge and skills necessary for this endeavour. This includes education focused on sustainable development, environmentally friendly lifestyles, human rights, gender equality, the promotion of a culture of peace and nonviolence, global citizenship, and an appreciation of cultural diversity. These educational components are vital to building a foundation for sustainable development, as outlined by the UNDP in 2022 (UNDP, 2022).

Achieving environmental sustainability necessitates recognizing and addressing the gender-environment nexus, maintaining policy coherence, adopting a well-being-centered approach, and transitioning toward inclusive and green growth. Women's multifaceted roles in society, the economy, and households have a direct impact on natural resource management. By adopting a gender equality perspective, we can observe that the 2030 Agenda's key objectives of "shifting the world onto a sustainable path" and "leaving no one behind" are intertwined with the gender-environment nexus. These insights are highlighted in a report by the Organization for Economic Co-operation and Development (OECD, 2021).

Increasing the representation of women in political leadership positions is pivotal to advancing gender equality. The removal of structural obstacles and the achievement of Goal 5 (gender equality) require political leadership, investments, and comprehensive policy reforms. This perspective on the importance of female political leadership was underscored by the United Nations in 2015. In order to achieve gender equality and sustainable development, women must have equitable access to property and land, sexual and reproductive health, technology, and the Internet. Implementing measures that provide women with equal access to financial resources, as well as ownership and authority over property, is essential. The UNDP works to advance gender equality in sustainable development and economic growth by putting into action programs that aim to give both men and women jobs, advocating for national legal frameworks and development strategies for women's economic empowerment, assisting governments and partners in the public and private sectors to acknowledge, redistribute, and lessen women's unpaid care and domestic work, and offering grants, business development, and other resources to women-led businesses (UNPD, 2022).

The relationship between gender equality and sustainable agriculture is gaining attention and significance. The many facets of this connection, the difficulties and possibilities it brings, and the evidence-based tactics that have been effective in advancing gender equality via sustainable agricultural development will all be explored in order to go deeper into this subject. In order to increase production, decrease poverty, and promote higher climate change resilience, the gender gap in agriculture needs to be addressed (FAO, 2011).

To shed light on the various ways that gender dynamics affect agricultural practices and how focused interventions might close the gender gap for the benefit of rural communities and the larger environment, we will rely on a variety of research literature and case studies. Unquestionably prevalent gender dynamics in agriculture highlight the critical responsibilities that both men and women play in a variety of aspects of agricultural productivity, resource management, and food security. The agricultural industry is a good example of the gendered division of labor since women are typically the backbone of agricultural operations while also being mostly excluded from resources, productive inputs, and decision-making positions. It impedes agricultural growth and contributes to the persistence of poverty and hunger in many rural areas (Kabeer, 2012). This gap in access to resources and opportunities is intricately tied to the larger discourse on gender inequality. The current as a springboard for a thorough investigation of the complex connections between gender equality and sustainable agriculture. We analyze the various ways in which gender disparities in agricultural systems manifest, clarify the structural barriers to gender equality, and assess the range of interventions and policy frameworks that have shown promise in reducing these disparities by drawing on a corpus of academic sources and empirical case studies. The need to support UN Sustainable Development Goal 5, which promotes gender equality and the empowerment of women and girls and is essential to global development, serves as a further justification for our thorough investigation (United Nations, 2015).

Literature review

The investigation into sustainable development through the lens of gender equality is a highly relevant and impactful topic in modern economies. Extensive literature has explored this subject, recognizing its significance in shaping sustainable economic and social frameworks. These studies emphasize the crucial role of gender equality in driving comprehensive and enduring sustainable development, advocating for inclusive policies that acknowledge and leverage the diverse contributions and needs of all genders. The growing body of research in this field underscores the need for comprehensive, gender-inclusive policies to foster equitable and enduring development across the globe. Duncan (2002) highlighted the emergence of economic rivalry, democratic processes, and the safeguarding of gender agreements as central themes at the inception of the EU's policy on gender. Nevertheless, ambiguity persists regarding the actual definition of gender mainstreaming, coupled with an inconsistent progression in the embrace of gender mainstreaming mechanisms. Rees (2005) aimed to enrich the discourse by pinpointing three core principles that seem to form the bedrock of gender mainstreaming in Europe: holistic individual consideration, democratic governance, and the principles of justice, equality, and fairness.

Also, Maier (2011) explored the repercussions of the prevailing economic downturn on gender dynamics within EU nations. Analyzes of employment trends, unemployment rates, and gender-specific evolutions led to the inference that the economic downturn, together with related fiscal policies, is unlikely to foster enhanced gender parity in the workforce or in societal realms. Debusscher (2012) scrutinized the integration of gender mainstreaming within EU's developmental assistance directed towards the European Neighbourhood, employing both quantitative and qualitative analyzes of policy documentation. Fagan and Rubery (2018) investigated the evolution of the EU's framework for gender equality, dissecting both the "hard" and "soft" legal aspects and noting the incomplete integration of gender mainstreaming within the European Employment Strategy.

Vinska and Tokar (2021) endeavored to identify categories and groupings of EU member states based on their economic progression and gender equality levels, aiming to augment the efficacy of the EU cohesion policy. Esteves (2018) underscored the significance of embedding a gender perspective within educational syllabuses, proposing recommendations to leverage gender equality as an instrument for educational enhancement and for fostering a more gender-balanced societal structure. Erum and Daud (2021) delved into the legal frameworks surrounding gender equality and mainstreaming in EU directives, with a particular focus on the EEC and the domestic policies of member states. Moreover, it underscored the universality of the Sustainable Development Goals (SDGs) under Agenda 2030, advocating for their implementation across all EU countries. Leal Filho et al. (2023) recognizes that gender issues are not confined to SDG5 (Gender Equality) alone but are integral to the achievement of all SDGs. Methodologically, Carlsen and Bruggemann (2021) recommended a specific partial order, the interval order, and provided a basic categorization placing Denmark at the forefront, Germany in an intermediate position, and the Czech Republic in a tier necessitating significant enhancements. Maranzano, Cerdeira Bento and Manera (2022) examine the role of education in influencing the relationship between pollution and income, factoring in income inequality across 17 European OECD countries from 1950 to 2015. Doğan and Kirikkaleli (2021) present a comprehensive macro-level framework that interconnects gender issues with energy policies and environmental quality, aligning these aspects with the Sustainable Development Goals (SDG). Each of these studies contributes to understanding the multifaceted nature of gender mainstreaming and gender equality within the EU, highlighting the progress, ongoing challenges, and the diverse methodologies and perspectives employed in tackling this complex issue.

Methodology

To explore the relationship between sustainable development and gender equality, it is important to have a comprehensive understanding of the complex dynamics involved. Therefore, we have adopted a descriptive analytical approach, concentrating on the data series related to the indicators presented in Figures 1–3. By using data series from Eurostat (2022) and Eurostat (2023), we aim to remain close to the central issues, ensuring that our analysis is relevant and accurately aligned with the nuances of the subject matter. This systematic approach allows us to analyze the implications of the connection between sustainable development and gender equality, presenting a clear and data-driven account that accurately reflects the situation on the ground.

Results and discussions

In the following subsections, we present and analyze the most important aspects of the concept of sustainable development through equality, as well as the most important aspects of the concept of sustainable development through equality.

Insights on gender equality in Europe

The European Union (EU) was founded on respecting and promoting equal opportunities. Therefore, discrimination based on sexual orientation, gender, age, race, nationality, disability, status, faith, and other characteristics is not tolerated. In addition to being established by law, equality must be properly implemented in all spheres of life, including politics, economics, society, and culture. In reality, there are still disparities: men and women do not have the same rights, for instance, when it comes to salary disparities, employment disparities, or the underrepresentation of women in leadership roles (Eurostat, 2022).

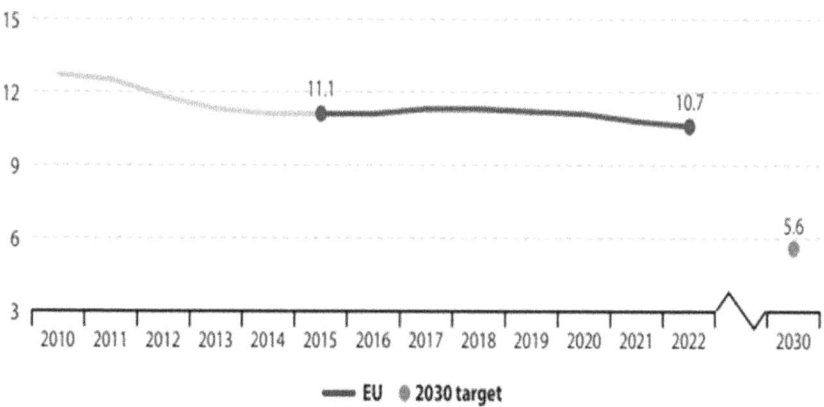

Figure. 1 Gender employment gap, EU, 2010–2022
Source: Eurostat, 2022.

As shown in Figure 1, only two metropolitan areas had higher employment rates for women than for men in 2021: Hildesheim in northern Germany and Limoges in south-west/central France. In these specific areas, women boasted employment rates of 84.1 % and 70.8 %, respectively, while men had employment rates of 71.1 % and 70.8 %, respectively. Conversely, eight metropolitan areas, all situated in western, Baltic, or Nordic Member States, displayed employment rate differences between men and women of no more than 1.5 percentage points, favouring men. In 2021, the gender employment gap will exceed 30.0 percentage points in Galați, Constanța and Taranto, which are all located in Romania, making them the three metropolitan areas with the highest gender employment gap in the EU. Similarly, various other metropolitan areas in Romania and Italy, along with Tarnów in eastern Poland, also exhibited significant gender employment gaps, as indicated in Figure 2. On a broader scale, numerous eastern and southern Member States in the EU continued to grapple with substantial gender disparities in employment rates. Based on a report published by Eurostat (2022), such differences may be influenced by regional economic structures and entrenched cultural norms that reinforce traditional gender roles due the fact that women are often assigned the role of career and housewife, contributing to the gender gap.

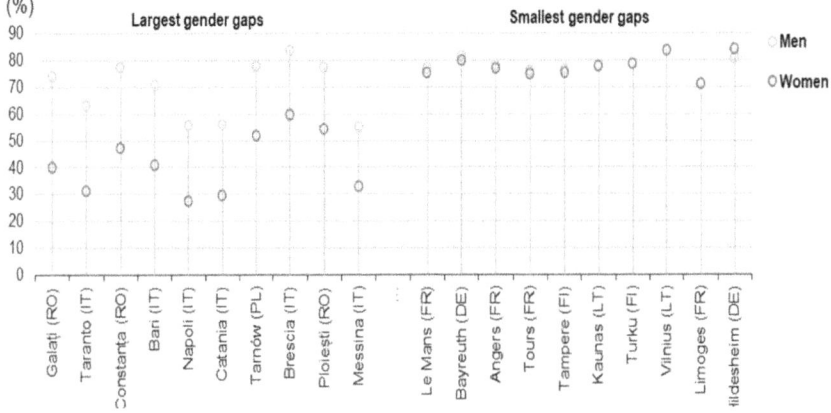

Figure 2. Employment rates (20–64 years) by sex, selected metropolitan regions, 2021 (%)
Source: Eurostat, 2022.

Within the European Union, it is a common occurrence for women to hold part-time, temporary, or precarious employment positions more frequently than men. This discrepancy plays a role in elevating the unemployment levels encountered by women, who also bear an uneven share of hardships during economic slumps, as evidenced in scenarios like the COVID-19 pandemic or the worldwide financial crisis. In urban zones within the EU, the 2021 data reveals that the unemployment figures for women aged 15 to 74 stands at 7.9 %, marginally surpassing the rate for their male counterparts, which is 7.8 %. When examining urban areas across the EU in 2021, the unemployment rate for women aged 15 to 74 was 7.9 %, slightly higher than the rate for men, which stood at 7.8 %. The most significant gender-based disparities in unemployment rates in the EU's metropolitan areas are illustrated in Figure 3.

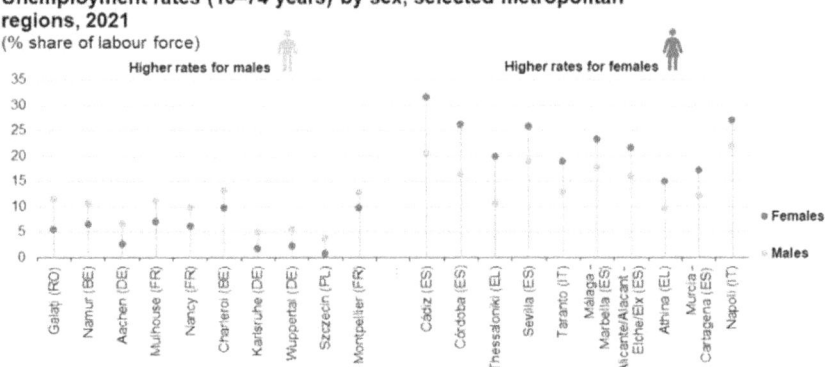

Figure 3. Unemployment rates (15–74 years) by sex, selected metropolitan regions, 2021 (% share of the labour force)
Source: Eurostat, 2022.

The figure on the left shows areas where male unemployment rates exceeded those of females. This pattern was generally observed across most metropolitan areas in western and northern EU Member States, as well as in Poland, where male unemployment rates exceeded female rates.

The importance of gender equality in agriculture

Sustainable development in agriculture depends on gender equality. Since they have less access to land, markets, farming methods, fertilizer, financing, and training, women farmers are less productive than men. Women might boost their yields and help feed more needy people around the globe if they had equal access to opportunities and resources. In reality, women could raise agricultural yields by 20–30 % and feed an extra 150 million people if they had equal access to productive resources as men (Sexsmith, Smaller and Speller, 2017). Firstly, enhancing productivity growth hinges on granting women equal access to vital resources, investments, and education. By eliminating gender-based barriers, women can not only produce more but also yield higher agricultural outputs. This, in turn, plays a crucial role in reducing global hunger and poverty, while simultaneously contributing to overall economic growth. Secondly, reducing the gender gap in agriculture holds the potential to safeguard the world's food supply, especially in the face of climate change and the ever-expanding, increasingly food-insecure global population. Prioritizing gender equality within the realm of agriculture is instrumental in securing food resources and ensuring their sustainability (Farhall and Rickards, 2021).

In addition to this, investments directed at addressing gender disparities within agricultural value chains and amplifying the yields and incomes of women-led farms carry profound implications for diminishing poverty worldwide. Such investments not only bolster overall economic growth but also mitigate hunger and poverty on a global scale. Lastly, sustainable agriculture necessitates the rectification of gender inequalities. The attainment of agricultural sustainability becomes a more complex endeavor when high levels of gender-based inequality persist. Thus, achieving sustainability in agriculture is intrinsically linked to rectifying these gender disparities, which are pivotal to global food security and economic well-being (Farhall and Rickards, 2021). To conclude, gender parity in agriculture is crucial for long-term growth. Women's productivity and yields can be increased, hunger and poverty throughout the world can be decreased, and total economic growth can be increased by giving women equal access to resources, investments, education and removing gender-specific barriers (Farhall and Rickards, 2021).

Gender issues in agriculture and rural development

Gender disparity in agriculture: For decades, agricultural policymakers have ignored the existence of women farmers because they see farmers as males, undermining women's right to participate in agriculture and other endeavors that have a direct impact on their lives. Because of this, rural women have consistently trailed behind men in terms of most social and economic indicators. Most agricultural policy makers perceive no need to address gender issues until an economic necessity, such as labor shortages, diminishing yields, or the realization of the need for the specific expertise that women have, requires a change (Farhall and Rickards, 2021). Feminization of agriculture: One significant problem is the "feminization of agriculture," which is a result of males moving off-farm in disproportionate numbers in search of paid jobs in cities. This has made women the primary food producers in many rural areas (Farhall and Rickards, 2021).

Gender inequality in the agricultural industry results in significant disparities between men and women. Women face obstacles to full mobility and political engagement, as well as limited access to opportunities and resources. This leads to a gender gap in agricultural production, which has a significant impact on countries where agriculture is a major source of both food and income. To secure women's access to and control over resources and assets, gender mainstreaming must be expanded in the major agricultural and rural development strategies. (FAO, 2019).

Conclusions

In conclusion sustainable development depends on gender equality to ensure food security, promote economic growth and reduce poverty. By removing gender-specific barriers and providing women with equal access to resources, investment and education, we can increase women's productivity and yields, reduce global hunger and poverty, and promote overall economic growth. The amount of hunger and poverty throughout the world may be decreased, and overall economic growth can be boosted, by making investments in methods that aim to close gender disparities in agricultural value chains and boost the yields and incomes of women-run farms. By eliminating the gender wage gap in agriculture, we can increase women's income, enhance children's health, nutrition, and educational outcomes, and encourage more environmentally friendly agricultural methods.

Ultimately, a fundamental element of sustainable development revolves around the intersection of gender and agricultural progress. In order to foster gender equality and empower women within the agricultural sector, enabling them to engage in and reap the benefits of agricultural growth, it becomes imperative to incorporate a gender-focused lens into agricultural practices. An indispensable facet of alleviating rural poverty and bolstering food security lies in the promotion of gender equality. The empowerment of women translates into advantages that extend beyond individuals and families to encompass entire communities and nations. By harnessing and furthering the contributions of women and dismantling the barriers they encounter, national governments and the global community can make significant strides in achieving their objectives in the realms of agricultural development, economic expansion, and food security.

References

Carlsen, L. and Bruggemann, R. 2021. Gender equality in Europe: The development of the sustainable development goal no. 5 illustrated by exemplary cases. *Social Indicators Research*, 158(3), pp. 1127–1151.

Debusscher, P. 2012. Mainstreaming gender in European Union development policy in the European neighborhood. *Journal of Women, Politics & Policy*, 33(4), pp. 322–344.

Doğan, N. and Kirikkaleli, D. 2021. Does gender equality in education matter for environmental sustainability in sub-Saharan Africa?. *Environmental Science and Pollution Research*, 28(29), pp. 39853–39865.

Doss, C. and Meinzen-Dick, R. 2015. Women in agriculture: Four myths. *Global Food Security*, 4, pp. 46–56.

Duncan, S. 2002. Policy discourses on 'reconciling work and life'in the EU. *Social Policy and Society*, 1(4), pp. 305–314.

Erum, R. and Daud, S. 2021. Gender parity and women empowerment in Europe efforts of European Union. *Pakistan Journal of International Affairs*, 4(2).

Esteves, M. 2018. Gender equality in education: A challenge for policy makers. *International Journal of Social Sciences*, 4(2), pp. 893–905.

Eurostat. 2022. *Urban-rural Europe – equality in cities*. [online] Available at: <https://ec.europa.eu/eurostat/statistics-explained/index.php?title=Urban-rural_Europe_-_equality_in_cities> [Accessed 13 October 2023].

Eurostat. 2023. *Sustainable development in the European Union – Statistical annex to the EU voluntary review – 2023 edition*. [online] Available at: <https://ec.europa.eu/eurostat/web/products-statistical-reports/w/ks-05-23-188> [Accessed 13 October 2023].

Fagan, C. and Rubery, J. 2018. Advancing gender equality through European employment policy: The impact of the UK's EU membership and the risks of Brexit. *Social Policy and Society*, 17(2), pp. 297–317.

Food and Agriculture Organization (FAO). 2011. *The state of food and agriculture 2010–11: Women in agriculture – Closing the gender gap for development*. [online] Available at: <https://www.fao.org/3/i2050e/i2050e.pdf> [Accessed 13 October 2023].

Food and Agriculture Organization (FAO). 2019. *Country gender assessment of agriculture and the rural sector in Indonesia*. [online] Available at: <https://www.fao.org/documents/card/zh?details=CA6110EN/> [Accessed 13 October 2023].

Farhall, K. and Rickards, L. 2021. The "gender agenda" in agriculture for development and its (lack of) alignment with feminist scholarship. *Frontiers in Sustainable Food Systems*, 5, 573424.

Kabeer, N. 2012. *Women's economic empowerment and inclusive growth: Labour markets and enterprise development*. SIG Working Paper 2012/1, Geneva: ILO.

Leal Filho, W., Kovaleva, M., Tsani, S., Țîrcă, D. M., Shiel, C., Dinis, M. A. P., ... and Tripathi, S. 2023. Promoting gender equality across the sustainable development goals. *Environment, Development and Sustainability*, 25(12), pp. 14177–14198.

Maier, F. 2011. Will the crisis change gender relations in labour markets and society?. *Journal of Contemporary European Studies*, 19(01), pp. 83–95.

Maranzano, P., Cerdeira Bento, J. P. and Manera, M. 2022. The role of education and income inequality on environmental quality: A panel data analysis of the EKC hypothesis on OECD countries. *Sustainability*, 14(3), p. 1622.

Mumtaz, K. 1995. *Gender issues in agricultural and rural development policy in Asia and the Pacific*. Bangkok: FAO.

OECD. 2021. *Gender and the environment: Building evidence and policies to achieve the SDGs*. Paris: OECD Publishing. https://doi.org/10.1787/3d32ca39-en.

OECD. 2022. *Gender, inclusiveness and the SDGs*. [online] Available at: <https://www.oecd.org/environment/gender-inclusiveness-and-sdg.htm> [Accessed 13 October 2023].

Quisumbing, A.R., Rubin, D., Manfre, C., Waithanji, E., van den Bold, M., Olney, D. K. and Meinzen-Dick, R. 2015. *Gender, agriculture, and assets project: Final research findings*. International Food Policy Research Institute (IFPRI).

Rees, T. 2005. Reflections on the uneven development of gender mainstreaming in Europe. *International Feminist Journal of Politics*, 7(4), pp. 555–574.

Sexsmith, K., Smaller, C. and Speller, W. 2017. *How to improve gender equality in agriculture*. International Institute for Sustainable Development (IISD).

United Nations. 2015. *Sustainable development goal 5: Achieve gender equality and empower all women and girls*. United Nations Sustainable Development Knowledge Platform.

UNDP Europe and Central Asia. 2022. *United nations development programme*. [online] Available at: <https://www.undp.org/eurasia> [Accessed 13 October 2023].

Vinska, O. and Tokar, V. 2021. Cluster analysis of the European Union gender equality and economic development. *Business, Management and Economics Engineering*, 19(2), pp. 373–388.

Andreea Adriana Petcov, Manuela Dora Orboi,
Ana Mariana Dincu, Andreia Sasu, Raul Pascalau

Study on the ammonium impact in the deep waters of some villages in Caras Severin county, Romania

Abstract: *Ammonium represents the stage of decomposition of organic substances containing nitrogen in their molecule and therefore indicates a recent pollution, hours-days, consequently very dangerous. It has a characteristic pungent smell, very soluble in water, with an alkaline character. In general, deep waters contain ammonium of telluric origin. Instead, surface waters are directly and immediately exposed to ammonia from the decomposition of organic substances or from manure. The precipitation regime is also of particular importance, which can have a double role. Following the ammonium analyzes carried out, on water samples collected from deep water in the studied villages, we can state that the deep waters of Caraş-Severin County above 100m are not affected in terms of ammonium content. This is far below the allowed limit of 0.50mg/l. However, it can be seen that the deeper the water, the lower the ammonium content is, up to 0.01mg/l.*

Keywords: water, ammonium, pollution.

Introduction

The consequences of pollution are becoming more and more felt, reflecting in the state of public health, in the biodiversity of ecosystems, in the balance of ecosystems, and not only that, jeopardizing their very existence. In this context, protecting the environment becomes an imperative. In natural conditions, water is not chemically pure, having in its composition different compounds in gaseous state (O_2, CO_2, H_2S), salts and a series of particles in suspension. Natural water is the result of the combination of different isotopes of hydrogen with different isotopes of oxygen (Bohlke, Smith and Miller, 2006). Due to its properties, water represents the environment microbiological quality of water is decreasing in most countries, and germs are conducive to the development of various physiological processes (Kamal, Md. Hashim and Zin, 2015). Small changes produce serious disturbances and the insufficiency of the water ratio is much less tolerated than the deficiency in other elements. By water pollution, we mean the alteration of the physical, chemical and biological characteristics of water (Butnariu, 2007). Water is an important factor in ecological balances, and its pollution is a topical issue (Bancila, Dima and Georgescu, 2014). The increasingly resistant to disinfectants. The decrease in the immunity of the population, mainly through the general improvement of hygiene, produced an increase in the susceptibility to water diseases (Jäntschi, 2005). Ammonium can also be of a mineral nature, coming from ores containing nitrogen. It is also formed by the decomposition of nitric acid in the presence of pyrite as well as through the decomposition process of organic substances in anaerobic conditions in the presence of bacteria. It can also be formed as a result of reactions that reduce nitrogen (Shafarina, Mangala and Aris, 2015). In natural water sources, ammonium concentrations are usually low, but due to industrial, agricultural and other human activities, its concentration is quite high. In water supply systems it leads to degradation of water quality and to formation of other by-products with effects on human health. Existing technologies used to reduce the ammonium concentration in water (Radu, Racoviteanu and Vulpasu, 2022) The surface water had been overwhelmed on account of the long-term discharge of domestic and industrial wastewater (Wang et al., 2023), Groundwater with high levels of ammonium not only threatens water supply security (Norman et al., 2015; Scheiber et al., 2015), but it may also have negative ecological effects because of groundwater-surface water interaction (Lingle Kehew and Krishnamurthy, 2017). The abundant ammonium in clayey sediment may serve as a (natural) source of groundwater ammonium (Liu, Chen and Ma, 2023).

Literature review

Ammonium is a common contaminant found in the soils and groundwater at former gasworks, associated with historical gas production and the storage and disposal of by-product (Thomas et al., 2022).

Ammonium is toxic to aquatic fauna even at very low concentrations and its allowable concentrations in the environment are controlled through several regulatory guidelines (Erskine 2000; Buss et al., 2004; Thomas et al., 2022). Ammonium (NH_4+) is listed in the EU Drinking Water Directive (2020/2184), with a threshold of 0.5 mg l^{-1}. Ammonium contents generally show a positive correlation with K, and increase with increasing salinity (Manning et al., 2004)

Meristematic tissues of plants generally show patterns of cytotoxic response similar with those of embryogenic tissues of vertebrates. Plants bioassays are most sensitive in detecting the environmental hazards in water and can serve as the first alert for their presence (Petcov et al., 2008).

Methodology

The stages necessary to obtain experimental data: sample preparation, sample preparation for analysis, actual analysis, calculation and expression of results.

The principle of UV-VIZ Spectrophotometry – the working technique is based on laws that measure the ability of certain substances to absorb or emit light energy at a characteristic wavelength. The equipment used in spectrometry includes: the radiation source, the monochromator, the sample tank, the detector (Figure 1).

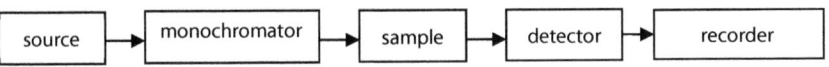

Figure 1. Spectroscopic analysis

The Spectroquant Ammonium 14752 test is used for the spectrophotometric determination of the ammonium ion (Figure 2). After alkalinization to pH=13, ammonium reacts with hypochlorite forming monochloramine.

The minimum detection limits according to the working method are 0.1 mg/l for ammonium.

Figure 2. Spectophotometer SQ – 118 MERCK with fixed wavelength
Photo: **https://www.sedgeochem.uni-bremen.de/po4_fotom.html**

From a qualitative point of view, the positions of absorption or emission lines and bands that appear in the electromagnetic spectrum indicate the presence of a certain substance. From a quantitative point of view, the intensity of emission or absorption lines or bands is measured for both standards and samples, then the concentration of the analyzed substances is determined. Quantitative spectroscopic analysis is based on two fundamental laws. These laws are applied in the case of the change in the radiant power of a monochromatic light radiation along with the change in the thickness of the layer traversed by the radiation and the change in concentration. Depth water sampling was carried out from 3 different localities in Caras Severin county, Romania, samples were taken from depths of 100 m, 150 m and 200 m as follows:

1. Sample 1 – locality Oravita, depth – 150 m (Figure 3);
2. Sample 2 – locality Grădinari, depth – 100 m (Figure 4);
3. Sample 3 – locality Calina, depth – 200 m (Figure 5).

Figure 3. Sample collecting point at Oravita location
Source: Petcov, 2023.

Figure 4. Sample collecting point at Gradinari location
Source: Petcov, 2023.

Figure 5. Sample collecting point at Calina location
Source: Petcov, 2023.

Water collection for physico-chemical analysis was carried out in glass or polyethylene bottles with hermetically sealed stoppers. Quantitative spectroscopic analysis is based on the laws that measure the ability of substances to absorb or emit light energy at a characteristic wavelength. Qualitatively, the positions of absorption or emission lines and bands that appear in the electromagnetic spectrum indicate the presence of a particular substance.

Located in the southwestern part of Romania, Caraș-Severin County borders Timis county to the northwest, Hunedoara County to the northeast, Gorj county to the east, Mehedinti county to the southeast, Mehedinti county to the west with the Republic of Serbia, for a length of 70 km, and in the southwestern part with the Danube River, for a length of 64 km. Although Caras-Severin County is predominantly mountainous, its relief is characterized by great diversity: 65.4 % of the surface is mountainous relief, 16.5 % depressional relief, 10.8 % hills and 7.3 % plains (National Institute of Statistics, Caraș-Severin County Statistics Directorate, 2023).

Figure 6. Geographical position of the localities under study
Source: **https://maps.app.goo.gl/9dpZ9XXsYgXFJeyYA**

Grădinari (Figure 6) village owns approx. 15 wells built with stone or brick and equipped with a wheel and bucket for removing water. There are also a number of wells in the residents' yards. The Grădinari village does not have a centralized rainwater collection system. Meteoric waters are collected in the street gutters inside the village, they are taken over by the ogasas with permanent and non-permanent flow and then they are discharged into the Caraș river. Domestic water drainage is carried out in the local system. They are collected in cesspools and individual septic tanks from where they are periodically emptied by their own care (Grădinari City Hall, 2023).

The town of Oravița (Figure 6) is located in the southwestern part of Caraș-Severin County, parallel 45 passes nearby. The town is located at a crossroads in the south-west of the county, where DN 57 connecting Reșita and Moldova Nouă branches off, crossing the town to connect with Anina and Bozovici on DN 57 B. (www.oravita.ro).

Calina locality (Figure 6) is part of Dognecea village and is located in the south-western extremity of the Dognecea mountains, at the base of the Moghila peak (496 m altitude), 12 km from the Reșita-Oravița road. There are stories in local folklore, that around the 1860s, the water from the Calina spring was prized by the imperial court in Vienna (Resita City Hall, 2023).

Results and discussions

The values obtained from the results of the analyzes are between 0.01 (Calina) and 0.22 mg/l (Grădinari), being within the normal limit. The values tend to remain within approximately the same limits during the five months of the study. (Table 1, Figure 7). It should be noted that these water samples were collected from quite deep depths (between 100 – 200 m depth), where the level of ammonium pollution is lower than in shallow water.

Table 1. Ammonium content in the samples collected from underground water

No.	Location/sample collection depth	Month				
		October 2022	December 2022	February 2022	May 2022	June 2022
		mg/l				
1	Oravița – 150 m	0.13	0.15	0.19	0.21	0.16
2	Grădinari – 100 m	0.16	0.14	0.19	0.22	0.17
3	Calina – 200 m	0.03	0.02	0.02	0.02	0.01

Source: Authors' own research.

On the basis of the obtained results, we can conclude that although negative values were observed in all the wells drilled at great depth, the maximum allowed limit of 0.50 mg/l was not exceeded; the lowest values were recorded in the locality of Calina, which is very famous, both in Romania and abroad for its mined and flat waters. The ammonium content is no longer determined by surface horizons or by anthropic activity specific to surface areas or economic activities carried out by man. The groundwater content in various salts is due to the nature of the soil layers through which the groundwater flows and, above all, to its salt content and solubility.

The analysis of the results of the ammonium – NH4+ content determinations allowed us to place it within the normal limits for drinking water for wells drilled at depth. Of particular importance is also the rainfall regime, which can play a double role.

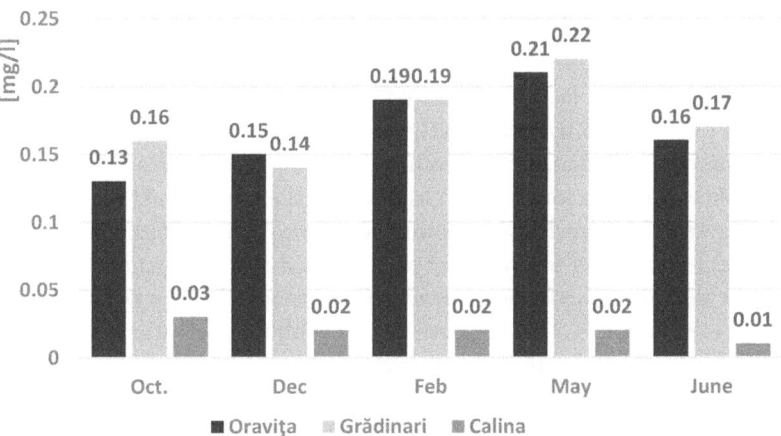

Figure 7. Representation of ammonium content in the samples collected from underground waters

As can be seen in Figure 7, lower values were recorded in Oravita, compared to Gradinari, a possible explanation could be that, the well from which the water was taken is in the city, being somewhat protected from floods, the exploitation of animals, intensive cultures, all of which are basic factors in the pollution of waters with ammonium. In the two localities (Gradinari, Oravita) the values tend to increase, but in the last month (June), the values decrease slightly.

Conclusions

We can conclude that the impact of pollution on the environment and implicitly on the waters, be they surface or deep, can be defined as a direct or indirect effect of a human activity that produces a change in the sense of evolution of the quality state of the ecosystems, a change that can affect health human, the integrity of the environment, cultural heritage or socio-economic conditions. In the case of the study of ammonium in the deep waters studied, there was no case of exceeding the allowed limit. Ammonium can also be of a mineral nature, coming from ores containing nitrogen. It is also formed by the decomposition of nitric acid in the presence of pyrite as well as through the decomposition process of organic substances in anaerobic conditions in the presence of bacteria. It can also be formed as a result of reactions that reduce nitrogen. The aim of the study was the analysis of ammonium and the presence of ammonium in wells for domestic use, it was observed that the maximum allowed limit was not exceeded, but, nevertheless, it was observed that the samples analyzed over the five months at the analysis point in Gradinari presented the highest values (0.19 mg/l respectively 0.22 mg/l) at a depth of 100m, followed by the collection point in Oravita at a depth of 150 m (0.19 mg/l respectively 0.21mg/it).

We can conclude that, as the depth of water sampling increases, the ammonium concentration is lower, which leads us to take into account the analysis of surface waters, with much lower depths because zootechnical farms were located in the respective areas whose activity determined that nitrogenous substances reach the shallow groundwater. A definite conclusion, that deep waters, from the Oravița – Gradinari – Calina area, can be used without reservations as sources of drinking water, provided that the other chemical content parameters fall within normal limits.

References

Bancila,V., Dima, V. and Georgescu, F. 2014. Fog as risk phenomenon during the cold season 2007–2008. *Romanian Journal of Meteorology*, 11(1–2), pp. 1–17. [online] Available at: <http://rjm.inmh.ro/articole/vol11-1-2/RJM-2014-01_Bancila.pdf f> [Acccesed 12 September 2023].

Bohlke, J.K., Smith, R.L. and Miller, D.N. 2006. Ammonium transport and reaction in contaminated groundwater: Application of isotope tracers and isotope fractionation studies. *Water Resources Research*, 42(5). https://doi.org/10.1029/2005WR004349.

Buss, S.R., Herbert, A.W., Morgan, P., Thornton, S.F. and Smith J.W.N. 2004. A review of ammonium attenuation in soil and groundwater. *Quarterly Journal of Engineering Geology and Hydrogeology.* https://doi.org/10.1144/1470-9236/04-005.

Butnariu, M. 2007. *Theoretical and practical notions of plant biochemistry.* Timisoara: Mirton.

Erskine, A.D. 2000. Transport of ammonium in aquifers: Retardation and degradation. *Quarterly Journal of Engineering Geology and Hydrogeology,* 33, pp. 161–170. https://doi.org/10.1144/qjegh.33.2.161

Google maps – satelit image. Available at: <https://maps.app.goo.gl/9dpZ9X XsYgXFJeyYA> [Acccesed 27 September 2023].

Jäntschi, L. 2005. *Microbiologie, Toxicologie și Studii Fitosanitare,* Cluj-Napoca: Academic Direct.

Kamal, M.Z., Md. Hashim, M. and Zin, M.S. 2015. The effect of the ammonium concentration in the groundwater. *PEOPLE: International Journal of Social Sciences,* 1(1), pp. 313–319. https://doi.org/10.20319/pijss.2015.s11.313319.

Lingle D.A., Kehew, A.E. and Krishnamurthy, R.V. 2017. Use of nitrogen isotopes and other geochemical tools to evaluate the source of ammonium in a confined glacial drift aquifer, Ottawa County. *Applied Geochemistry,* 78, pp. 334–342. https://doi.org/10.1016/j.apgeochem.2017.01.004.

Liu, R., Chen, J. and Ma, T. 2023. Releasing mechanism of ammonium during clayey sediments compaction and its impact on groundwater environment. *Science of The Total Environment,* 898, p. 165579. https://doi.org/10.1016/j.scitotenv.2023.165579.

Manning, D.A.C. and Hutcheon, I.E. 2004. Distribution and mineralogical controls on ammonium in deep groundwaters. *Applied Geochemistry,* 19, pp. 1495–1503, https://doi.org/10.1016/j.apgeochem.2004.01.019.

Marine Geochemistry – Laboratory methods. *Phosphate detection with Merck Photometer.* [photograph] [online] Available at: <https://www.sedgeochem.uni-bremen.de/po4_fotom.html> [Acccesed 27 September 2023].

National Institute of Statistics, Caraș-Severin County Statistics Directorate. 2023. *Despre județul Caraș-Severin* [About Caraș-Severin county]. [online] Available at: <https://carasseverin.insse.ro/despre-noi/despre-judetul-caras severin/> [Acccesed 27 September 2023].

Norman, J., Sparrenbon C.J., Berg, M., Nhan, D.D., Jacks, G., Harms Ringdahl, P., Nhan, P.Q. and Rosqvist, H. 2015. Tracing sources of ammonium in reducing groundwater in a well field in Hanoi (Vietnam) by means of stable nitrogen isotope ($\delta 15N$) values. *Applied Geochemistry,* 61, pp. 248–258. https://doi.org/10.1016/j.apgeochem.2015.06.009.

Oravița – orașul premierelor [Oravița, the city of premieres]. [online] Available at: <https://oravita.ro/> [Acccesed 27 September 2023].

Petcov, A., Botos, A., Corneanu, M., Butnaru, G. and Lazureanu, A. 2008. Studies on the enviromental hazards in drinking water evaluation from Caras-Severin district by *Allium sativum* L., *Scientific Agriculture – Agriculture of the Future*, pp. 448–453.

Petcov, A. 2023. *[photograph] own private collection*.

Primăria Comunei Grădinari, județul Caraș-Severin (Grădinari City Hall, Caraș-Severin County). *Informații generale* [General Information]. [online] Available at: <https://gradinari-cs.ro/comuna-gradinari/monografia-comunei/> [Acccesed 27 September 2023].

Primăria Reșița (Resita City Hall). *Informații generale* [General Information]. [online] Available at: <https://www.primariaresita.ro/portal/cs/resita/portal.nsf/AllByUNID/informatii-generale-00004422?OpenDocument> [Acccesed 27 September 2023].

Radu, G., Racoviteanu, G. and Vulpasu, E. 2022. Biofilters efficiency in removing ammonium from water intended for human consumption. *Romanian Journal of Civil Engineering* https://doi.org/10.37789/rjce.2022.13.2.4.

Scheiber, L., Ayora, C., Vázquez-Suñé, E., Cendón, D.I., Sole, A. and Baquero, J.C. 2015. Origin of high ammonium, arsenic and boron concentrations in the proximity of a mine: Natural vs. anthropogenic processes. *Science of The Total Environment*, 541, pp. 655–666, https://doi.org/10.1016/j.scitotenv.2015.09.098.

Shafarina, N.R., Mangala, P.S. and Aris, A.Z. 2015. Drinking water assessment on ammonia exposure through tap water in Kampung Sungai Sekamat, Kajang. *Procedia Environmental Sciences*, 30, pp. 354–357. https://doi.org/10.1016/j.proenv.2015.10.063.

Thomas, R.A.P., Riding, M.J., Robinson, J.D.F, Brown, S.J.A. and Taylor, C. 2022. Distinguishing sources of ammonium in groundwater at former gasworks sites using nitrogen isotopes, 55(4). https://doi.org/10.1144/qjegh2021-139.

Wang, Y., Dongpeng, L., Xin, C., Xinshan, S., Chenteng, G., Yuhui, W., Zhongshuo, X. and Wei, H. 2023. Integrating a manganese ores-filled module with a submerged plants cathode sediment fuel cell: In-situ remediation of ammonium pollution in surface water, *Journal of Water Process Engineering*, 53, p. 103696. https://doi.org/10.1016/j.jwpe.2023.103696.

Veronica Prisacaru, Alina Caradja

Modeling the relationship between the performance of agricultural vocational education and the sustainable development of the rural environment

Abstract: *Since the significance of sustainable rural development is widely recognized, an important issue is to identify and mobilize the factors that can contribute to achieving the assumed objectives. In this context, we highlight the role of agricultural professional education. Thus, being a good post-experience, the effects of education are not immediately visible, but its role is extremely important in achieving the goals of sustainable development. Therefore, it is certain that agricultural vocational education is a "factory" for training young specialists qualified in the required fields of activity in the rural areas of the Republic of Moldova.*

The purpose of the research consisted in quantifying the contribution of the performance of agricultural vocational education in the sustainable development of the rural environment. In order to achieve the assumed goal, an opinion poll was conducted on a sample of 194 managers and specialists who graduated from agricultural study programs and are working in agricultural enterprises. Based on the respondents' perception of the importance and role of the competences obtained in professional education, the econometric model was created that elucidates the impact of the professional education performance on the sustainable development of the agricultural enterprises. By generalizing the findings, the directions of intervention in the agricultural vocational education programs were identified and exposed in order to adapt the educational offer to the requirements of the sustainable rural development.

Keywords: sustainable rural development, vocational education, performance, competence, enterprise.

Introduction

The sustainable development of the rural environment in the Republic of Moldova is an objective of major significance. Over time, the approach to the subject has become more intense both globally and nationally. An indispensable condition for achieving sustainable development is the improvement of the vocational education process with an agricultural profile and the extension of the area of its performance indicators and the impact that they have on the economic and social development of the rural environment.

Being a post-experience thing, the effects of education are not immediately visible, but its role is extremely important in achieving the goals of sustainable development. Therefore, it is certain that agricultural vocational education is a "factory" for training young specialists qualified in the required fields of activity in the rural areas of the Republic of Moldova.

Agriculture, as a basic branch of the analyzed environment, has wide ecological, economic and social implications. In this sense, the sustainable development of agriculture is a national priority, on which in the last decade more attention has been drawn, finding its expression in the developed policies and strategies.

At the moment, the sustainable development of the rural environment is a major concern, having as a reference the environmental problems and the crisis of natural resources, the urgent need to approach them, being aware and public declared about five decades ago. At the same time, more and more efforts have been directed towards highlighting the role and place of education at all levels in achieving the goals of sustainable development. The international community's concerns in education for sustainable development are reflected in a number of international events.

The Republic of Moldova has, in turn, assumed the objectives of sustainable development, by nationalizing the respective indicators, as well as reflecting the concern for sustainable development in a series of strategic documents. Implicitly, there are a number of shortcomings both in the content of the related strategies and in the extent to which the performance indicators of vocational education, but also the performance itself, contribute to achieving the objectives of sustainable development.

Literature review

Originating in the private sector, performance management later penetrated the public sector, including universities, as a mechanism for improving the public services' performance, productivity, accountability and transparency (Forrester, 2011). Thus, universities have had to adhere to the new public management, the latter representing a set of broadly similar administrative doctrines that have dominated the bureaucratic reform agenda in many OECD countries since the late 1970s (Hood, 1991).

Even though performance management has entered academic activity very quickly, there are many opinions regarding its definition and use in a practical context. Thus, while some authors place a particular emphasis on its procedural side (Armstrong, 2003; Forrester, 2011; Pulakos, 2009; Samoilenco and Sverdlic, 2010; Verweire and Berghe, 2004), others particularly highlight its systemic character (Avasilicăi, 2001; Briscoe and Claus, 2008; Shields, 2016). The strategic nature of the performance management is also undeniable, this being clearly reflected in the approaches of Bădescu, Mirci and Bögre (2008), Mathis and Jackson (2010), Armstrong (2014).

Returning to the real reasons that determined the new orientations of the university management, we deduce, as a significant problem, the evaluation of the impact of the actions undertaken in this regard in achieving the most important objectives for the society, the sustainable development goals being a priority at the current stage.

Elucidating and, in particular, quantifying the contribution of vocational education in achieving the goals of sustainable development is both important and challenging. On the one hand, the professional education system tends towards performance and, in this sense, it imposes itself through a vast series of indicators through which the objective of continuously increasing the quality of the educational offer is pursued. In this context, the importance of performance measurement can be highlighted which, according to Lebas (1995), refers to a system of key indicators designed to provide contextual and specific information. For a good operation, it is considered that the performance indicators must have the following characteristics: to have the function of monitoring, being defined as a series of information collected at equal time intervals to track the performance of a system (Fitz-Gibbon, 1996); to be quantitative (Guenin, 1987); to correspond to the predetermined objectives (CVCP, 1986).

And if the presence of indicators in measuring performance is perceived by the academic communities as something natural, the change in the nature of these indicators, with the increasingly insistent penetration of marketization in higher education (Brown, 2010, 2015; Engwall, 2007; Hemsley-Brown, 2011; Kalio et al., 2016), has sparked heated discussions. As a result, researchers and practitioners have divided into two camps: allies of marketing processes – those who believe in the virtues of marketing, as a rational device for determining the allocation of insufficient resources for securing "efficiency" (Barner, 2011; Furedi, 2010; Kalio et al., 2016) and opponents – those who believe that the transition to marketized universities will have a damaging impact on universities (Brown, 2015; Engwall, 2007; Knights and Clarke, 2014).

Regardless of the discussions that take place with reference to how we measure the performance of a vocational education institution, the need for high performance qualifications is indisputable. Moreover, if the initial efforts to change the approach to performance were focused on the need to ensure a higher efficiency in the use of public money (Fitz-Gibbon, 1996; Gherghina, Vaduva and Postole, 2009), today we can speak with certainty about the need to evaluate the performance of vocational education in terms of its impact on the economic and social performance of the sectors where the workforce trained by vocational education institutions can be found (Prisacaru and Caradja, 2018, p. 367). Implicitly we mention the objectives of sustainable development as targets which, being reached, create real premises to optimize these impacts. Moreover, mankind's concern for achieving the sustainable development goals has led to the introduction of the term "Education for Sustainable Development" (ESD). According to UNESCO, ESD "empowers learners to make informed decisions and take responsible actions for environmental integrity, economic viability and a fair society for present and future generations, respecting cultural diversity" (UNESCO, 2023). It is obvious that the exercise of these powers takes place within each economic entity, and personal performances will be found in the sustainable development indicators of the enterprise.

With reference to agricultural enterprises, we note that a series of indicators of sustainable development have been identified and stated, these being exposed by fields and subfields, as follows (Lazar, Mortan and Vereş, 2007, p. 56):

I. The field *Sustainable development of the production environment (ecological)*
   1.1 Subfield *Diversity of activities*: the diversity of annual crops; the diversity of perennial crops; the diversity of other associated vegetable crops; the diversity of livestock; the evaluation and improvement of genetic heritage; the development of non-agricultural activities.
   1.2 Subfield *Territorial organization*: the type of crop rotation; the sole size; the preservation of the natural environment; the fodder areas management.
   1.3 Subfield *Production technology*: the amount of chemical fertilizers; the level of pesticides and veterinary products; the soil protection; the water resources management; the degree of energy autonomy.

II. The field *Sustainable development of the social environment*
   2.1 Subfield *Quality of products and rural space*: the quality of food produced; the treatment of the agricultural waste; the capitalization of buildings and landscape; the quality of the roads.
   2.2 Subfield *Development of rural services*: the complex services; the direct capitalization; the association of producers.
   2.3 Subfield *Human development*: the staff training; the employment rate; the life quality; the hygiene and work safety.

III. The field *Sustainable economic development*
   3.1 Subfield *Income level*: the income level.
   3.2 Subfield *Economic independence*: the financial autonomy; the influence of direct aid.
   3.3 Subfield *Efficiency*: the profit rate.

It is obvious that the level of the enterprise's sustainable development indicators depends on a whole series of subjective and objective factors. An alternative for evaluating the contribution of the competences obtained within the professional training is through the prism of the own perception by the graduates.

Methodology

In order to determine the impact of the competencies obtained in agricultural vocational education institutions on the performance in sustainable development, there was carried out a selective research of the managers and specialists of the agricultural enterprises from the rural area, with specialized studies completed in the agricultural field.

In order to ensure an acceptable representativeness of the selective research results, the sample (194 respondents) was taken on the basis of a mixed selection, with a limit error of representativeness of 3.2 employees/enterprise, and a significance of $\alpha=0.05$.

In order to optimize the results of the selective research and to evaluate the quantitative and qualitative aspects of enterprise performance, correlated with educational variables, the survey questionnaire consisted of a set of 37 questions structured in three blocks:

I. Questions to evaluate the contribution of the competences obtained in the vocational education institution on the development of the production environment;
II. Questions on the contribution of the stated competences in the sustainable development of the social environment;
III. Questions to quantify the extent to which the stated competences (professional and transversal) have contributed to increasing labor productivity.

Six variables included in this section (q1) referred to the activities involving:

- q1_1 – Diversification of agricultural crops;
- q1_2 – Diversification of livestock;
- q1_3 – Evaluation and improving genetic heritage;
- q1_4 – Maintaining and increasing soil fertility;
- q1_5 – Rational water management;
- q1_6 – Application of ecological production technologies.

The research was carried with the support of the project "Predictive approaches to increasing the quality of competences in agricultural higher education based on the partnership with the business environment", No. 20.80009.0807.41, contracting authority – the National Agency for Research and Development.

Results and discussions

Having interpreted the data, we found out that the most favorable rating was obtained by "Diversification of agricultural crops", the average level tending to the maximum value of satisfaction (3.27), while the lowest value is "Diversification of livestock" (2.10). Small difficulties (values below the average level) can also be noticed in the case of the variables "Assessment and improvement of genetic heritage" and "Rational water management" (Figure 1.).

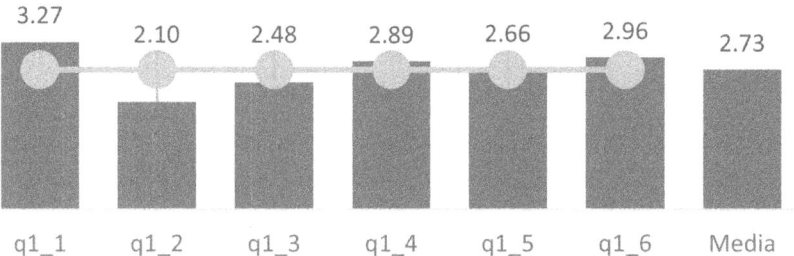

Figure 1. Media on satisfaction with competencies in the activities from q1 section
Source: Authors' own research.

In general, the most common value selected was 3, which corresponds to the grade "good" (Figure 2). At the same time, more than 1/5 of the respondents opted on the average for the "weak" option, although, a good thing, this extreme was more than 2 times (25 %) advanced by the opposite option ("very good").

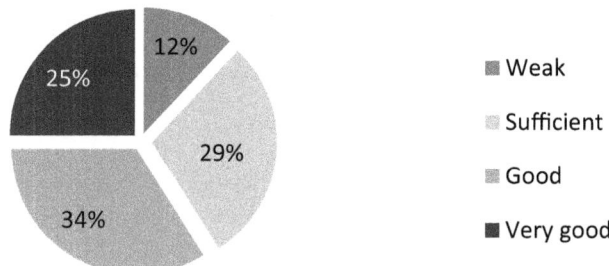

Figure 2. The average structure of respondents by satisfaction level in the activities from q1 section
Source: Authors' own research.

The results of estimating the partial linear correlation indicate that only the variable q1_2 manifests itself independently of the other variables from the set of variables q1.

The eight variables included in section q2 were oriented towards a series of activities:

- q2_1 – Increasing the quality of agricultural products obtained;
- q2_2 – Processing of agricultural waste;
- q2_3 – Optimal use of buildings and landscape;
- q2_4 – Increasing the quality of roads;
- q2_5 – Development of rural services;
- q2_6 – Human resources development under your subordination;
- q2_7 – Ensuring adequate working conditions of the employed staff;
- q2_8 – Ensuring decent salaries of the employed staff.

And in the case of this set of variables, the most favorable qualifier obtained the first variable from the list "Increasing the quality of agricultural products obtained", which is characterized by an average score of 3.39, which determines this variable as a goal in itself in the preparation agricultural specialists (Figure 3). The maximum grade was given to this variable by more than half of the respondents, and cumulatively grade 3 and 4 represent almost 90 % of the options.

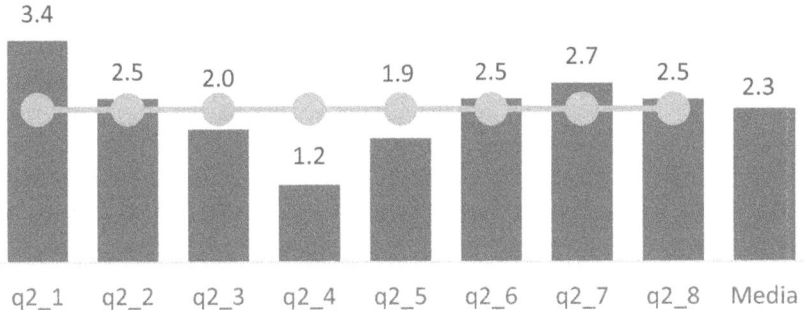

Figure 3. Media on satisfaction with competencies in the activities from q2 section
Source: Authors' own research.

In the case of this set of variables, the options were distributed more evenly, highlighting the "sufficient" rating, by 33 %, followed by the "good" rating (27 %). In the competition of extreme values, the " very good" rating gave way to the "weak" rating (Figure 4).

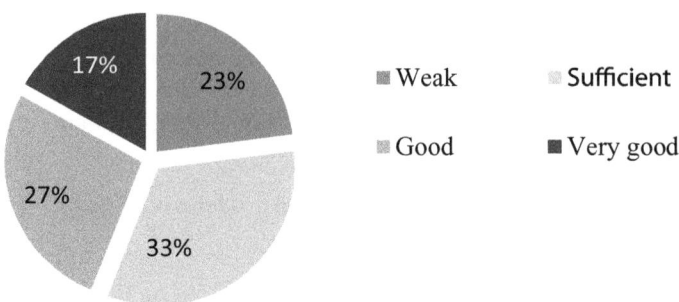

Figure 4. The average structure of respondents by satisfaction level in the activities from q2 section
Source: Authors' own research.

The hypothesis of the existing link between the quantitative variables "age" and the variables expressing the level of satisfaction with the skills formed in the training process necessary for sustainable development of the rural environment (verified by determining the Pearson correlation coefficients) refutes the hypothesis of the link between correlated variables. Regarding "seniority", no significant interdependencies are found with the variables in set q2, except in the case of variable q2_8, where a weak dependence is reflected.

Analyzing the aspects of interdependence between the two sets of variables, we found out a slightly different view of respondents on basic agricultural activities and those for sustainable development, although in some cases significant links, but of low intensity can be seen. The described aspects may cause some difficulties in the case of separate employment of these variables in econometric models, because the existence of an interdependence between factorial variables may prejudice the quality of the estimated model. In this sense, procedures will be undertaken to transform and include these variables in the developed models.

A logical extension of the quantitative analysis process will materialize in elaboration of an econometric model, through which the analyzes performed previously will be deepened, so that the behavior of the entities described through the 194 respondents (survey subjects) will be completed with new quantitative approaches (estimated by econometric models). Likewise, the elaboration of an econometric model will allow a more thorough knowledge of the extent to which the resultant variable (the effect being expressed by the average annual increase in labor productivity [variable q3]) was determined by the sets of factorial variables (cause), through which the level of satisfaction with the competencies obtained in the process of vocational education is characterized (variables of section q1 and q2).

Starting from the assumption that a higher level of satisfaction would encourage higher results (higher average labor productivity increase), we will try to estimate the model of dependence between the sets of variables q1 and q2. The big difference between the values of the dependent variable (q3) and the values of the independent variables (data sets q1 and q2), but also the different character of the variables (different scales: q3 – of proportional type, while q1 and q2 – of ordinal type), will involve the operation of transformations of the resultant variable by logarithm. In this case, the specified model will have a semi logarithmic analytical expression, of the form:

$$\log y = b_0 + \sum_{j=1}^{k} b_j x_j + e \qquad (1)$$

$\log y$ - Logarithmic values of the variable average productivity increase (q3);

b_0 - Free term of the model – indicates the value of the variable under conditions of zero values of the factorial variables;

b_j - Regression parameters, which indicate by how many units the logarithm of the value of the resultant variable will increase with an increase of the factorial variable by one unit;

x_j - Factorial variable j;

e - Error/residual value of the model (difference between empirical and adjusted resultant value).

The activities of specifying the model and estimating it were performed using the EViews econometric analysis software, and the results were expressed in the graphic style of that software, as follows:

$$LOG(Q3) = C(1)*Q1\_1 + C(2)*Q1\_2 + C(3)*Q1\_3 + C(4)*Q1\_4 \\ + C(5)*Q1\_5 + C(6)*Q1\_6 + C(7) \qquad (2)$$

The value of C (free term) shows that if the influence of the factors included in the model is zero, then the average increase in labor productivity for the last three years will be 31.8 thousand lei per employee.

Increasing the appreciation of the level of appreciation compared to the skills obtained in the process of vocational education, in order to diversify agricultural crops, will lead to an increase in labor productivity by about 1.21 thousand lei. The estimation results show that two of the estimators of the model parameters (C and q1_2) have a higher significance, while the significance of the other estimators of the model parameters are characterized by significance levels from low to moderate.

The multiple correlation coefficient indicates a weak to medium interdependence ($R = 0.34$), which results in a determination ratio ($R^2 = 0.15$), or about 15 % of the variation of the resulting variable is determined by the variation of the factorial variables employed in the model. The results of estimating the overall quality of the model speak of its acceptable quality (F-statistic=5.09, Probe F-statistic=0.000077).

A combination of the factorial variables from the two sections (q1 and q2) could contribute to a qualitative strengthening of the model, but also to an associated explanation of the influence of the factors to establish their ranking. In this case, the estimated regression model will take the form:

$$\begin{aligned}LOG(Q3) = &\ C(1)^*Q1\_1 + C(2)^*Q1\_2 + C(3)^*Q1\_3 + C(4)^*Q1\_4 \\ &+ C(5)^*Q1\_5 + C(6)^*Q1\_6 + C(7)^*Q2\_1 + C(8)^*Q2\_2 \\ &+ C(9)^*Q2\_3 + C(10)^*Q2\_4 + C(11)^*Q2\_5 + C(12)^*Q2\_6 \\ &+ C(13)^*Q2\_7 + C(14)^*Q2\_8 + C(15)\end{aligned} \qquad (3)$$

From the estimated results we can notice that the most influential factor of increasing labor productivity is q2_1 – "Increasing the quality of agricultural products obtained", and changing the opinion of respondents to this factor causes a change in the same direction of the resulting variable by about 2.01 thousand lei, following q2_5 – "Development of rural services" (+1,56) and q1_4 – "Maintaining and increasing soil fertility" (+1.26). At the other pole are the variables with a lower influence, the least influential being q1_2 – "Diversification of livestock", q2_6 – "Development of human resources" and, respectively, q2_2 – "Agricultural waste processing" (Figure 5).

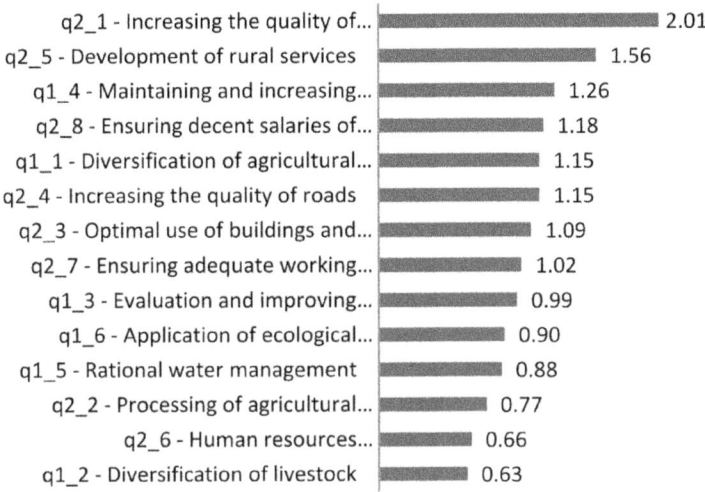

Figure 5. Ranking of factors according to the level of influence on the labour productivity increase
Source: Authors' own research.

It is obvious that this model presents more favorable results than the two previously presented, in terms of creditworthiness indicators. The model errors are characterized by a little significant positive autocorrelation. It is equally obvious the normality of the model errors distribution (Figure 6), determined by an approximate lack of asymmetry and excess.

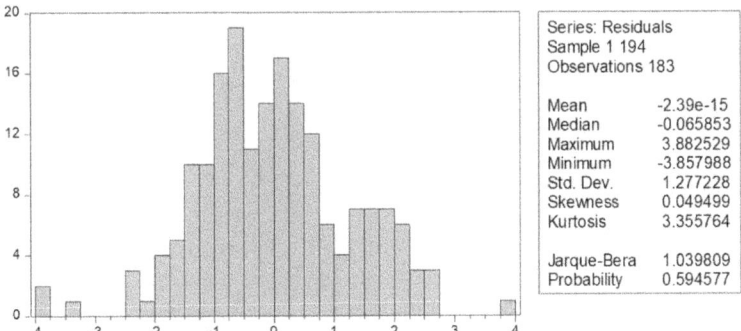

Figure 6. Distribution of model 3 errors
Source: Authors' own research.

At the next stage there was developed a fictitious econometric model, only to make a detail of the influence of structural factors that contribute to the formation of labor productivity increase. For this purpose, certain transformations of the structural variables q4 set were made in contributions to the formation of the productivity increase in the form of absolute values. The results of the model estimation are as follows:

$$\begin{aligned} q3 = {} & C(1)*P4\_01 + C(2)*P4\_02 + C(3)*P4\_03 + C(4)*P4\_04 \\ & + C(5)*P4\_05 + C(6)*P4\_06 + C(7)*P4\_07 + C(8)*P4\_08 \\ & + C(9)*P4\_09 + C(10)*P4\_10 + C(11)*P4\_11 \\ & + C(12)*P4\_12 + C(13) \end{aligned} \qquad (4)$$

The parameter estimators, which in this case have an absolute significance, denote that the most influential factorial variable is the variable q4_7 – "Quality of problem-solving skills", and its change by a monetary unit determines the change of the resulting variable by about 1.33 units. It is followed by the variable q4_3 – "Assurance of the enterprise with labor force" (+1.23), the whole classification of the factorial variables being presented in Figure 7.

Unfortunately, quite important operational factors are placed on the last three positions, including one related to the involvement of information technologies, the change in value of which has the least influence. It cannot be omitted the fact that, as a whole, the transversal competencies obtained in the process of carrying out specialized studies (q4_6, q4_7, q4_8, q4_9, q4_10), compared to each of the other factors analyzed, would have the greatest influence on the increase of labor productivity, the modification of which by one point would lead to an increase of productivity by about 3.98 thousand lei.

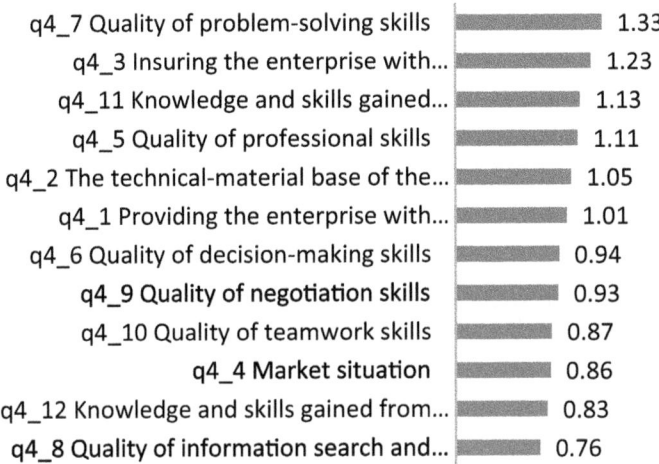

Figure 7. The influence of q4 factors on the average increase of labor productivity
Source: Authors' own research.

By generalizing the above said, we can deduce that the quality of competences provided by the training process influences the indicators of sustainable development at the level of agricultural enterprise along with other factors (such as, for example, the level of remuneration or the quality of rural services). At the same time, we noticed the needs to prioritize the competences of rational soil management and to increase the quality of products, which are identified as having a greater impact on the economic performance of enterprises.

In order to improve the quality of the educational offer with agricultural profile, in accordance with the objectives of sustainable development, we propose an updated mechanism for its evaluation, which is has the following distinct aspects: updated performance indicators; continuous process of improving performance indicators and evaluation methods and procedures; active and continuous involvement of several actors: educational institutions, employers, pupils/students and graduates, the relevant ministry (Figure 8).

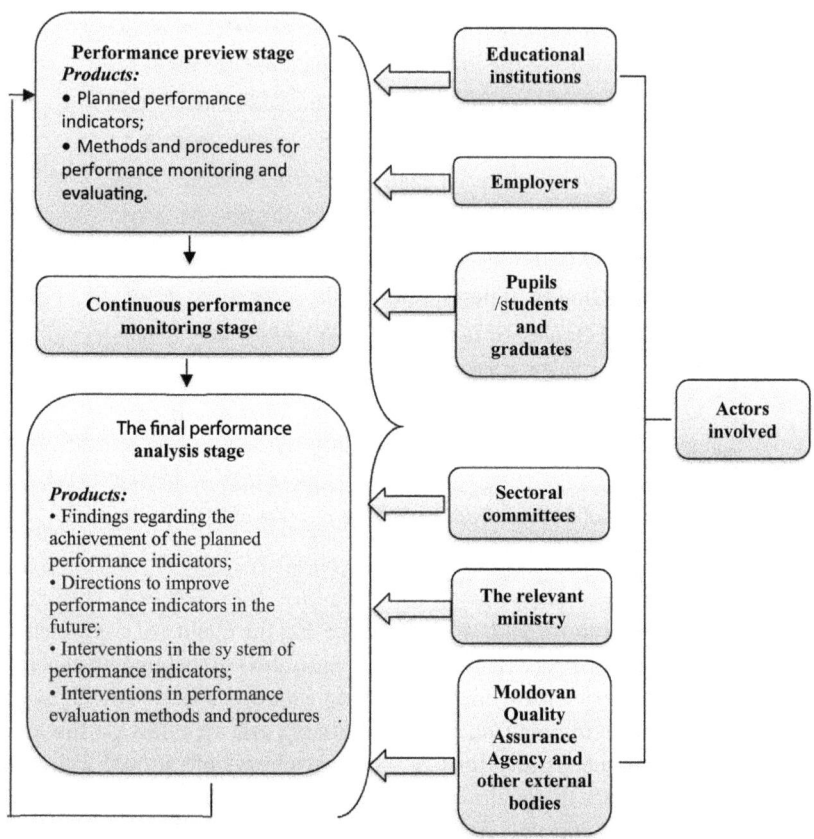

Figure 8. The updated mechanism to evaluate the quality of vocational education with agrarian profile
Source: Authors' own research.

Conclusions

As a result of quantifying the contribution of the performance of vocational education with agricultural profile to the sustainable development of the rural environment by modeling the impact of educational performance on indicators of sustainable development of agricultural enterprises, it was deduced that the quality of competences offered in the training process has an influence on indicators of sustainable development at the level of agricultural enterprise, along with other factors (level of remuneration, quality of rural services, etc.).

Based on the above, we propose to use the econometric model developed in order to determine the impact of the performance management indicators of the agricultural vocational education on the sustainable development of the rural environment of the Republic of Moldova by vocational education institutions, employers and other stakeholders, including the process of developing policies related to vocational education and rural labor force management.

We also propose to implement the updated mechanism to evaluate the quality of vocational education with agrarian profile in which each actor involved, directly or indirectly, in the process of vocational education, come with a maximum contribution in the functioning of the performance management system, as follows:

- educational institutions – by exercising the management responsibilities of the system;
- pupils/students – by declaring preferences, exposing satisfaction with the quality of the process, offering suggestions for its improvement;
- graduates – by expressing satisfaction with the quality of the competences obtained and offering suggestions for improving their quality;
- employers – by assessing the quality of the prepared workforce, providing information on the development trends of the sector and, implicitly, by the contribution in predicting the necessary competences on different time horizons;
- the relevant ministry (in this case – the Ministry of Agriculture and Food Industry) – by contributing to highlighting the trends of the sector, as well as assuming the role of moderator of the communication of educational institutions with the business environment;
- sectoral committees – by contributing to the connection of the educational offer in a qualitative aspect (within the competences corresponding to the qualifications) to the requirements of each sector;
- National Agency for Quality Assurance in Education and Research and other external bodies (quality certification bodies, professional unions, etc.) – by contributing to the rationalization of the entire quality management system and connection to the needs of the real sector.

References

Armstrong, M. 2003. *Managementul resurselor umane: Manual de practică*. București: Codecs.

Armstrong, M. 2014. *Armstrong's handbook of performance management. An evidence-based guide to delivering high performance.* 5th ed. London: Kogan Page.

Avasilicăi, S. 2001. *Managementul performanței organizaționale.* Iași: Tehnopress.

Barner, R. 2011. *The marketised university: Defending the undefensible.* London and New York: Routledge.

Bădescu, A., Mirci, C. and Bögre, D. 2008. *Managementul Resurselor Umane: manualul profesionistului.* Timișoara: Brumar.

Briscoe, D. and Claus, L. 2008. Employee performance management: Policies and practicies in multinational entreprises. In: A. Varma, P. Budhvar, and A. DeNisi, eds. *Performance management systems. A global perspective.* London and New York: Routledge, pp. 15–39.

Brown, R. 2010. The march of the market. In: M. Molesworth, R. Scullion, and E. Nixon, eds. *The marketisation of higher education and the student as consumer.* London and New York: Routledge, pp. 11–24.

Brown, R. 2015. The marketisation of higher education: Issues and ironies. *New Vistas.* UK: Liverpool Hope University, 1(1), pp. 4–9.

CVCP. 1986. *Performance indicators in Universities: A first statement by joint CVCP/UGC Working Group.* London: CVCP/UGC.

Engwall, L. 2007. Universities, the state and the market. *Higher Education Management and Policy.* USA: OECD. 19(3), pp. 87–104.

Fitz-Gibbon, C. 1996. *Monitoring Education – Indicators, quality and Effectiveness.* London: Continuum Intl Pub Group.

Forrester, G. 2011. Performance management in education: Milestone or millstone? *Management in Education.* London: SAGE. 25(1), pp. 5–9.

Furedi, F. 2010. Introduction to the marketisation of higher education and the student as consumer. In: M. Molesworth, R. Scullion, and E. Nixon, eds. *The marketisation of higher education and the student as consumer.* London and New York: Routledge, pp. 1–7.

Gherghina, R., Vaduva, F. and Postole, M. 2009. The performance management in public institutions of higher education and the economic crisis. *Annales Universitatis Apulensis. Series Oeconomica,* 11(2), pp. 639–645.

Guenin, S. 1987. The use of performance indicators in Universities: An international survey. *International Journal of Institutional Management,* 11(2), pp. 117–139.

Hemsley-Brown, J. 2011. Market heal thyself: the challenges of a free marketing in higher education. *Journal of Marketing for Higher Education,* 21(2), pp. 115–132.

Hood, C. 1991. A public management for all seasons. *Public Administration*, 69(1), pp. 3–19.

Kalio, K., Kalio, T., Tienari, J. and Hyvonen, T. 2016. Ethos at stake: Performance management and academic work in universities. *Human relation*, 69(3), pp. 685–709.

Knights, D. and Clarke, C. 2014. It's a bittersweet symphony, this life: Fragile academic selves and insecure identities at work. *Organization Studies*, 35(3), pp. 335–357.

Lazar, I., Mortan, M. and Vereș, V. 2007. Un posibil model de evaluare a durabilității exploatațiilor agricole din zona de Nord Vest a României. *Revista Transilvana de Stiinte Administrative*, 20, pp. 52–67.

Lebas, M. 1995. Performance measurement and performance management *International Journal of Production Economics*, 41(1), pp. 23–36.

Mathis, R. and Jackson, J. 2010. *Human Resource Management*. 13en ed. USA: South-Western Cengage Learning.

Prisacaru, V. and Caradja, A. 2018. Factors of competitiveness of the higher education institutions of the Republic of Moldova under conditions of marketization of professional education. *Scientific Papers. Series Management, Economic Engineering in Agriculture and Rural Development*, 18(2), pp. 367–374.

Pulakos, E. 2009. *Performance management. A new approach for driving business results*. Singapore: Wiley-Blackwell.

Samoilenco, F. and Sverdlic, V. 2010. *Managementul performanțelor rețelei naționale aeriene*. Chişinău: Ed. TEHNICA-INFO.

Shields, J. 2016. *Managing employee performance and reward: Concepts, practices, strategies*. 2rd ed. Cambridge: Cambridge University Press.

UNESCO. 2023. *What you need to know about education for sustainable development*. [online] Available at: <https://www.unesco.org/en/education-sustainable-development/need-know> [Accessed 9 October 2023]

Verweire, K. and Berghe, L. 2004. *Integrated performance management. A guide to strategy implementation*. London: SAGE Publications Ltd.

Valentina Ofelia Robescu, Valentina Nicoleta Florea,
Gabriel Croitoru, Vasile Cumpănașu

Theoretical aspects regarding the use of biomass for a sustainable economy

Abstract: *Agriculture generally produces raw materials and substances for the food industry. But in the secondary, especially in wetter and warmer areas, it produces biogas and solid fuels using straw and other waste. The food industry arises from the need to process and make food. Still, part of the raw material is used to manufacture fuel oils, biodiesel, ethanol, and biogas, which are also used as fuels. From the forest area, the industry uses wood for furniture, paper, supplies, chemical industry, etc. We are talking about biomass as a mass of substances that come from the living world, which is intended for various needs. In this paper, we will analyze the role and importance of biomass as a pillar of the development of the bio-economy in a constantly changing society. We will identify the main ways of using biomass for a sustainable economy (bioeconomy) and try to determine its impact on society.*

Keywords: biodiversity, bio-economy, sustainable food use, sustainable development.

Introduction

In the European Union's view, the term or notion of the bio-economy presupposes an economy that uses soil and sea biological resources, as well as waste as food raw materials, for animal feed and industrial and energy production. It also includes the use of green processes for sustainable industrial sectors. For example, bio-waste has considerable potential as an alternative to chemical fertilizers or for conversion to bioenergy and can contribute to 2 % of the European Union's target on energy from renewable sources (Jennings and Wcislo, 2012).

Nicholas Georgescu-Roegen was one of the greatest Romanian scientists in mathematical statistics, economics, and the environment. By the end of the 1960s, he noted that the economy and ecology had already reached irreconcilable conflicts, from which the economy and human society were losing concern. At that time, the world of research, like economic practice, had drawn a barrier between the industrial revolution, technical progress on the one hand, and the evolutionism of the world on the other. In his work, "The Entropy Law and The Economic Process" published in 1971, Georgescu-Roegen (Georgescu-Roegen, 1971) notices the rapid increase in entropy in biological systems and formulates several restrictions on weather research. He calls for the integration of the economy and the environment, stating that "solving decisive environmental problems are also closely linked to the scientific, technological and IT progress of human society, but at the same time, only the existence of the widespread progress of the human race, in itself, cannot automatically solve the ecological problems that all people and the accelerated development of mankind, starting with the industrial revolution, have created".

Lester Brown, an agronomist, became a famous environmental analyst who wrote a lot and campaigned for the construction of a new economic system, which he defined in 2001, with the establishment of the Earth Policy Institute (EPI) – Institute of World Policy, which also dissolved immediately after his retirement. Lester Brown's theory is based on research on economic-environmental relations by many scientists. He said that "economic and environmental deficits now shape not only our future but also the present" (Brown, 2001, 2006, 2008); as such, we can say that the evolution of the model from N. Georgescu-Roegen to L. Brown is dramatic.

N. Georgescu-Roegen used the law of entropy to calculate a module of economic development that would lead to "the joy of living, which is the main purpose of the economy". In that case, Lester Brown proposes modeling an economic system strongly coupled with the environment, with respect to it and its resources, leading to the salvation of human civilization, with all its good and evil.

As such, the bio-economy presupposes a new way of life in which the order of nature is found in the economy, the two functioning as a unitary whole in favor of man and nature. Respect for life will become an equal phenomenon in importance, both for man and for the rest of non-anthropic living things.

In this paper, we will analyze the role and importance of biomass as a pillar of the development of the bio-economy in a constantly changing society. We will identify the main ways to use biomass for a sustainable economy, and we will try to determine its impact on society.

Methodology

The methodological approach will be based on a revision of the literature. We will largely analyze some documents of the European Commission, and on the other hand, we will start from the findings of the recently published theoretical and empirical research, which we will analyze and develop from our point of view.

In the vision of the European Commission and the OECD, the bio-economy is presented as "a set of economic activities related to innovation, production development, and the use of biochemical products and processes". The new strategy, developed in the case of the project "Bio-economy", aims, in the vision of these institutions, to use research and innovation to make the transition of our current economy, based on carbon and other fossil energies, to a green economy, without fossil and sustainable carbon (Berca et al., 2019).

Biodiversity is the most widely used term and has the role of generically replacing terms such as species diversity but also their wealth (Walker, 1992). A widely used definition of biodiversity is that biodiversity includes "all the genes, species and ecosystems of a region" (Larsson, 2001). This definition takes a unified view of the traditional names of biological varieties previously identified as taxonomic diversity. Sahney, Benton and Ferry (2010), mention ecological diversity, which is viewed from the perspective of ecosystem diversity. At Campbell (2003), we find the concept of morphological diversity, which comes from genetic diversity and molecular diversity. Lefcheck (2014) defines functional diversity as a measure of the number of functionally disparate species in a population.

Food and Agriculture Organization of the United Nations (FAO) defined biodiversity in 2019 as "the variability that exists between living organisms (both within and between species) and between the ecosystems of which they belong" (Bélanger and Pilling, 2019).

A strategy for the protection of the European Union's biodiversity was presented in 2021. Forest and maritime management, environmental protection, and addressing the issue of loss of species and ecosystems are all aspects of this target area (European Commission, 2020a, 2020b). The restoration of affected ecosystems can be done by implementing ecological farming methods, managing pollination processes, restoring watercourses, using less pesticides, reforestation. Within the climate change mitigation strategy of the European Union, an important place is occupied by the strategy for biodiversity. Thus, from the European budget intended for climate change, a large part is directed towards the restoration of biodiversity (European Commission, 2020a, 2020b).

Starting from the objectives set by the EU which include the protection of 30 % of the territory of the sea and 30 % of the land, in particular primary forests and mature forests, planting 3 billion trees by 2030, restoring about 25,000 kilometers of rivers, so that they will become flowing, as well as reducing the use of pesticides by 50 % by 2030 to support the increase in the share of organic farming but also the increase in biodiversity in agriculture, at least €20 billion per year is expected to implement these goals and support the transformation of these goals business in the field of sustainable business, practically supporting the development of the bio-economy. Knowing that about half of global GDP depends on nature. In Europe, many parts of the economy that generate trillions € per year depend on nature (European Commission, 2020a, 2020b).

People depend on biodiversity for survival, such as the food we eat, the drugs we use to stay healthy, and the materials we wear or use to build our homes. These services are the tangible products or articles we and other species are subject to survive. Human dependence on biodiversity extends beyond the food we eat, the air we breathe, and the water we drink (Buttke, Allen and Higgins, 2014).

Results and discussions

A team of researchers led by Stephen Shennan (2013) from the Institute of Archeology at College-London University, who was working with radioactive isotopes immediately after their appearance in 1950, proved that in the period 10,000 BC. – In 4,000 BC, there was an extremely flourishing civilization in the current territory of Western and Central Europe (Wikipedia, 2023a).

The development of agriculture began on the current territories of Romania and Hungary in the year 8,000 B.C. It extended to France (7,000 B.C.) and England (5,000–6,000 B.C.), according to Hamuda and Patkó (2010). Agriculture developed rapidly and according to Alfred J. Lotka and Vito Volterra equation (Wikipedia, 2023b) developed exponentially in the area (Figure 1).

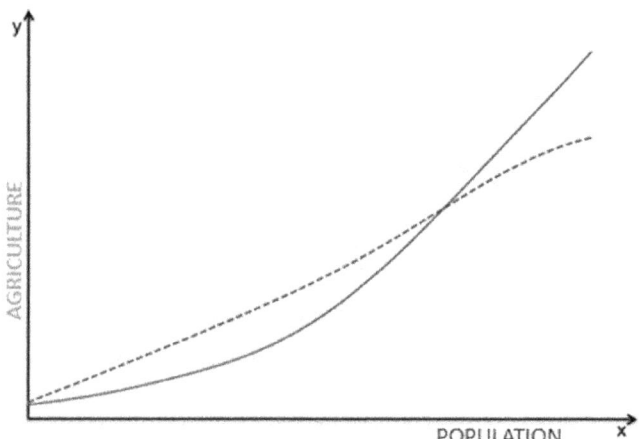

Figure 1. The exponential development model of agriculture and population with 5,000–10,000 years B.C.

Source: Authors' own research.

The same studies show that although food production has increased four times, the population has increased eight times. With small exceptions (game), the population feeds on agriculture. The collapse was possible due to famine and collateral disease, which trained the entire population. In the years 3,500–4,000 BC, extinction was total, and civilization was compromised. The collapse left no room for recovery of the population (Figure 2).

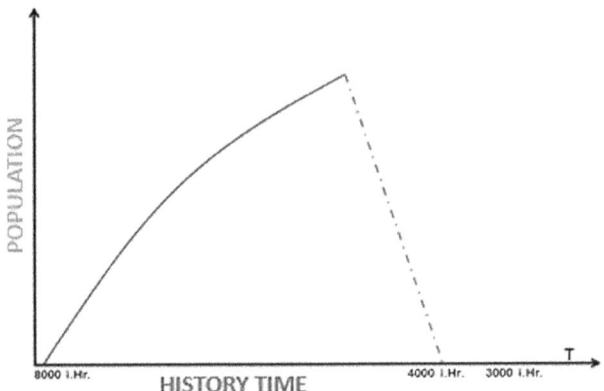

Figure 2. The collapse model of the oldest European civilization
Source: Authors' own research.

Archaeological studies conducted by Joseph A. Tainter (1988) also demonstrate that the Mayan civilizations and Mesopotamian, large and deep, collapsed due to food crises. The knowledgeable reader has already deduced that the collapse is due to the decline of marginal yields in agriculture. If much more is placed at the table than those who make the production, then the marginal yields are minus, and the production system collapses (Table 1).

Table 1. The relationship between total production, labor productivity, and marginal productivity – consumption of 200 kg wheat/individual

Nr. crt.	Capital	Work	Production (kg/ha)	Average productivity (kg/individual)	Medium productivity (kg/individual)	Feed people
1.	1	1	1000	1000	1000	5
2.	1	2	2400	1200	1400	7
3.	1	3	3600	1200	1200	6
4.	1	4	4200	1050	600	3
5.	1	5	4200	840	0	0
6.	1	6	3900	650	– 300	Hunger 1.5
7.	1	7	3500	500	– 400	Hunger 2
8.	1	8	3000	375	– 500	Hunger 2.5

Source: Authors' own research based on Tainter, 1988.

The production system will feed both those who work and the others. In the last three positions, the number of hungry people increases from 1.5 for position 6 to 2.5 for position 8, i.e., from 26 % to 31 %. Committing labor resources has the same effect as other resources in establishing marginal productivity. Most of the time, the entry into food collapse is due to the unlimited degradation of the main means of production – soil. This happened in Mesopotamia, and the same happened with the great empires – Roman and Ottoman, but also with others. As a tsar, synthesizing the main causes that led to agricultural and food collapses, determinants for all crises and collapses, and we list some of them, namely: loss of humus by erosion or oxidation; labor degradation (education, health); increasing bureaucracy and bureaus; complication of thinking and the system.

Human food costs financial and environmental resources, sometimes greater than we imagine. Both the lack of food (hunger) and their surplus (obesity) are associated with very high financial and energy costs required by the social status of patients of human individuals. Maintaining balance, i.e., placing people at a BMI between 18–25, would be ideal for minimal financial and energy costs. But maintaining a balance is always the most difficult problem of humanity. Deviations from the right path of balance accelerate the advance of society to collapse.

In the same study, FAO demonstrates that acute hunger is already settling at a deficit of 500 kcal/day, but there are several intermediate stages of the social phenomenon. Depending on the spread of hunger worldwide and the depth of the phenomenon, five groups of countries can be defined (Table 2).

Table 2. Spreading and severity of hunger in the world between 1996 and 1998

Hunger spread – percentage of undernourished population	Hunger severity – food deficit of underpowered persons (Kcal/day)			
	<200 (poor)	200–s300 (average)	>300 (high)	Total
	Number of countries			
<5 % (poor spread)	52	0	0	52
5–19 % (average spread)	17	29	0	46
≥20 % (high spread)	0	31	23	54
Total	69	60	23	152

Source: Authors' own research.

One of the forms shown to us by this table is that over 1 billion inhabitants are in a starvation state and live in 152 countries. More than 800 million are affected by medium and severe hunger. They live in 100 countries around the world, on all continents. They also exist in Romania too, because some of them are living in houses without direct access to a source of water, and electricity, and without regular access to basic food baskets.

Mankind is partly hungry and needs food. Isn't there enough food? Without going into details, we only present here the conclusion that food is now sufficient on Earth, but some limitations make it inaccessible and very easy to waste (Figure 3). The largest food loss per flow takes place between industry and marketing. They exceed 14 % of total agricultural production for food. Large losses are also in the private area (15 % after some authors, while after others, they can reach over 30 %).

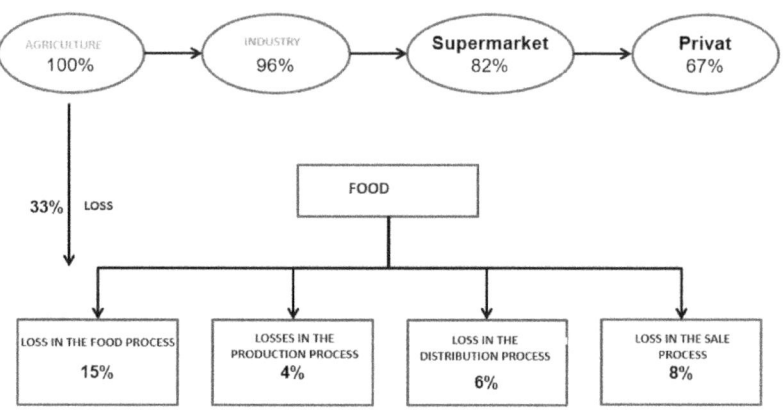

Figure 3. The flow of food losses from agriculture to the final consumer

Only foods that reach household waste amount to 43.2 million tons (Kreczi, 2012). At the level of the European Union, each citizen throws away about 180 kg of food, over half of which is in the original packaging. The figure leads to a total volume of over 90 million tons. Or, that means capital, water, energy, and labor, evolution towards unsustainable and collapse.

Similar statistics can be found in Austria, the United States, etc. With food thrown into the world yearly, more than 3 billion people can be fed for a year, i.e., 40 % more than at present, and hunger could be eradicated. Without further global action, food saving could feed a planet with more than 9 billion inhabitants and delay regional and global collapse by at least 50 years.

According to our calculations, the energy value of food lost and dumped is just over $9 \times 1,012$ kcal or $3.8 \times 1,013$ kJ (1 kcal = 4.184 kilojoules/kJ). We allow ourselves to inform that 1 kg oil = 4.6×10^3 kJ = 46,000 kJ. This means that the energy value of discarded, lost food equals

8.2×10^9 kg oil = 8.2×10^6 tons = 8,200,000 tons oil.

The energy and environmental resources used to obtain the food needed by mankind are much higher than those necessary to ensure the metabolic requirements of the human species and the adjacent waste at the current stage. The human species has become the largest waste of resources, a real danger to ecosystems and the life of the planet, and an imminent danger to its own existence.

The way in which the current population of the globe feeds, especially in the industrial areas of the Earth, is a generator of crisis.

Food swallows a lot of energy, and the amount used is all the higher as the food used is more elaborate and complexly manufactured. On the other hand, the amount of energy used is positively correlated with the amount of pollutant emissions and hazardous to the greenhouse effect and climate change.

If we start from FAO data (2012), which shows that globally, every person consumes, on average, 39 kg of meat/year, means that 273 million tons of meat are produced and consumed annually, consuming 2.73 x 1011 liters of oil, i.e., over 30 % of all world energy. Such energy consumption will no longer be sustained in the near future. In the not-too-distant future, which we have presented before, we will be significantly or totally concerned with the appetite for steak and cheese, in which place to use the vegetarian menus.

In Figure 4, we performed an analysis of the correlation between the amount of energy used to obtain one kg of food and the amount of CO_2 emitted into the atmosphere with greenhouse gases. The authors performed the analysis using primary data published by Agenda 21 Treffpunkt (2015) and by UN Development Programs – FAO. Figures and functions work in favor of primary, photosynthetic food. The analysis is actual even if the data are based on previous statistical data.

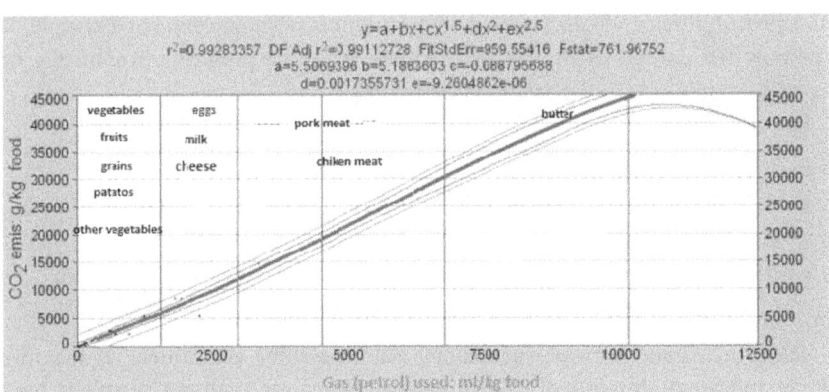

Figure 4. Correlation between the amount of petrol used in food production and related gas emissions, CO2 equivalent (original data processing)

The smallest amount of fossil energy, in oil gas equivalent, is required for vegetables – vegetables, fruits, cereals, potatoes, etc. (column 1 of the graph), consuming up to 1 liter of oil gas/kg product and removing in the air between 150 g/kg CO2 (fresh vegetables) and almost 1500 g/kg CO2 (sugar). The food obtained in step 2 on the food chain consumes an average of 2.5 liters of oil gas equivalent to eggs, milk, and cheese, 3.5 liters of oil gas for poultry and pork, 4.5 – 7.0 liters of oil gas in cheeses and sausages and over 7.5 – 8.0 liters of oil gas for beef.

Butter, this animal fat so requested for breakfast, has the highest energy consumption of up to 8.00–10.00 liters of oil gas and eliminates up to 45 kg of CO_2/ into the atmosphere/kg of food, as well as the prepared beef and other food dishes, at a digestive energy of 500 kcal/portion 140 g.

The large amount of energy required by steak and butter is due to the long maintenance period of cattle, a lower conversion of feed, and high energy maintenance costs in the stables.

In Figure 4 we have described an exponential function of the type:

$$y = a + bx + cx^{1,5} + dx^2 + ex^{2,5}$$

where y = amount of emissions, equivalent CO2, in g/kg food;

x = The amount of energy used, calculated in oil gas equivalent, in ml/kg of food.

The function and curve are calculated for 95 % and 99 % (confidence intervals) probabilities and can be used as calculation nomograms. For example, if we know the energy used, we can calculate potential pollution graphically or through evaluations in the calculated function.

For example:

to x = 100 ml petrol; y = 452 g CO2/kg food;
to x = 1,000 ml petrol; y = 3,828 g CO2/kg food = 3.8 kg CO2/kg food;
to x = 5,000 ml petrol; y = 21,572 g CO2/kg food = 21.6 kg CO2/kg food;
to x = 10,000 ml petrol; y = 44,045 g CO2/kg food= 44.04 kg CO2/kg food.

Calculations are statistically insured for a 99 % probability or a 1 % risk.

The conclusion of most researchers, but also some consumers, is that the planet, in today's situation of using conventional energy, will not be able to bear the great meat eaters for a long time, butter and cheese, and that a rethinking of the feeding pattern of mankind will become an equally important concern of the new bioeconomic science, even if energy becomes ecological. This is because animals have other important sources of pollution and dirt, such as gastric methane or their manure. Returning to the 1st link (vegetable – vegan) of the nutrient chain, which leads to humans, could save nine times the 30 % dedicated to the food industry, creating significant breathing in human evolution towards the bioeconomic era.

By stimulating sustainable activity in key sectors and supporting new business opportunities, we can unlock economic growth and influence the bioeconomic potential at both micro and macro levels. These elements help us to achieve a correct design of the bioeconomic future based on the correct development of strategies based on innovative approaches to the production of plastics and chemicals, also with regard to the re-use in the economic process of food waste, construction waste, critical raw materials, as well as industrial and mining waste. This approach allows us to support and promote the implementation of sustainable economic measures at every stage of the value chain – from production to consumption, repair and re-manufacturing, waste management, and secondary raw materials brought back into the economy.

Conclusions

If we continue to use resources at the current rate, by 2050, we will need, in total, the equivalent of more than two planets to support us, and aspirations for a better quality of life will not be achieved. In order to avoid this unsustainable perspective, a resource-efficient economy is needed, close to the concept of an ideal economy, a bio-economy in this case, to promote a systemic transformation in how resources are used in the economic system and by society.

The global dimension of the economy and supply chains is particularly important in areas such as sustainable supply, food waste, and an increasingly globalized market for secondary raw materials. In implementing this action plan, there must be global cooperation to achieve sustainable development goals. The adoption of solutions to reduce the impact on the environment and the generation of waste in the manufacturing process and during the use of products should be encouraged from the design phase of the product. Encourage the production and marketing of multipurpose products, technically sustainable products which, once they become waste, can be properly exploited and whose disposal is compatible with the principles of environmental protection.

Important elements for the sustainable implementation of measures to support the promotion of the bio-economy are based on directions such as the reuse of goods, recycling of materials, and design of goods, which allows circulation without loss of own quality. At the same time, the production of substances that make the circulation and accumulation of pollutants more present should be avoided. The life of the goods should be preserved as long as possible, and their return to circulation at the end of their useful life should be rapid. Economic actors, such as businesses and consumers, are essential for leading this process. Local, regional, and national authorities allow for the transition, but the EU also has a key role in supporting it. The aim is to ensure that an appropriate regulatory framework is in place for the development of the bio-economy in the single market and give clear signals to economic operators and society at large on the way forward with long-term targets, as well as a set of concrete, comprehensive and ambitious actions, to be carried out before 2030.

References

Agenda 21 Treffpunkt. 2015. Treibhausgase (THgG), Daten/Statistiken/Infografiken. [online] Available at: <http://www.agenda21-treffpunkt.de/daten/treibhausgase.htm> [Accessed 10 September 2023].

Bélanger, J. and Pilling, D. 2019. *The state of the world's biodiversity for food and agriculture* (PDF). Rome: FAO. [online] Available at: <https://www.fao.org/3/ca3129en/CA3129EN.pdf> [Accessed 5 September 2023].

Berca, M., Robescu, V.O. and Horoiaș, R. 2019. *Bioeconomy: Return to natural models*. Bucharest: House Bren.

Brown, L.R. 2001. *Eco-economy: Building an economy for the earth*. New York: W.W. Norton & Co.

Brown, L.R. 2006. *Plan B 2.0: Rescuing a planet under stress and a civilization in trouble*. New York: W.W. Norton & Co.

Brown, L.R. 2008. *Plan B 3.0: Mobilizing to save civilization*. New York: W.W. Norton & Co.

Buttke, D., Allen D. and C. Higgins. 2014. Benefits of biodiversity to human health and well-being. *Park Science*, 31(1).

Campbell, A.K. 2003. Save those molecules: Molecular biodiversity and life. *Journal of Applied Ecology*, 40(2), pp. 193–203. https://doi.org/10.1046/j.1365-2664.2003.00803.x.

European Commission. 2020a. *EU biodiversity strategy for 2030*. [online] Available at: <https://environment.ec.europa.eu/topics/nature-and-biodiversity/habitats-directive_en> [Accessed 5 September 2023].

European Commission. 2020b. *The essentials of the "Green Deal, of the European Commission"*. Green Facts. [online] Available at: <https://www.greenfacts.org/en/europe-green-deal-2019/index.htm> [Accessed 5 September 2023].

FAO. 2012. *The State of Food and Agriculture*. [online] Available at: <http://www.fao.org/docrep/017/i3028e/i3028e.pdf> [Accessed 5 September 2023].

Georgescu-Roegen, N. 1971. *The entropy law and the economic process*. Harvard University Press.

Hamuda, H.E.A.F.B. and Patkó, I. 2010. Relationship between environmental impacts and modern agriculture. *Óbuda University e-Bulletin*, Vl(1), pp.

87–98. [online] Available at: <http://uni-obuda.hu/e-bulletin/Hamuda_Patko_1.pdf> [Accessed 5 September 2023].

Jennings, M. and Wcislo, M. 2012. *Commission proposes a strategy for sustainable bioeconomy in Europe*. European Commission – Press Release. [online] Available at: <http://europa.eu/rapid/press-release_IP-12-124_en.html> [Accessed 5 September 2023].

Kreczi, F. 2012. *Lebensmittelverschwendung*. RESET – Digital for Good. [online] Available at: <https://reset.org/knowledge/lebensmittelverschwendung> [Accessed 5 September 2023].

Larsson, T-B. 2001. *Biodiversity evaluation tools for European forests*. Oxford: Wiley-Blackwell.

Lefcheck, J. 2014. *What is functional diversity, and why do we care?*. Sample (Ecology). [online] Available at: <https://jonlefcheck.net/2014/10/20/what-is-functional-diversity-and-why-do-we-care-2/> [Accessed 5 September 2023].

Sahney, S., Benton, M.J. and Ferry, P. 2010. Links between global taxonomic diversity, ecological diversity and the expansion of vertebrates on land. *Biology Letters*, 6(4), pp. 544–547. https://doi.org/10.1098/rsbl.2009.1024.

Shennan, S., Downey, S.S., Timpson, A., Edinborough, K., Colledge, S., Kerig, T., Manning, T. and Thomas, M.G. 2013. Regional population collapse followed initial agriculture booms in mid-Holocene Europe. *Nature Communications*, 4. https://doi.org/10.1038/ncomms3486.

Tainter, J.A. 1988. *The collapse of complex societies*. Cambridge: Cambridge University Press. [online] Available at: <https://risk.princeton.edu/img/Historical_Collapse_Resources/Tainter_The_Collapse_of_Complex_Societies_ch_1_2_5_6.pdf> [Accessed 10 September 2023].

Walker, B.H. 1992. Biodiversity and ecological redundancy. *Conservation Biology*, 6(1), pp. 18–23. https://doi.org/10.1046/j.1523-1739.1992.610018.x.

Wikipedia. 2023a. *History of Europe*. [online] Available at: <https://en.wikipedia.org/w/index.php?title=History_of_Europe&oldid=1194572590> [Accessed 6 September 2023].

Wikipedia. 2023b. *Lotka–Volterra equations*. [online] Available at: <https://en.wikipedia.org/w/index.php?title=Lotka%E2%80%93Volterra_equations&oldid=1192851784> [Accessed 6 September 2023].

Srdjan Šljukić, Milovan Mitrović

Sociological aspects of sustainable development: Cultural capital in rural settlements in Serbia

Abstract: *In this paper the authors focus on cultural capital in the rural settlements in Serbia, seen as an aspect of the social dimension of sustainable development. Among the three dimensions of sustainable development (environmental, economic and social), the social dimension attracted less attention than the first two, and has been seen as difficult to operationalize. Since there cannot exist any social phenomena without a cultural side or cultural traits, the authors start from the concept of cultural capital (Bourdieu). By using the results from the last Census of Population, Households and Dwellings in Serbia (2022), they present the data for three indicators of cultural capital in the rural settlements: educational attainment, literacy and computer literacy. The presented data are analyzed and compared with the same sort of data for urban settlements, as well as with the data from the previous Census (2011).*

Key words: rural settlements, Serbia, sustainable development, cultural capital, literacy, computer literacy, educational attainment, Census.

Introduction

Almost a hundred years ago, the Russian-American sociologist Sorokin (1931) argued that, despite the historically and empirically confirmed dominance of the city over the countryside, culminating in the time beginning with the Industrial Revolution, these two forms of human settlements remain mutually dependent economically, politically, demographically and culturally. The rural and urban world represent, each in their own way, positive values, and they need each other; their simultaneous presence increases the value of both. Some kind of "pure" urban society would have no chance to survive. When we connect Sorokin's idea of interdependence with the concept of sustainable development, which emerged in the 70s of the 20th century as a compromise between development and conservation (Du Pisani, 2006), then it is perfectly clear that the sustainability of rural areas is a necessary precondition for the survival of society as a whole.

Among the three "pillars" (dimensions) of sustainable development (ecological, economic and social), the third one has so far attracted the least attention and is considered the most difficult to operationalize (Bostrom, 2012). However, no matter how the social dimension of sustainable development is operationalized, it must certainly include education and various skills and knowledge. This is also because there are no social phenomena that do not have their own cultural side. Bearing this in mind, in this text we start from the concept of cultural capital (Bourdieu, 1986), with the intention of presenting and analyzing three indicators of cultural capital in rural settlements in Serbia: literacy, computer literacy and educational attainment. Using the just published statistical data from the Census of Population, Households and Dwellings (2022), we are trying to establish whether, according to these three indicators, in the eleven-year period between the two censuses (2011 and 2022), any changes took place and in what direction. In doing so, we started from the assumption that the number of illiterates will decrease between the two censuses, for the simple reason of a change of generations. Namely, while the generations whose parts could have been left out of the process of compulsory education leave the scene of life, the generations to come are, almost without exception, included in the process of education. The second assumption we started from is that between the two censuses, the number of computer literates in rural settlements increased as a result of the combination of the education process, the evident accelerated spread of wireless internet and the increase in the necessity of knowing how to work with a computer in everyday work and life activities. At the same time, we looked at the changes in the structure of educational attainment in rural settlements as something we cannot make any assumption we can be sure about; several factors are at work in that field, both those that improve that structure (educational system) and those that worsen it (migrations). We also wondered whether the gap between rural and urban settlements in terms of literacy, computer literacy and schooling had narrowed between the two censuses. Given that Bourdieu's theory of forms of capital (economic, social, cultural and symbolic capital) implies the possibility of transforming one form into another, the confirmation of our hypotheses and possible positive answers to the research questions would mean that the rural component of society in this segment has strengthened its sustainability, which can have positive consequences for the sustainability (sustainable development) of the entire Serbian society.

Literature review

Understanding the concept of sustainable development is not possible without understanding its historical roots, primarily the idea of progress. In ancient times, the concept of progress was almost unknown; it appears with Jewish and Christian theology and its linear conception of time and history. With the rise of Western civilization and its modernity, the idea of progress gets an additional impulse. It was first formulated by the French scientist Fontenel in 1683, and it reached its peak in the Age of Enlightenment and its offshoots and successors such as Thirgot, Condorcet, Saint-Simon, Comte, Hegel, Marx and Spencer (1750–1900). With the advent of the industrial revolution and the changes it brought, the idea of progress became associated with economic growth and material progress. Belief in progress has become universal in intellectual circles in the West (Du Pisani, 2006).

At the same time, since ancient times, processes of deforestation, loss of soil fertility, etc., have been observed, concern for the environment has been expressed and measures for its preservation have been recommended (Plato, Strabo, Columella, Pliny the Elder). The term "sustainability" was first used by Hans von Karlowitz in 1713, recommending the constant renewal of forests, so that wood resources would not be depleted. In the 19th century, the focus shifted to concern for the depletion of coal reserves, and in the 20th to the issue of oil. Theories of social development (the theory of modernization and the theory of dependent development) represented the basis for discussions about growth, but also about sustainability. At the end of the 60s and the beginning of the 70s of the 20th century, the understanding that the Enlightenment did not fulfill its promises about the constant progress of the human race (world wars, colonial exploitation, etc.), as well as the awareness of the increasing threat to the environment through uncontrolled economic growth came to the fore. In 1972, a group of scientists gathered in the so-called "Club of Rome" published a report called *The Limits of Growth*, in which they warned of the limitation of global resources and gave apocalyptic forecasts, which sparked discussions about the need to limit and control economic growth. From the beginning of the 80s, the new paradigm of sustainable development was increasingly in use, so that the United Nations Organization (UN) finally formed a commission of 22 members called the World Commission on Environment and Development (WCED), better known as the "Bruntland Commission". In 1987, this commission published a report called *Our Common Future*. The commission established that there is a tension between economic growth and environmental protection, recommending a transition to sustainable development (Du Pisani, 2006), at the same time defining it: "Sustainable development is development that meets the need of the present without compromising the ability of future generations to meet their own needs" (WCED, 1987, p. 43).

This definition was and remains the starting point for all subsequent considerations of the issue of sustainable development, and it was soon joined by the definition according to which sustainable development rests on three "pillars" (Bostrom, 2012), i.e. that it has three dimensions: ecological, economic and social (Nasirzadeh et al., 2020; Orboi et al, 2010). The effort to list more dimensions, even up to seven (Pawlowski, 2008), turned out to be more like a breakdown of the social dimension of sustainable development, than a different definition of it. Apart from the fact that the social dimension has attracted less attention and has proven to be rather difficult to operationalize, it is often seen as fluid, vague, subjective, ideological and the least explicit (Bostrom, 2012). The social dimension of sustainable development is connected and intertwined with numerous other concepts in social sciences, such as social capital, moral economy, identity, democracy, participation, social cohesion, etc.

A large number of attempts have been made to operationalize the social dimension of sustainable development and to determine specific goals that should be pursued within it. Regardless of the numerous differences among these efforts, almost all of them mention education, learning, knowledge, skills, etc. as aspects of the social dimension of sustainable development (Bostrom, 2012; Burja and Burja, 2014; Dempsey et al., 2011; Galdeano-Gomez, 2016; Murphy, 2012; Lopez-Penabad et al., 2022). This practically means that in addition to intertwining and connecting with the concept of social capital, the social dimension of sustainable development can also be viewed from the point of view of cultural capital, because all the terms mentioned in the previous sentence represent precisely this form of capital. According to Bourdeiu (1986), cultural capital takes three forms: embodied (enduring dispositions of mind and body), objectified (cultural goods), and institutionalized (educational qualifications) (Bourdieu, 1986; Brown, 2016; Huang, 2019). The importance of cultural capital is seen in its ability to transform (under certain conditions) into economic capital, which not only significantly affects the reproduction of the social structure (Bourdieu, 1986; Kim and Kim, 2009), but can also seriously contribute to social (sustainable) development. There is no doubt that the indicators we look at in this text, namely literacy, computer literacy and educational attainment can be seen as expressions of cultural capital.

There are not many studies devoted to cultural capital (regularly without using this term) in the form of education and literacy in rural areas in Serbia. At the end of the 1980s, Trnavac (1992) researched small rural schools in central Serbia, a hilly-mountainous region characterized by a large distance between settlements and a fragmented type of village, with a small number of inhabitants. The sample included about 300 rural schools, 60 % of which were located more than 15 kilometers from the municipal center, and half did not have a local bus connection. In these schools, the teachers changed quickly, waiting for the first opportunity to leave the village; at the same time, migration to cities led to the closure of a number of rural schools. With the help of statistical data from the

Census, Mitrović (2015) dealt with these issues, and in this paper we actually continue this research. Illiteracy in the rural Serbia, which for a long time was very high in mountain villages, especially among older peasants (and especially peasant women), has been significantly reduced in the last half century, and the educational structure of the rural population has been improved. However, according to the data from the Census held in 2011, significant differences in the level of education between residents of urban and other (rural) settlements remained: in the latter there were more people with incomplete elementary school and complete elementary school, while in urban areas there were four times more residents with high and higher education. Jelić and Kolarević (2016) state that a large number of rural children travel long distances to school, which often leads to irregular schooling and/or early school leaving. Negative deviations from the average are very high in rural settlements, especially when it comes to the number of children enrolled in kindergartens (−43 %) and those attending university (−34 %). Also, far fewer people in rural settlements know how to work with a computer (19.84 %) than in cities (44.09 %). Stojić (2023) also provides data on computer literacy and educational attainment of residents of rural settlements compared to cities from the 2011 Census.

Methodology

The fact that the experiment, as the basic method of the natural sciences, is difficult to perform in the social sciences, including in sociology, has long been established. The reasons are of both ethical and methodological nature. Namely, there are questions that are very sensitive for those who should be observed in some kind of experiment, and researchers must strictly take care not to injure the physical and psychological integrity of the participants in an experiment. This means that the participants in the experiment must not be harmed in any way, including that they must not be deceived, i.e. that they must know that they will be watched. Then, however, a methodological problem arises: those who know that they are being observed will not behave in the experimental situation as they would otherwise behave in similar situations, but will manage their behavior in relation to what they think is expected of them. This necessarily leads to distorted and unusable results. For these reasons, the comparative method appears as a substitute for the experiment and is therefore widely used in the social sciences and humanities (Griffiths, 2017). The importance of the comparative method was particularly emphasized by Durkheim, emphasizing that it is "the only one suitable for sociology" (Durkheim, 2012). The basic method of this paper is precisely the comparative method, whereby phenomena are compared over time (literacy, computer literacy and educational attainment in two different censuses), while at the same time the same phenomenon is observed in two different social environments (urban and rural), and sometimes in different countries as well.

The main source of data that we use is statistics, especially the results of the Censuses of Population, Households and Dwellings conducted in Serbia in 2011 and 2022. The data from the mentioned Censuses were obtained in the same way, which is a necessary condition for a meaningful comparison. Creating one's own records in sociology can often be a very difficult thing, requiring a lot of energy, money and time, while statistical data provide ready-made material for research. Of course, we are aware that the reliability of statistical data has its limitations and that is why they must not be taken for granted, but in this case we believe that their usability cannot be seriously questioned. Simple percentage calculations in the tables are largely the work of the authors.

We must mention one more difficulty. The types of settlements used by Censuses in Serbia ("urban settlements and other settlements") are distinguished from each other in an administrative manner, i.e. a settlement is considered urban if there is a decision on its status by the appropriate authority (Živanović, 2018). From a sociological point of view, this criterion can hardly be considered sufficient and it certainly represents the most serious objection to the validity of the data used. The whole matter is complicated by the fact that in this dichotomous typology, rural settlements are not even mentioned directly (although it is clear that they are being discussed), but rather as "other settlements". The problem is that there is no other data, so we can only rely on these if we want to say something about these issues.

Results and discussions

After the fall of the Serbian medieval states under the rule of the Ottoman Turks (14th and 15th centuries), the upper social classes were either physically destroyed or became Islamized. Cities, as the centers of every government, became places where mainly Turks and Islamized population lived, while Serbs lived in villages. However, from the end of the 17th century, in the neighboring Habsburg Monarchy, Serbs formed part of the city's population in some areas. The creation of the modern Serbian state, which began in 1804, marked the beginning of the process of population change in the cities, which grew slowly during the 19th and the first half of the 20th century. Thus, in 1859, the share of the urban population in the total population was 8.1 %, 9.5 % in 1866, 10.2 %

in 1873, 12.4 % in 1884, and 13.3 % in 1895 (Statistical Office of the Republic of Serbia, 2008; Šljukić and Janković, 2015).

Table 1. Share of the urban population in Serbia 1953–2022 (%)

Year	1953	1961	1971	1981	1991	2002	2011	2022
Percentage of urban population	22.5	29.8	40.6	46.6	50.7	56.4	59.44	62.0

Source: Authors' own research based on Statistical Office of the Republic of Serbia, 2008, 2012a, 2023a.

As it can see in Table 1, a turning point in the process of increasing the share of the urban population occurred after the Second World War. In less than 20 years (1953–1971), it almost doubled, and in 1991, for the first time, the number of inhabitants living in cities exceeded the number of the rural population. After 2002, the growth of the urban population slowed down, so almost two-fifths (38 %) of the total number of inhabitants of the Republic of Serbia live in the countryside today, which means that Serbia still has a very strong rural component, indispensable when it comes to (sustainable) development. including the presence of cultural capital.

During the centuries under Turkish rule, literacy was preserved only within the confines of the church. Long after the acquisition of the status of a vassal principality (1830), there were very few literate people in Serbia (the situation was somewhat better among the Serbs in the Habsburg Monarchy). According to the census data from 1866, only 4.2 % were literate, with the percentage being 27.9 % in cities and only 1.7 % in villages. The number of literates grew very slowly, so eight years later (1874) there were 6.7 % literates (33.6 % in cities, 3.7 % in villages) (Statistical Office of the Republic of Serbia, 2008). The tenfold higher percentage of literates in the cities was a consequence of the fact that for the construction of state institutions, as well as for the market economy (both were concentrated in the cities), it was necessary to have educated people (Šljukić and Janković, 2015).

Table 2. Share of illiterate population (10 years and older) in Serbia in 2011 and 2022 by type of settlement (%)

Year	2011	2022
Republic of Serbia	1.96	0.63
Urban settlements	1.03	0.40
Rural ("other") settlements	3.31	1.01

Source: Authors' own research based on Statistical Office of the Republic of Serbia, 2012b, 2023b.

After free and compulsory primary education was proclaimed (1882), the number of illiterates began to decrease rapidly (Šljukić and Janković, 2015). The data from Table 2 testify that today illiteracy has been reduced to a minimum, both in cities and in villages. In 2011, there were three times more illiterates in the villages than in the cities, but these percentages are very small. The gap between the city and the countryside was further reduced in the following eleven years, as was the share of illiterates in the total population, which fell below 1 %. We can state that in Serbia, illiteracy has been reduced so much that it has ceased to be a factor that can seriously affect development processes, and that the gap between rural and urban areas has been maintained, but the percentages have been reduced so much that the gap itself has become irrelevant.

Table 3. Population aged 15 and over by computer literacy, by type of settlement in 2011 (%)

	Computer literate persons	Persons with partial computer skills	Computer illiterate persons
Republic of Serbia	34.21	14.78	51.01
Urban settlements	44.09	15.11	40.80
Rural ("other") settlements	19.84	14.29	65.87

Source: Authors' own research based on Statistical Office of the Republic of Serbia, 2012b.

Table 4. Population aged 15 and over by computer literacy, by type of settlement in 2022 (%)

	Computer literate persons	Persons with partial computer skills	Computer illiterate persons	Unknown
Republic of Serbia	45.73	29.62	24.19	0.46
Urban settlements	55.47	25.57	18.35	0.61
Rural ("other") settlements	30.08	36.11	33.56	0.25

Source: Authors' own research based on Statistical Office of the Republic of Serbia, 2023b.

If the issue of literacy is "taken off the agenda", the issue of computer literacy comes to the fore, even when it comes to (sustainable) development. Determining the number and share of computer literates in Serbia was carried out for the first time in the 2011 Census. At that time, slightly more than half of the population were computer illiterate, with two thirds of the rural population being computer illiterate, and two fifths of the urban population also computer illiterate; in urban settlements there were more than twice as many computer literates as in rural ones (Table 3). In the next census (Table 4), the share of computer literates (completely and partially) increased significantly; only one quarter of the population of Serbia was computer illiterate. The gap between the countryside and the city remained, but it was somewhat reduced: in the city, a fifth of the inhabitants were computer illiterate, and in the countryside, one third. It is worth noting that even in the European Union (EU) there is a gap between rural and urban areas when it comes to computer literacy (Popescu et al., 2022).

While literacy and computer literacy are a matter of having a certain ability, i.e. skill, not necessarily with any formal certification, educational attainment includes certain certificates, diplomas, etc., which are guaranteed and supported by the state as an organization. Mass formal education is a recent phenomenon, about two and a half centuries old, although it sometimes seems that it has always existed. Schooling among Serbs in Austria was initiated by pressure from the Catholic Church to convert them from Orthodoxy to Catholicism, after which the Serbs asked for help from Russia, which sent teachers who founded schools for the Orthodox population. The first teacher training school was founded in Sombor in 1778, and the first high school in 1791 in Sremski Karlovci. In liberated Serbia, Velika škola was founded (1808), while formal education at all levels would be regulated by laws during the 40s, 50s and 60s of the XIX century. Elementary schools multiplied rapidly, and education in them lasted four years, which continued between the two world wars. The first secondary school was founded in 1830, and at the end of the century (1890) there were 29 secondary schools, with 7230 students. The University of Belgrade (first called Velika škola) was founded in 1863, with three faculties; in 1855, there were about 200 people with a university degree in Serbia, 50 of whom obtained their degree abroad. Mass formal education gained momentum especially after the Second World War (Šljukić and Janković, 2015; Statistical Office of the Republic of Serbia, 2008).

The rural population was not among the first to be drawn into the organized schooling system, but it could not be completely left aside, given the needs of modern society to integrate each individual into the central value system and train them for work outside of agriculture. Nevertheless, the education of the rural population has always been neglected, and in the first decade of this century it was further aggravated by the so-called "rationalization of the school network" by recommendation of the International Monetary Fund (IMF). Along with the notorious fact of the concentration of schools and universities in cities, as well as the concentration of various institutions and businesses also in cities, this has led to a very pronounced gap in the level of education between urban and rural populations.

Table 5. Population aged 15 and over by educational attainment, by type of settlements in 2011 (%)

	Without educational attainment	Incomplete primary education	Primary education	Secondary education	High education	Higher education	Unknown
Republic of Serbia	2.68	11	20.86	48.94	5.65	10.59	0.28
Urban settlements	1.48	5.49	16.01	53.44	7.52	15.69	0.37
Rural ("other") settlements	4.41	19.01	27.68	42.37	2.94	3.16	0.42

Source: Authors' own research based on Statistical Office of the Republic of Serbia, 2012b.

Table 6. Population aged 15 and over by educational attainment, by type of settlements in 2022 (%)

	Without educational attainment	Incomplete primary education	Primary education	Secondary education	High education	Higher education	Unknown
Republic of Serbia	1.01	5.27	17.80	53.08	6.05	16.40	0.40
Urban settlements	0.67	2.41	12.81	53.40	7.60	22.61	0.51
Rural ("other") settlements	1.55	9.85	25.80	52.56	3.56	6.44	0.22

Source: Authors' own research based on Statistical Office of the Republic of Serbia, 2023b.

In Table 5, we see that in 2011 there were three times as many residents in rural settlements without any formal education, as well as almost four times as many people with incomplete primary education; almost a quarter of the villagers did not even finish primary school. The most common highest completed level of education both in the city and in the countryside was high school (in the city, however, by some percentage more). The educational gap between the city and the countryside was greatest at the highest levels of education: five times as many residents of Gras settlements had a university degree. In the last period (2011–2022), the number of rural residents without formal education decreased significantly, even three times (Table 6). The number of those who did not complete elementary school decreased (significant decrease), as did the number of those who stopped at completed elementary school (minimum decrease). The number of rural residents with a high school diploma increased by 10%, while those with a university degree were twice as many as in 2011. However, the educational gap between the city and the countryside remained and is still the largest when it comes to having a university degree. As in the case of computer literacy, in the EU the gap is the largest when it comes to the third level of education (high and higher education) (Popescu et al., 2022).

Conclusions

Both assumptions from which we started were confirmed. The level of illiteracy has decreased in rural areas between the two Censuses (2011 and 2022), even to the level that it can be considered negligible, that is, without a serious impact on development processes. The percentage of computer literate population in the countryside has increased significantly, which can be a valuable resource when it comes to (sustainable) development of rural areas. We also received answers to our research questions: the educational level of rural residents improved in the examined period, and the gap between rural and urban areas in terms of computer literacy and educational level remained, although somewhat reduced. Overall, the level of cultural capital according to these indicators in the rural settlements of Serbia can be considered satisfactory, except when it comes to educational attainment, where, apart from the relatively low level of education, there is still a very noticeable gap between the countryside and the city.

If we were to apply the Organization for Economic Cooperation and Development (OECD) criteria to Serbia, according to which rural areas are defined as those with a population density of less than 150 inhabitants per square kilometer, rurality would be 10% higher than that shown by official statistics (Vlada Republike Srbije, 2012). Thus, it turns out that Serbia, with close to 50 % of the population living in villages, is more rural than all EU member countries, where the average is 22.3 %, with the most rural countries being Lithuania (44.9), Romania (44.7) and Slovakia (43.9) (Popescu et al., 2022). This fact additionally increases the importance of the potential of rural areas when it comes to cultural capital (as well as all other forms of capital) for sustainable development, as well as for rural-urban balance (Sorokin) as a precondition for the stability of Serbian society. The future development of events on these issues, i.e. whether the level of education in the countryside will increase and the educational gap between the countryside and the city will be reduced, which would slow down rural-urban migration and increase the stability of Serbian society, depends to a large extent on the policy that will be pursued towards rural areas.

References

Bostrom, M. 2012. A missing pillar? Challenges in theorizing and practicing social sustainability: Introduction to the special issue. *Sustainability: Science, Practice and Policy,* 8(1), pp. 3–14. https://doi.org/10.1080/

Bourdieu, P. 1986. The forms of capital. In: J. Richardson, ed. *Handbook of theory and research for the sociology of education.* Westport CT: Greenwood, pp. 241–258.

Brown, P., Power, S., Tholen, G. and Allouch, A. 2016. Credentials, talent and cultural capital: A comparative study of educational elites in England and France. *British Journal of Sociology of Education,* 37(2), pp. 191–211. https://doi.org/10.1080/

Burja, C. and Burja, V. 2014. Sustainable development of rural areas: A challenge for Romania. *Environmental Engineering and Management Journal*, 13(8), pp. 1861–1871.

Dempsey, N., Bramley, G., Power, S. and Brown, C. 2011. The social dimension of sustainable development: Defining urban social sustainability. *Sustainable Development*, 19, pp. 289–300. https://doi.org/10.1002/sd.417.

Durkheim, E. 2012. *The Rules of the Sociological Method*. Novi Sad: Mediterran Publishing. [in Serbian]

Du Pisani, J.A. 2006. Sustainable development – Historical roots of the concept. *Environmental Sciences*, 3(2), pp. 83–96. https://doi.org/10.1080/15693430600688831

Galdeano-Gomez, E., Perez-Mesa, J.C. and Godoy-Duran, A. 2016. The social dimension as a driver of sustainable development: the case of family farms in southeast Spain. *Sustain Science*, 11, pp. 349–362. https://doi.org/10.1007/s11625-015-0318-4.

Griffiths, D. 2017. The comparative method and the history of modern humanities. *History of Humanities* 2 (2), pp. 473–505. http://dx.doi.org/10.1086/693325.

Huang, X. 2019. Understanding Bourdieu – cultural capital and habitus. *Review of European Studies* 11(3), pp. 45–49. https://doi.org/10.5539/res.v11n3p45

Jelić, S. and Kolarević, V. 2016. O socijalnoj isključenosti tokom perioda tranzicije u Srbiji [On social exclusion during the period of transition in Serbia]. *Sociološki pregled*, 50(2), pp. 209–228.

Kim, S. and Kim, H. 2009. Does cultural capital matter?: Cultural divide and quality of life. *Social Indicators Research*, 93, pp. 295–313. https://doi.org/10.1007/s11205-008-9318-4.

Lopez-Penabad, M. C., Iglesias-Casal, A. and Rey-Ares, L. 2022. Proposal for sustainable development index for rural municipalities. *Journal for Cleaner Production*, 357. https://doi.org/10.1016/j.jclepro.2022.131876.

Mitrović, M. 2015. *Sela u Srbiji. Strukturne promene i problem održivog razvoja [Villages in Serbia. Structural Changes and the Problems of Sustainable Development]*. Beograd: Republički zavod za statistiku.

Murphy, K. 2012. The social pillar of sustainable development: A literature review and framework for policy analysis. *Sustainability: Science, Practice and Policy*, 8(1), pp. 15–29. https://doi.org/10.1080/15487733.2012.11908081

Nasirzadeh, F., Ghayouminan, M., Khanzadi, M. and Cherati, R.N. 2020. Modeling the social dimension of sustainable development using fuzzy cognitive maps. *International Journal of Construction Management*, 20(3), pp. 223–236. https://doi.org/10.1080/15623599.2018.1484847

Orboi, M. D., Banes, A., Petroman, I., Monea, M. and Balan, I. 2010. Sociological dimension of sustainable development. *Research Journal of Agricultural Science,* 42(3), pp. 749–755.

Pawlowski, A. 2008. How many dimensions does sustainable development have? *Sustainable Development,* 16, pp. 81–90.

Popescu, A., Tinceche, C., Marcuta, A., Marcuta, L., Hontus, A. and Angelescu, C. 2022. Gaps in the education level between rural and urban areas in European Union. *Scientific Papers Series Management, Economic Engineering in Agriculture and Rural Development,* 22 (3), pp. 531–546.

Sorokin, P. and Zimmerman, C. 1931. *Principles of rural-urban sociology.* New York: Henry Holt and company.

Statistical Office of the Republic of Serbia. 2008. *Two centuries of Serbian development.* Belgrade: Statistical Office of the Republic of Serbia.

Statistical Office of the Republic of Serbia. 2012a. *Ethnicity.* Belgrade: Statistical Office of the Republic of Serbia. [online] Available at: <Microsoft Word – tekst, REV.GN.doc (stat.gov.rs)> [Accessed 9 October 2023].

Statistical Office of the Republic of Serbia. 2012b. *Educational attainment, literacy and computer literacy.* Belgrade: Statistical Office of the Republic of Serbia. [online] Available at: <G20134001.pdf (stat.gov.rs)> [Accessed 9 October 2023].

Statistical Office of the Republic of Serbia. 2023a. *Ethnicity.* Belgrade: Statistical Office of the Republic of Serbia. [online] Available at: <G20234001.pdf (stat.gov.rs)> [Accessed 9 October 2023].

Statistical Office of the Republic of Serbia. 2023b. *Educational attainment, literacy and computer literacy.* Belgrade: Statistical Office of the Republic of Serbia. [online] Available at: <G20234006.pdf (stat.gov.rs)> [Accessed 9 October 2023].

Stojić, G. 2023. Information and communication technologies and rural development of Serbia. *Ekonomika,* 69(1), pp. 69–80. https://doi.org/10.5937/ekonomika2301069S.

Šljukić, S. and Janković, D. 2015. *Selo u sociološkom ogledalu* [Rural in the Sociological Mirror]. Novi Sad: Mediterran Publishing.

Trnavac, N. 1992. *Male seoske škole: šanse za preživljavanje i dalji razvoj* [Small Rural Schools: Chances for Survival and Further Development]. Beograd: Institut za pedagogiju i andragogiju Filozofskog fakulteta.

Vlada Republike Srbije. 2012. *Budućnost ruralnih područja u Srbiji* [*Government of the Republic of Serbia, 2012. The Future of Rural Areas in Serbia*]. Beograd: Vlada Republike Srbije.

WCED (World Commission on Environment and Development). 1987. *Our common future*. Oxford: Oxford University Press.

Živanović, Z. 2018. Contribution to the discussion on typology of the settlements of Serbia. *Demografija*, 15, pp. 33–49. doi:10.5937/demografija1815033Z 10.5937/demografija1815033Z.

Cosmina-Simona Toader, Ciprian Ioan Rujescu,
Andrea Ana Feher, Małgorzata Zajdel,
Małgorzata Michalcewicz-Kaniowska, Iveta Ubrežiová,
Levente Komarek

Investigating tourists' interest in sustainable travel when using online booking accommodation – A multinational approach

Abstract: *The factors that underlie tourists' decisions are in a constant state of change, determined by the evolution of society as a whole. It is precisely the different speed of transformation at the socio-economic level that characterizes certain states that induces different attitudes, interests and perceptions of consumers of tourism services in different geographical areas. Sustainability is starting to become a criterion taken into account in the decision-making process regarding the choice of tourist destinations. The level of sustainability achieved by some locations for tourist purposes, indicated by the online platform Booking.com, constituted in the present work an indicator used to test the differences of opinion. The data that is the subject of statistical comparisons comes from questionnaires applied to people who use the Booking.com platform for travel planning. The sample under study, consisting of 221 respondents, was divided into two categories. The first category is represented by respondents from Romania (127), and the second category includes respondents from Hungary, Poland and Slovakia (94). Among the respondents from the HU/PL/SK group, there was a higher average of answers (3.27) compared to the average of the answers from Romania (2.69) when it was tested to what extent they take into account the "Sustainable Travel" attribute in the reservations they make they did them. Considering the future bookings and the intention to take into account the "Sustainable Travel" label, the respondents from Romania indicated a higher average response than the respondents from the HU/PL/SK group. Even if Romanian respondents are currently showing a less responsible behavior towards the environment when they book a place to stay, the attitude shown by this study regarding their intention to have a sustainable behavior in the future is encouraging.*

Keywords: sustainable travel, travel behavior, booking, tourism.

Introduction

Today, travel websites have become the starting point for travelers to search for information about their destinations and make reservations (Chaw and Tang, 2019). Online accommodation booking platforms like Booking.com,

Tripadvisor, Agoda and Expedia offer a wide range of accommodation options and relevant information such as rates, star ratings, customer review ratings, booking policies, amenities and so on (Chaw and Tang, 2019; Zhang, Lu and Lu, 2023; Xia et al., 2022). However, in recent times we are witnessing a change in the behavior of tourists regarding how they select their accommodation and the factors they consider when booking a stay (Assaker and O'Connor, 2023). Growing concerns about the environment have led travelers to look for more sustainable, "green" accommodation options when planning their trips (Assaker, 2020).

Hotels are perceived as having a negative effect on the environment because they use a large amount of water, generate large amounts of waste and a significant percentage of carbon dioxide emissions (Popescu et al., 2022; Robin et al., 2016; Salama and Abdelsalam, 2021). Under these conditions, customers increasingly want to be perceived as responsible people who play an important role in the field of environmental protection. Due to customers' increasing awareness of the importance of sustainability, environmental concerns have become a competitive advantage in the hospitality industry (Salama and Abdelsalam, 2021). Moreover, the increasing awareness of customers towards the environment forces hoteliers to develop and implement increasingly diverse environmental practices (Bohdanowicz, 2005; Han, 2021; Stanciu et al., 2022).

In light of these trends and to better respond to the environmental aspirations of travelers when booking accommodation online, Booking.com has developed the "Travel Sustainable" program (Booking.com), through which it labels "green" properties on 3 levels (1, 2 and 3), respectively without a label, depending on the steps they take towards sustainability. Through this program Booking.com helps customers in providing information about the level achieved by each property in terms of the implementation of environmental protection measures. A study carried out in the USA (Assaker and O'Connor, 2023) shows that about 40 % of American respondents are willing to pay more for a hotel with ecological certification.

There are also other travel platforms that have implemented similar systems. For example, Tripadvisor has developed the "GreenLeaders" program (Tripadvisor) according to which the accepted properties are classified into four statuses: Bronze, Silver, Gold and Platinum. The higher the status, the higher the impact of a property's green practices on the environment.

Although online platforms are actively implementing green certification labels/badges, existing academic studies on the determinants of online accommodation booking have, to date, largely failed to integrate this construct into the selection attributes investigated. Assaker and O'Connor (2023) conducted an analysis on seven online hotel attributes (including eco-certification labels/badges) in two specific scenarios (pre- and post-COVID 19) and concluded that eco-certification labels/badges influence hotel choice online travel booking, but not as much as cancellation policies, hotel rating, price and location.

Doing so, the present article aims to investigate tourists' interest in sustainable travel when they use the Booking.com platform for accommodation and highlight the differences in opinion between Romanian travelers and travelers from three other European countries (Hungary, Poland and Slovakia).

The article contains a review of the specialized literature, followed by the description of the research methodology, the analysis of the results obtained and, finally, conclusions and discussion of the findings.

Literature review

The impact of global tourism growth on the environment has become a vital topic in the hospitality industry (Lee and Cheng, 2018). More and more customers prefer green products/services and environmentally responsible companies that satisfy the green needs of customers (Cuc et al., 2022; Moise, Gil-Saura and Ruiz-Molina, 2021a). With the depletion of global resources and the increasing awareness of environmental protection and ecological conservation, the management of many tourism establishments has set the goal of implementing environmentally friendly practices (Agag and Colmekcioglu, 2020), such as: rooms with refillable shampoo dispenser, energy efficient light bulbs, towel and linen reuse policies, waste reduction and recycling, key card to control room energy consumption, environmental education programs (Assaker, 2020). A study conducted by Millar and Baloglu (2011) on a sample of 571 business and agreement travelers concluded that the most influential attribute on hotel room preference was the hotel's environmental certification (such as LEED – Leadership in Energy and Environmental Design). Thus, more hotels are making efforts to implement ecological practices and adopt sustainable guidelines in their activities, with the aim of attracting more environmentally conscious customers (Abdou, Hassan and Dief, 2020; Barakagira and Paapa, 2023; Kim et al., 2017).

Most researchers have identified three reasons for the adoption of green practices by accommodation facilities: financial benefits, customer needs and desires, and stakeholder relations (Abdou et al., 2020; Moise, Gil-Saura and Ruiz Molina, 2021b; Singal, 2014). On the other hand, Buunk and van der Werf (2019) concluded that the main reasons for the adoption of eco-labelling criteria consisted of increasing the hotel's image and relationship with the environment. Although green practices were primarily introduced as a means of reducing costs, recently they have focused on gaining a larger market share by improving relationships with stakeholders (Abdou, Hassan and Dief, 2020; Khatter et al., 2019).

Previous academic research clearly highlights the positive influence of both hotel green practices and green certifications on travelers' hotel attitudes, level of satisfaction, intention to return, and willingness to pay more (Kim, Barber and Kim, 2019; Nelson et al., 2021; Olya et al., 2021). However, most studies have been conducted in a post-purchase context, typically using reviews from respondents who have already stayed and experienced the hotel and its green practices (Assaker and O'Connor, 2023). Therefore, while existing research establishes the importance of green practices and certification in the post-purchase and booking context, the present article aims to establish the importance of these aspects in influencing the initial booking, in the context of online booking on the Booking.com platform.

Methodology

In order to attain the purpose of the article the authors followed the research design presented in Figure 1.

Literature review	**Data collection method:** document analysis
	Sources: articles, published works, reports
	Analyzed aspects: sustainable travel, travel behaviour, sustainable practices, ecological certification labels

Investigating tourists' interest in sustainable travel when using online booking accomodation	**Data collection method:** survey (questionnaire)
	Sources: Booking.com platform users from Romania, Hungary, Poland and Slovakia
	Analyzed aspects:
	Q1 Have the Booking.com users noticed the Sustainable Travel label displayed on some of the properties available on Booking.com platform?
	Q2 Did Booking.com users noticed that there is a possibility to include in the search filters on the Booking.com platform also the criterion of Sustainable Travel?
	Q3 Do Booking.com users know the meaning of Sustainable Travel levels on Booking.com platform?
	Q4 To what Booking.com users extent do they also consider Sustainable Travel criteria when booking accommodation through the platform?
	Q5 Do Booking.com users plan to take into account the Sustainable Travel criteria when booking accommodation through the Booking.com platform in the future?
	Q6 To what extent do Booking.com users consider that choosing a place to stay also taking into account the levels of Sustainable Travel has a beneficial effect on the transformation of society?

Figure 1. Research design

The questionnaire, designed to investigate tourists' interest in sustainable travel when booking accommodation through the Booking.com platform, was developed using the Google Forms application. The questionnaire was applied online because it does not involve costs, the answers are obtained quickly, and the data (answers) can be easily organized and processed from a statistical point of view.

Before applying the questionnaire, it was decided to pre-test it. Thus, the questionnaire was sent to a number of 20 people in order to complete it, identify uncertainties and offer suggestions. Following the feedback provided, the questionnaire was reorganized and simplified.

The link or QR code of the questionnaire was distributed to various groups through social networks (Facebook, WhatsApp). The questionnaire application period was September 19, 2023 – October 4, 2023.

The questionnaire reached a total of 302 people, of which 300 people responded positively to the invitation to participate in the study. We note that out of the total of 300 responses, 221 responses were considered valid. The 221 respondents who declared that they used the Booking.com platform at least once to book a place to stay (apartment, apartment, villa, hotel room, etc.) had access to the questionnaire questions. The distribution of answers according to the country of origin of the respondents can be seen in Table 1.

Table 1. Distribution of responses by country

	Total responses	Consent answers *I have read the introduction and I agree to take part in the survey*	Valid answers *I have used Booking.com platform at least once to book accommodation*
	302	300	221
Hungary	115	113	64
Slovakia	22	22	16
Poland	23	23	14
Romania	142	142	127

Source: Authors' own research.

The questionnaire contains a number of 6 questions, distributed in 4 sections. Questionnaire design and item coding are presented in Table 2.

Table 2. Questionnaire design and item coding

Section	Item code	Question	No of responses
Respondents' consent		I have read the introduction and I agree to take part in the survey	300
Respondents' validation		Have you used Booking.com platform at least once to book accommodation (e.g. apartment, hotel room, etc.)?	221
1	Q1	Have you noticed the Sustainable Travel label displayed on some of the properties available on Booking.com platform?	221
2	Q2	Did you notice that there is a possibility to include in the search filters on the Booking.com platform also the criterion of Sustainable Travel?	221
3	Q3	Do you know the meaning of Sustainable Travel levels on Booking.com platform?	143
4	Q4	To what extent do you also take into account the Sustainable Travel criteria when booking accommodation through Booking.com platform?	143
	Q5	Do you intend to take into account the Sustainable Travel criteria when booking accommodation through the Booking.com platform in the future?	143
	Q6	To what extent do you consider that choosing a place to stay also taking into account the levels of Sustainable Travel has a beneficial effect on the transformation of society?	143
General information		Gender/Age group/Educational level/Place of residence/Country	221

Source: Authors' own research.

The first two sections include single-choice questions (Q1, Q2) and are accessible to all respondents who stated that they used the Booking.com platform at least once to book an accommodation. The third section also includes a single-choice question (Q3), this is accessible only to respondents who had at least one positive answer to the questions from the previous sections. The 4th section includes 3 rating scale questions (Q4, Q5, Q6). The evaluation of the 3 items was done using an interval scale (1 – not at all; 2 – in small measure; 3 – in some wise; 4 – largely; 5 – very much).

At the end of the questionnaire, study participants accessed a separate section dedicated to general information (gender, age group, educational level, place of residence, country). The socio-demographic profile of the respondents is presented in Table 3.

Table 3. Respondents' socio-demographic profile

	Total	Country			
		Hungary	Poland	Slovakia	Romania
Valid answers	221	64	16	14	127
Female	129	31	11	6	81
Male	89	31	5	8	45
No answer	3	2	–	–	1
under 25 years old	64	38	3	7	16
25–50 years old	138	25	9	4	100
over 50 years old	19	1	4	3	11
Secondary school	6	6	–	–	–
High school	41	36	1	1	3
University	174	22	15	13	124
Urban	157	40	14	11	92
Rural	64	24	2	3	35

Source: Authors' own research.

The sample was divided into two categories. The first category is represented by respondents from Romania, and the second category by respondents from Hungary, Poland and Slovakia. The grouping of answers from the 3 countries was chosen because they can be seen as a homogeneous group, as they present certain common features from a geographical, economic and social point of view. The 3 countries are located in Central Europe, are developed countries (World Population Review, 2023), show interest in practicing responsible tourism (Mamula et al., 2021; Nga, Erdélyi and Formádi, 2018; Ernszt and Marton, 2021), are concerned about sustainability (Ingaldi and Dziuba, 2021), the application of measures aimed at protecting the environment (Echeverría, Giménez-Nadal and Molina, 2022) and show interest in adopting a sustainable lifestyle (Eurostat, 2023b, 2023c). Table 4 shows key figures of the countries selected for this study.

Table 4. Key figures of the countries selected for study

	HU	PL	SK	RO
Location				
Sub region of Europe	Central			South East
Economy & Society				
GDP per capita, 2021 (% of EU average)**	48.7	46.4	56.6	38.8
Inflation rate, 2022 (% change compared to previous year)*	15.3	13.2	12.1	12.0
Unemployment rate, 2022 (% of the active population aged 15–74 years)*	3.6	2.9	6.1	5.6
Minimum wage, 2023-S2 (euro per month)*	623.8	811.0	700.0	604.4
People at risk of poverty or social exclusion, 2022 (% of the population)*	18.4	15.9	16.5	34.4
Human Development Index 2021 (HDI)****	0.846	0.876	0.848	0.821
Environment				
Generation of municipal waste, 2020 (kg per capita)**	403	346	478	290
Recycling of municipal waste, 2020 (% of total waste treatment)**	35.9	34.1	38.5	11.5
Digitalization & Internet activities				
People with basic or above basic digital skills, 2021 (% of people)***	49.1	42.9	55.2	27.8
People finding information about goods and services, 2022 (% of people who used internet in the last 3 months)***	87.2	85.4	75.3	57.6
Goods and services bought online-rented accommodation, 2022 (% of people who purchased online in the last 3 months)***	24.8	14.0	25.7	18.2

Source: Authors' own research based on *Eurostat, 2022a; **Eurostat, 2022b; ***Eurostat, 2023a; ****World Population Review, 2023.

Therefore, the two groups were compared, looking at the interest in sustainable travel when booking accommodation through the Booking.com platform. The information obtained through the application of the questionnaire was organized in a database in order to analyze the results through the application of statistical tests.

The statistical data resulting from the answers to questions Q4, Q5 and Q6, which contain integer values from 1 to 5, represented the inputs for the t-test and the non-parametric Wilcoxon rank-sum test, respectively. SAS Studio was used to perform these calculations, as well as the related graphic representations.

Results and discussions

According to the answers given by the study participants to the question Q1 – Have you noticed the Sustainable Travel label displayed on some of the properties available on Booking.com platform?, 60.64 % of the respondents belonging to the HU/PL/SK group declared that they had noticed the Sustainable Travel label displayed on some of the properties available on Booking.com platform, while the respondents from Romania who noticed the labels represent 48.03 %.

Regarding the possibility of including the "Sustainable Travel" criterion in the search filters, 56.4 % of the respondents who belong to the HU/PL/SK group observed the existence of the criterion, and the percentage of respondents from Romania was 40.94 %.

Regarding knowing the meaning of Sustainable Travel levels on Booking.com platform (Q3), 61.76 % of the respondents belonging to the HU/PL/SK group stated that they know the meaning, while only 45.33 % of the respondents from Romania know the meaning of the labels.

Table 5. Distribution of responses to Q1, Q2, Q3

		HU/PL/SK	RO	Total
Q1	No	37	66	103
	Yes	57	61	118
	Total	94	127	221
Q2	No	41	75	116
	Yes	53	52	105
	Total	94	127	221
Q3	No	26	41	67
	Yes	42	34	76
	Total	68	75	143

Source: Authors' own research.

The results of the comparisons between the two analyzed groups are discussed in the following for the questions for which the evaluation of the items was carried out using the interval type evaluation scale (Q4, Q5 and Q6).

To the question Q4 – To what extent do you also take into account the Sustainable Travel criteria when booking accommodation through Booking.com platform? 143 responses were analyzed, 68 responses coming from people from the HU/PL/SK group and 75 from people from Romania. The average value of the responses related to the HU/PL/SK group was 3.27, while the average value of the responses from Romania was 2.69. The data are indicated in the statistical summary in Table 6. The differences are statistically significant, according to t-test with p=0.003 and t=3.02. The use of the non-parametric Wilcoxon rank-sum test confirms differences between the two groups, with p=0.007.

Table 6. Statistical summary of responses to Q4

Country	N	Mean	Std Dev	Std Err	Minimum	Maximum
HU/PL/SK	68	3.27	0.959	0.116	1	5
RO	75	2.69	1.345	0.155	1	5
Differences		0.58	1.177	0.197		

Source: Authors' own research.

The size of the differences can also be observed graphically, in the diagrams in Figure 2, where the histograms of the two statistical series for the two groups are shown, as well as the boxplot diagrams containing, for comparison, the minimum, maximum, average and median values for the series RO, respectively HU/PL/SK.

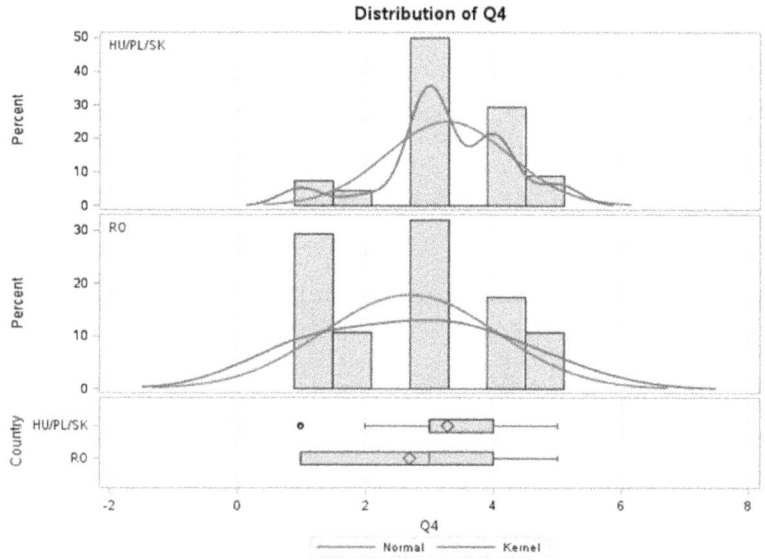

Figure 2. Histograms and boxplot diagrams for answers to Q4
Source: Authors' own statistical processing using SAS Studio.

It is observed that when they are asked to what extent they take into account the criteria of sustainable travel, the respondents from Romania indicate a lower average answer compared to the people in the sample made up of respondents from Hungary, Slovakia and Poland.

For the question Q5 – Do you intend to take into account the Sustainable Travel criteria when booking accommodation through the Booking.com platform in the future?, the statistical summary, respectively the comparative diagrams are reproduced in Table 7 and Figure 3.

Table 7. Statistical summary of responses to Q5

Country	N	Mean	Std Dev	Std Err	Minimum	Maximum
HU/PL/SK	68	3.5	0.872	0.105	1	5
RO	75	3.89	0.894	0.103	1	5
Differences		-0.39	0.883	0.148		

Source: Authors' own statistical processing using SAS Studio.

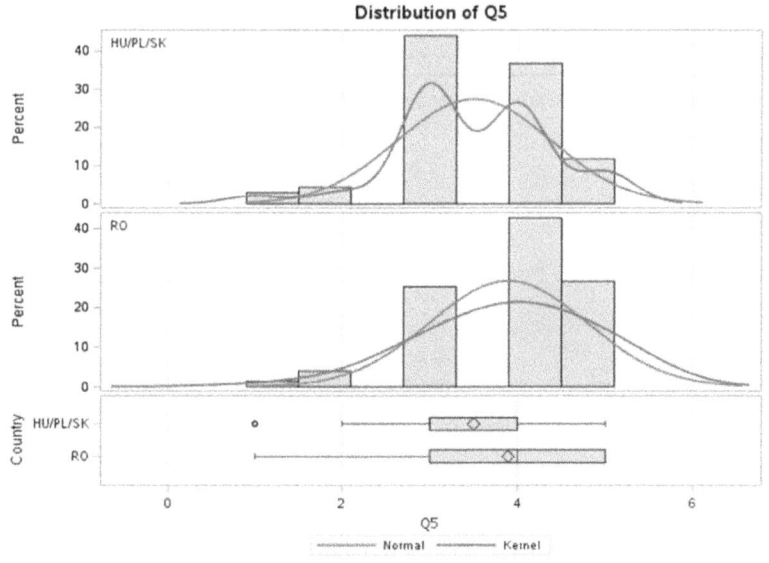

Figure 3. Histograms and boxplot diagrams for answers to Q5
Source: Authors' own statistical processing using SAS Studio.

When asked if in the future they intend to take into account the sustainable travel criteria when booking accommodation through the Booking.com platform, respondents from Romania indicated a higher average score than the respondents from the other group. Thus, for the RO group, the average value was 3.89 compared to the average value of 3.5 for the HU/PL/SK group. The t-test indicated statistically significant differences with p=0.008, t=−2.66 also confirmed by the non-parametric Wilcoxon rank-sum test, p=0.006. It is likely that the previous question has determined the respondents' interest in sustainable travel, and possibly an information in this regard could have an important effect in changing the attitude of some tourists.

For the question Q6 – To what extent do you consider that choosing a place to stay also taking into account the levels of Sustainable Travel has a beneficial effect on the transformation of society?, the statistical summary, respectively the comparative diagrams are reproduced in Table 8 and in Figure 4.

Table 8. Statistical summary of responses to Q6

Country	N	Mean	Std Dev	Std Err	Minimum	Maximum
HU/PL/SK	68	3.5147	0.9541	0.1157	1	5
RO	75	3.8667	0.9054	0.1046	1	5
Differences		-0.352	0.9289	0.1555		

Source: Authors' own statistical processing using SAS Studio.

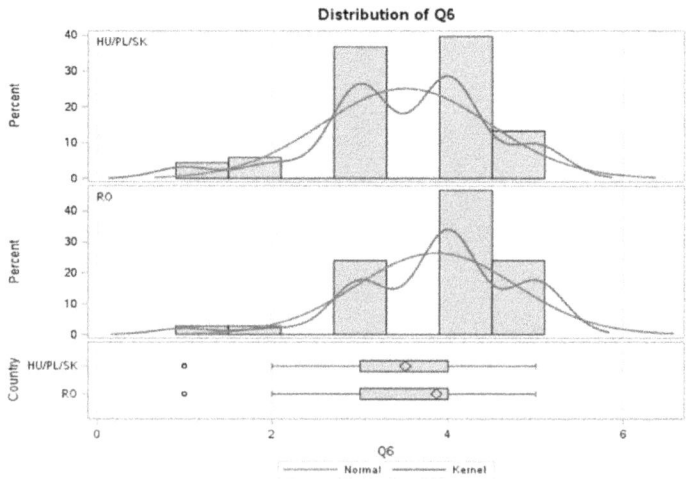

Figure 4. Histograms and boxplot diagrams for answers to Q6
Source: Authors' own statistical processing using SAS Studio.

The answers to question Q6 reinforce the opinions formed after analyzing the previous questions. Choices in the tourist reservation process taking into account the criterion of sustainability have, in the opinion of respondents from Romania, a beneficial effect on the transformation of society with an average measure of 3.86 on a value scale from 1 to 5. At the same time, respondents from HU/PL/SK indicate an average response of 3.51. The results are statistically significant (t-test, p=0.025, t=−2.26, respectively p=0.019 in the non-parametric Wilcoxon rank-sum test).

Conclusions

Of the 221 people who participated in the study, 53.40 % declared that they had noticed the existence of the "Sustainable Travel" label displayed on some of the properties available on the Booking.com platform. This attitude of the respondents is in line with the worldwide trends of increasing awareness of sustainable travel practices. In order to respond to the environmental aspirations of tourists, online booking platforms (Booking.com, Tripadvisor, Agoda, Expedia) have developed programs through which they label "green" or "sustainable" properties. Thus, travelers have the opportunity to choose the accommodation option in accordance with their aspirations in terms of sustainable travel.

The possibility of including the "Sustainable Travel" criterion in the search filters is known to 47.51 % of the respondents. Although this construct is included in search filters, a relatively small percentage of users have noticed the existence of this possibility, because they still place more importance on price, cancellation policies or property valuation.

A little over half (53.14 %) of the respondents who noticed the existence of the "Sustainable Travel" label or the option to include the label in search filters resorted to looking for additional information about the meaning of this attribute, being interested in practicing responsible tourism.

The tourism industry is among the actors that have a negative impact on the environment (large water consumers, waste generators, carbon dioxide emitters). At the level of the HU/PL/SK group, greater attention is paid to waste recycling compared to the group of Romanian respondents, recording an average of 36 % of total waste treatment, compared to Romania, where the percentage is 11.5. This behavior towards waste recycling, and implicitly towards the environment, is also reflected in the different answers given by the respondents of the 2 groups. When asked to what extent they take into account the "Sustainable Travel" attribute, the respondents who belong to the HU/PL/SK group recorded a higher average of answers (3.27) compared to the average of answers from

Romania (2.69). Considering the future bookings and the intention to take into account the "Sustainable Travel" label, the respondents from Romania indicated a higher average response than the respondents from the HU/PL/SK group. Even if Romanian respondents are currently showing a less responsible behavior towards the environment when they book a place to stay, the attitude shown by this study regarding their intention to have a sustainable behavior in the future is encouraging.

In order to better respond to environmental aspirations, respondents from Romania believe to a significant extent (3.86) that choosing a place to stay while also taking into account the Sustainable Travel levels has a beneficial effect on the transformation of society. The same attitude is found among respondents from the HU/PL/SK group.

The results of the study can serve online accommodation booking platforms as it clearly highlights the positive influence of sustainable property practices on future accommodation intention. Platforms that adopt such policies tend to gain ground and acquire a larger market share among environmentally responsible customers.

It is important to note that there are some limitations of this study that should be evaluated in future research. First, the study sample can be extended to a larger number of respondents and could include other homogenous country groups to collect cross-national and even multicultural data to examine differences in preferences and opinions on sustainable travel. Second, this study sampled respondents who used the Booking.com platform only. Future studies could include respondents who have used other online accommodation platforms as well, in order to have a more comprehensive picture of travelers' perception of green (sustainable) labels/badges of tourist facilities.

References

Abdou, A.H, Hassan, T.H. and El Dief, M.M. 2020. A description of green hotel practices and their role in achieving sustainable development. *Sustainability*, 12, p. 9624. https://doi.org/10.3390/su12229624.

Agag, G. and Colmekcioglu, N. 2020. Understanding guests' behavior to visit green hotels: The role of ethical ideology and religiosity. *International Journal of Hospitality Management*, 91, 102679. https://doi.org/10.1016/j.ijhm.2020.102679

Assaker, G. 2020. The effects of hotel green business practices on consumers' loyalty intentions: an expanded multidimensional service model in the upscale segmen. *International Journal of Contemporary Hospitality Management*, 32(12), pp. 3787–3807. https://doi.org/10.1108/IJCHM-05-2020-0461.

Assaker, G. and O'Connor, P. 2023. The importance of green certification labels/badges in online hotel booking choice: A conjoint investigation of consumers' preferences pre- and post-COVID-19. *Cornell Hospitality Quarterly*, 64(4). https://0e10pahfu-y-https-journals-sagepub-com.z.e-nformation.ro/doi/10.1177/19389655231184474.

Barakagira, A. and Paapa, C. 2023. Green practices implementation for environmental sustainability by five-star hotels in Kampala, Uganda. *Environ Dev Sustain*, 2389. https://doi.org/10.1007/s10668-023-03101-7.

Bohdanowicz, P. 2005. European hoteliers environmental beliefs: Greening the business. *The Cornell Hotel and Restaurant Administration Quarterly*, 46(2), pp. 188–204. https://doi.org/10.1177/0010880404273891

Booking.com, *Research Report: Sustainability in Tourism: Challenges, opportunities and role of Booking.com*. [online] Available at: <https://www.sustainability.booking.com/_files/ugd/6b9913_ae27941687d3425d9a7e27672dc38d41.pdf> [Accessed 3 October 2023]

Buunk, E. and van der Werf, E. 2019. Adopters versus non-adopters of the green key ecolabel in the Dutch accommodation sector. *Sustainability*, 11, 3563. https://doi.org/10.3390/su11133563.

Chaw, L.Y. and Tang, C.M. 2019. Online accommodation booking: What information matters the most to users?. *Inf Technol Tourism*, 21, pp. 369–390. https://doi.org/10.1007/s40558-019-00146-1.

Cuc, L.D., Pelau, C., Szentesi, S.G. and Sanda, G. 2022. The impact of green marketing on the consumers' intention to buy green products in the context of the Green Deal. *Amfiteatru Economic*, 24(60), pp. 330–345. https://doi.org/10.24818/EA/2022/60/330.

Echeverría, L., Giménez-Nadal, J.I. and Molina, J.A. 2022. Who uses green mobility? Exploring profiles in developed countries. *Transportation Research Part A: Policy and Practice*, 163, pp. 247–265. https://doi.org/10.1016/j.tra.2022.07.008.

Ernszt, I. and Marton, Z. 2021. An emerging trend of slow tourism: Perceptions of Hungarian Citizens. *Interdisciplinary Description of Complex Systems*, 19(2), pp. 295–307. https://doi.org/10.7906/indecs.19.2.8.

Eurostat. 2022a. *Country facts*. [online] Available at: <https://ec.europa.eu/eurostat/cache/countryfacts/#> [Accessed 10 October 2023].

Eurostat. 2022b. *Key figures on Europe*. [online] Available at: <https://ec.europa.eu/eurostat/cache/digpub/keyfigures/> [Accessed 10 October 2023].

Eurostat. 2023a. *Digitalisation in Europe – 2023 edition*, interactive publication. [online] Available at: <https://ec.europa.eu/eurostat/web/interactive-publications/digitalisation-2023> [Accessed 10 October 2023].

Eurostat. 2023b. *Sustainable development in the European Union – Monitoring report on progress towards the SDGs in an EU context – 2023 edition*. [online] Available at: <https://ec.europa.eu/eurostat/web/products-flagship-publications/w/ks-04-23-184> [Accessed 12 October 2023].

Eurostat. 2023c. *Sustainable development in the European Union – Statistical annex to the EU voluntary review – 2023 edition*. [online] Available at: <https://ec.europa.eu/eurostat/web/products-statistical-reports/w/ks-05-23-188> [Accessed October 2023].

Han, H. 2021. Consumer behavior and environmental sustainability in tourism and hospitality: a review of theories, concepts, and latest research. *Journal of Sustainable Tourism*, 29(7), pp. 1021–1042. https://doi.org/10.1080/09669582.2021.1903019

Ingaldi, M. and Dziuba, S.T. 2021. Sustainable tourism: Tourists' behaviour and their impact on the place visited. *Visions for Sustainability*, 17, 5828, pp. 8–38. http://dx.doi.org/10.13135/2384-8677/5828.

Khatter, A., McGrath, M., Pyke J., White L. and Lockstone-Binney L. 2019. Analysis of hotels' environmentally sustainable policies and practices: Sustainability and corporate social responsibility in hospitality and tourism. *International Journal of Contemporary Hospitality Management*, 31(6), pp. 2394–2410. https://doi.org/10.1108/IJCHM-08-2018-0670.

Kim, W.G., Li, J., Han, J.S. and Kim, Y. 2017. The influence of recent hotel amenities and green practices on guests' price premium and revisit intention. *Tourism Economics*, 23(3), pp. 577–593. https://doi.org/10.5367/te.2015.0531

Kim, Y.H., Barber, N. and Kim, D.K. 2019. Sustainability research in the hotel industry: Past, present, and future. *Journal of Hospitality Marketing & Management*, 28(5), pp. 576–620. https://doi.org/10.1080/19368623.2019.1533907

Lee, W.H and Cheng, C. 2018. Less is more: A new insight for measuring service quality of green hotels. *International Journal of Hospitality Management*, 68, pp. 32–40, https://doi.org/10.1016/j.ijhm.2017.09.005.

Mamula Nikolic, T., Pantic, S.P., Paunovic, I. and Filipovic, S. 2021. Sustainable travel decision-making of Europeans: Insights from a household survey. *Sustainability*, 13, 1960. https://doi.org/10.3390/su13041960

Millar, M. and Baloglu, S. 2011. Hotel Guests' preferences for green guest room attributes. *Cornell Hospitality Quarterly*, 52(3). https://0e10pamr3-y-https-doi-org.z.e-nformation.ro/10.1177/193896551140

Moise, M.S., Gil-Saura I. and Ruiz-Molina, M-E. 2021a. "Green" practices as antecedents of functional value, guest satisfaction and loyalty. *Journal of Hospitality and Tourism Insights*, 4(5), pp. 722–738. https://doi.org/10.1108/JHTI-07-2020-0130.

Moise, M.S., Gil-Saura, I. and Ruiz Molina M.E. 2021b. The importance of green practices for hotel guests: Does gender matter?. *Economic Research-Ekonomska Istraživanja*, 34(1), pp. 3508–3529. https://doi.org/10.1080/1331677X.2021.1875863

Nelson, K.M., Partelow, S., Stäbler, M., Graci, S. and Fujitani M. 2021. Tourist willingness to pay for local green hotel certification. *PLOS ONE*, 16(2). https://doi.org/10.1371/journal.pone.0245953.

Nga, N.T.T., Erdélyi, Év. and Formádi, K. 2018. Investigation into responsible tourism tours in Budapest, Hungary. *WIT Transactions on Ecology and the Environment*. WITPress. 227, pp. 141–150. https://doi.org/10.2495/ST180141.

Olya, H., Altinay, L., Farmaki, A., Kenebayeva, A. and Gursoy, D. 2021. Hotels' sustainability practices and guests' familiarity, attitudes and behaviours. *Journal of Sustainable Tourism*, 29(7), pp. 1063–1081. https://doi.org/10.1080/09669582.2020.1775622

Popescu, D., Coroș, M.M., Pop, I. and Bolog, C. 2022. The Green deal – Dynamizer of digitalization in tourism: The case of Cluj-Napoca Smart City. *Amfiteatru Economic*, 24(59), pp. 110–127. https://doi.org/10.24818/EA/2022/59/110.

Robin, C.F., María, U.T.F.S., Valencia, J.C., Muñoz, G.J., Astorga, P.S. and Martínez, D.Y. 2016. Attitude and behavior on hotel choice in function of the perception of sustainable practices. *Tourism & Management Studies*, 12(1), pp. 60–66. https://doi.org/10.18089/tms.2016.12106.

Salama, W. and Abdelsalam, E. 2021. Impact of Hotel Guests' trends to recycle food waste to obtain bioenergy. *Sustainability*, 13, p. 3094. https://doi.org/10.3390/su13063094

Singal, M. 2014. The link between firm financial performance and investment in sustainability initiatives. *Cornell Hospitality Quarterly*, 55(1), pp. 19–30. https://doi.org/10.1177/1938965513505700.

Stanciu, M., Popescu, A., Antonie, I., Sava, C. and Nistoreanu, B.G. 2022. Good Practices on Reducing Food Waste Throughout the Food Supply Chain. *Amfiteatru Economic*, 24(60), pp. 566–582. https://doi.org/10.24818/EA/2022/60/566.

Tripadvisor. *Green Leaders*. [online] Available at: <https://www.tripadvisor.com/GreenLeaders> [Accessed 3 October 2023].

World Population Review. 2023. *Developed countries 2023*. [online] Available at: <https://worldpopulationreview.com/country-rankings/developed-countries> [Accessed 10 October 2023].

Xia, Y., Chan, H.K., Zhong L. and Xu S. 2022. Enhancing hotel knowledge management: The influencing factors of online hotel reviews on travellers' booking intention. *Knowledge Management Research & Practice*, 20(1), pp. 34–45, https://www.tandfonline.com/doi/full/10.1080/14778238.2021.1967214

Zhang, S., Lu, Y. and Lu, B. 2023. Shared accommodation services in the sharing economy: understanding the effects of psychological distance on booking behavior. *Journal of Theoretical and Applied Electronic Commerce Research*, 18, pp. 311–332. https://doi.org/10.3390/jtaer18010017.

Sanyam Varma, Manish Sen, Anush Jain,
Laura Iosefina Smuleac, Sorin Mihai Stanciu

Cross cultural management: Challenges and strategies for managing a global workforce

Abstract: Cross-cultural management is the adept handling of a diverse workforce comprising individuals from varied cultural backgrounds within an organization. This entails acknowledging and valuing cultural differences, comprehending their impact on behavior and decision-making and formulating strategies to harness these distinctions for the betterment of the organization. In essence, cross-cultural management is about fostering an inclusive and harmonious workplace that maximizes the potential inherent in cultural diversity. This research paper delves into the intricacies of managing a global workforce, shedding light on the challenges encountered by organizations and proposing effective strategies to navigate the complexities of cross-cultural management. The focal point of the study is the pivotal role played by cultural intelligence and intercultural communication in fostering a workplace that is both harmonious and conducive to productivity. A thorough examination of existing literature forms the basis for identifying key challenges, and the paper explores a range of approaches to address these issues in the realm of cross-cultural management. In terms of research methodology, the study adopts a multifaceted approach that includes the analysis of case studies, interviews, and surveys to gather pertinent data. The results underscore the critical importance of cultural sensitivity, cross-cultural training initiatives, and adept leadership practices in successfully managing a global workforce. The ensuing discussion section delves deeper into the implications of the research findings, offering practical recommendations tailored for organizations seeking to enhance their cross-cultural management capabilities. In conclusion, the paper provides a summary of the challenges inherent in cross-cultural management, insistence the inescapable for continuous attainment and malleability in the frame of contemporary business landscape. It accentuates the exigency for organizations to remain agile and proactive in order to thrive amidst the dynamic interplay of diverse cultural elements.

Keywords: Cross Cultural management, global workforce, intercultural communication, cultural intelligence, challenges, strategies.

Introduction

In today's globalized business landscape, organizations increasingly operate in diverse cultural contexts and manage a workforce that spans across national boundaries. The growth of international trade, advancements in technology, and the mobility of talent have all contributed to the need for effective cross-cultural management.

In the contemporary global business environment, organizations are navigating diverse cultural landscapes and overseeing workforces that transcend national borders, studied by Stoermer, Hildisch and Froese (2015). The expansion of international trade, advancements in technology and the fluidity of talent have collectively underscored the essential requirement for proficient cross-cultural management (CCM)

It cites to the dexterity of organizations to maneuver the intricacy of cultural differences and leverage them as a source of competitive advantage. It involves understanding and adapting to distinct cultural benchmark within the workplace reveals by Javidan et al. (2005). The connotation of CCM lies in its budding to foment collusion, contrivance, and productivity in a multicultural environment. IT is the effective management of employees from various cultural backgrounds within an organization. It involves recognizing and appreciating cultural diversity, understanding how it influences behavior and decision-making, and developing strategies to clout this divergence for organizational success. Failure to adequately wield cross-cultural challenges can output in miscommunication, conflicts, drained employee spirit, and diminished productivity.

Cross-cultural management holds a crucial role in cultivating collaboration, innovation, and productivity within a multicultural setting. It entails the adept management of individuals from diverse cultural backgrounds within an organization, scrolled by Chang and Lin (2015). This involves acknowledging and valuing cultural diversity, comprehending its impact on behavior and decision-making, and formulating strategies to harness these differences for organizational success. Failing to navigate cross-cultural challenges effectively can lead to miscommunication, conflicts, lower employee morale, and decreased productivity. In essence, the essence of cross-cultural management lies in its power to capitalize on diversity, fostering a harmonious and thriving workplace.

One of the primary reasons why cross-cultural management has gained importance is the increasing trend of globalization has explained by Meyer (2014). Businesses are expanding their operations to new markets, necessitating the recruitment and management of employees from vibrant cultural backdrops. potent CCM acquiesce organizations to overpass cultural disparity, augment cross-cultural communication, and build relationships based on collective perceptive and respect.

The soaring eminence of cross-cultural management is intricately tied to the accelerating trend of globalization, as elucidated by Søderberg and Holden (2002). With businesses extending their operations to new markets, there arises a need for recruiting and managing employees from diverse cultural backgrounds. Successful cross-cultural management serves as a crucial tool for organizations, enabling them to overcome cultural disparities, improve cross-cultural communication, and foster relationships grounded in mutual understanding and respect. In essence, the growing global landscape necessitates effective cross-cultural management as a means to navigate the complexities of diverse work environments.

Cultural differences can have a profound impact on various aspects of organizational dynamics, including decision-making processes, leadership styles, employee motivation, and conflict resolution. Without proper management strategies, these cultural differences can lead to misunderstandings, conflicts, and reduced productivity. Consequently, organizations recognize the need to develop cross-cultural management competencies to create inclusive work environments and optimize the potential of their diverse workforce mentioned by Moran, Abramson and Moran (2014).

Cultural disparities significantly influence key elements of organizational functioning, such as decision-making procedures, leadership approaches, employee motivation, and conflict resolution. In the absence of effective management strategies, these cultural variations can give rise to misunderstandings, conflicts, and a decline in productivity. As emphasized by Diamantidis and Chatzoglou (2018), organizations acknowledge the imperative of cultivating cross-cultural management competencies. This recognition stems from the understanding that such competencies are essential for establishing inclusive work environments and unlocking the full potential of a diverse workforce.

Moreover, cross-cultural management is essential for organizations seeking to build global brands and maintain a competitive edge in international bazaar. Compassionate cultural subtlety and marketing strategies to specific cultural contexts is critical for success in global business operations, mentioned by Elsbach and Stigliani (2018).

Furthermore, cross-cultural management is indispensable for organizations aiming to construct global brands and sustain a competitive advantage in international markets. The ability to comprehend cultural nuances and customize products, services, and marketing strategies to align with specific cultural contexts is pivotal for success in the realm of global business operations, described by Bortolotti, Boscari and Danese (2015).

The background and significance of cross-cultural management stem from the growing need to effectively manage cultural diversity in organizations operating in global markets. By recognizing and addressing the challenges associated with cultural differences, organizations can create a supportive and inclusive work environment, improve employee satisfaction and retention, and achieve sustainable growth in today's interconnected world, stated by Mills (2017).

The foundation and importance of cross-cultural management arise from the increasing necessity to adeptly handle cultural diversity within organizations that operate in global markets. Acknowledging and proactively addressing the challenges linked to cultural differences enable organizations to establish a nurturing and inclusive work environment, published by Terjesen, Hessels and Li (2016). This, in turn, contributes to enhanced employee satisfaction and retention, fostering sustainable growth in our interconnected world.

Research objectives

The research objectives of a cross-cultural management study serve as a roadmap, offering clear guidance and focus for the investigation into essential aspects of overseeing a global workforce within a cross-cultural framework. These specific objectives may encompass:

1. Pinpoint the primary challenges encountered by organizations in overseeing a global workforce characterized by diverse cultural backgrounds. This involves delving into the specific hurdles arising from cultural distinctions and examining how they influence the dynamics within the organization.
2. Investigate the significance of cultural dexterity in the canvas of CCM. This equitable dragnet to understand how it plays a pivotal role in skillfully navigating cultural diversity and utilizing it as a potential source of competitive advantage for organizations.
3. Assess the consequences of intercultural communication on cross-cultural management, examining how effective communication strategies and intercultural competence influence the management of interactions and relationships across diverse cultures within the organization.
4. Investigate the tactics and optimal methodologies implemented by organizations for the effective management of a global workforce. This objective aims to understand the diverse strategies organizations employ to surmount challenges in cross-cultural management, emphasizing the creation of a productive and inclusive work environment.
5. Evaluate the efficiency of training and development programs in agreeable potentialities for CCM. The primitive intent is to appraise how cross-cultural training initiatives influence the comprehension of cultural differences among managers and employees, and their ability to adapt and thrive in multicultural settings.

6. Explore the impact of leadership on enabling successful cross-cultural management. This objective involves scrutinizing leadership practices, behaviors, and qualities that play a pivotal role in effective cross-cultural management and the establishment of a culturally inclusive organizational culture
7. Furnish practical recommendations for organizations to bolster their cross-cultural management capabilities. The objective is to offer actionable insights and guidance on strategies, policies, and interventions that can empower organizations to adeptly manage a global workforce

By pursuing these research objectives, the study seeks to augment the current knowledge base in the field of cross-cultural management. The intent is to offer practical insights that can benefit organizations, fostering the cultivation of culturally competent managers and organizations equipped to thrive in the complexities of a globalized world.

Literature review

Adler and Gundersen (2007) examines cross-cultural management in this context encompasses key dimensions such as cultural values, communication styles, leadership, and negotiation. The focus of this exploration is to offer practical insights for handling cultural diversity and formulating strategies that contribute to successful and effective cross-cultural management. Earley and Mosakowski (2004) provides an extensive overview of cultural savvy and its purport in the ambience of CCM. It delves into the components of cultural intelligence, explores how it is measured, and examines its influence on diverse organizational outcomes.

Hofstede (1980) in his study established the cultural dimensions theory, which identifies fundamental aspects of national culture and their influence on work-related values and behavior. It furnishes scarce acumen into understanding cultural differences and their implications for cross-cultural management.

Lee and Gyamfi (2023) explore the concept of conceptualizing domestic employees within the framework of globalization, addressing the challenges associated with cross-cultural management in this era. It emphasizes the imperative for cultural sensitivity, proficient communication, and adaptability when overseeing multicultural teams. Furthermore, the paper suggests potential avenues for future research aimed at improving cross-cultural management practices. Tarique and Schuler (2010) center on talent management within a cross-cultural context, delving into the difficulties of recognizing, attracting, and nurturing talent across diverse cultural landscapes. It offers recommendations for devising successful global talent management strategies.

Thomas and Peterson (2017) investigate the concept of cultural intelligence (CQ) and its pivotal role in navigating cross-cultural interactions. It underscores the significance of CQ in effective leadership, decision-making, and relationship-building within multicultural environments. Trompenaars and Hampden-Turner (2011) in their study highlight the observable influence of cultural diversity on business practices and provides valuable insights into the effective management of cultural differences. The paper offers practical tools and strategies applicable to diverse organizational contexts for navigating cross-cultural management successfully.

The examined research papers significantly enrich our comprehension of cross-cultural management, offering valuable insights into the hurdles and tactics linked to overseeing a global workforce. These studies underscore the critical role of cultural intelligence, proficient communication, leadership, and talent management in attaining success within cross-cultural contexts. Through the synthesis of these findings, organizations can formulate effective strategies for managing cultural diversity, ultimately harnessing it as a means to gain a competitive advantage.

Methodology

The methodology for this research paper will adopt a mixed methods approach, combining qualitative research review, secondary data & case studies. This approach allows for a comprehensive understanding of the challenges faced by organizations in managing a global workforce and the strategies employed to address these challenges. The research design will be exploratory in nature, aiming to gather rich insights and qualitative data on cross-cultural management challenges and strategies. This research paper will facilitate a holistic understanding of the topic and provide a basis for practical recommendations.

Results and discussions

1. Achieving success in cross-cultural management necessitates prioritizing cultural sensitivity and awareness. Organizations can enhance their capabilities by focusing on cultural sensitivity training, improving intercultural communication, fostering an understanding of cultural norms, and cultivating an inclusive organizational culture. These measures enable organizations

to navigate cross-cultural challenges more effectively and capitalize on the advantages that cultural multifariousness brings to their labor pool.

   i. Significance of Cultural Delicacy: Acknowledging the importance of cultural sensitivity is compelling for rewarding CCM. This entails recognizing, respecting, and valuing cultural distinctions within individuals and groups. Cultural sensitivity not only contributes to a positive work environment but also nurtures mutual understanding, mitigating the likelihood of fallacy and strife.
  ii. Tutelage in Cultural Realization: To harbor a deeper consideration of many cultural measures, values, and behaviors, organizations should contrivance cultural awareness training programs for both employees and managers. Such training initiatives aid individuals in cultivating empathy, appreciation, and pliancy, essential qualities for adequately agreeing with compeers from divergent refined backgrounds.
 iii. Remodeling Intercultural Communication: It relies on effective intercultural communication. Training programs should concentrate on honing both verbal and non-verbal communication skills, encompassing active listening, clarity in speech, and awareness of cultural distinctions in communication styles. Underlining the significance of empathy and steering clear of cultural assumptions further reinforce the effectiveness of communication in cross-cultural contexts.
  iv. Presumptions with Cultural Norms and Standards: Cultural sensitivity training should impart knowledge about the distinct cultural norms, practices, and etiquette pertinent to the target regions. Grasping cultural nuances equips employees to steer clear of misunderstandings and promotes respectful interactions in cross-cultural settings.
   v. Conflict Resolution in Cross-Cultural Contexts: In cultural sensitivity training, incorporating conflict resolution strategies customized for cross-cultural contexts is imperative. Employees and managers should cultivate skills to navigate conflicts stemming from cultural differences, including diverse communication styles, decision-making approaches, and preferences in conflict resolution.
  vi. Fostering Inclusive Organizational Culture: Organizations should actively work towards establishing an inclusive culture that appreciates and embraces diversity. Encouraging open dialogue, acknowledging and celebrating cultural differences, and ensuring equal opportunities for growth and development all contribute to cultivating an inclusive work environment.

vii. Continuous Learning and Feedback: Cultural sensitivity and awareness are dynamic processes that necessitate ongoing learning and feedback. Organizations should endorse an environment where employees feel encouraged to share their experiences, challenges, and insights regarding cross-cultural interactions. Establishing platforms for dialogue and learning from diverse perspectives fosters a culture of uninterrupted preferment and enhances cultural competence.

2. Calibrating Leadership Practices with Cross-Cultural Training: Assimilation cross-cultural training and development strategies with leadership practices, as highlighted by Oberg (1960), enables organizations to cultivate a culturally competent workforce and encourage inclusive leadership. These strategies not only enhance employees' cross-cultural skills but also improve their proficiency in navigating diverse work environments, fostering effective collaboration across cultural boundaries. Additionally, leadership practices emphasizing cross-cultural competence contribute significantly to nurturing an comprehensive and culturally different organizational culture.
    i. Thorough Cross-Cultural Training Programs: Organizations should enact detailed cross-cultural training initiatives aimed at lifting employees' cultural expertise and understanding, as explained by Xing (1995). These programs may cover topics such as cultural norms, values, communication styles, and cultural intelligence. To ensure effectiveness, training sessions should be interactive, integrating case studies, replicas, and real-life examples to offer practical insights into the dynamics of synergy.
    ii. Leadership Evolution: Organizations should implement leadership development programs with a specific emphasis on competence, aligning with insights from Cullen and Parboteeah (2013). These programs should concentrate on enhancing leaders' cultural intelligence, encompassing their awareness, perceptive, and compliancy in diverse cultural contexts. Modules within pilotage can environ sensitivity, and intercultural communication skills.
    iii. Advisership and Indoctrination: Endowing relationships between employees from colored backgrounds can speed the learning and development. These relationships offer valuable opportunities for individuals to receive guidance, feedback, and support in navigating challenges presented by cross-cultural contexts. Mentors and coaches play a pivotal role in aiding individuals to develop cultural awareness, viewpoint-taking skills, and artifice for competent concert in settings.

iv. Empiric Learning: Employee participation in memoir like international assignments, cross-functional projects, or job rotations. Encouraging experiential learning enables individuals to immerse themselves in diverse cultural settings, draw insights from firsthand experiences, and cultivate a profound understanding of cross-cultural dynamics. Organizations should offer support and resources to ensure that employees can maximize the benefits of these opportunities, as elucidated by House et al. (2004).

v. Fostering Diversity in Leadership Roles: Advocate for diversity in leadership positions by incorporating inclusive talent management practices, as discussed by Pudelko and Harzing (2007). Actively identify and create opportunities for individuals from diverse cultural backgrounds to assume leadership roles. Cultivating diversity in leadership across different organizational levels enriches cross-cultural perspectives, promotes inclusive decision-making, and communicates a robust commitment to diversity and cross-cultural management within the organization.

vi. Role Modeling and Leading by Example: Leader's ought to function as exemplars of cross-cultural competence, showcasing inclusive leadership practices. They should exemplify cultural sensitivity, demonstrate respect for diverse perspectives, and exhibit open-mindedness, as stated by Yeganeh and Su (2006). Active engagement in cross-cultural interactions, encouragement of collaboration across cultural boundaries, and the proactive addressing of any instances of bias or discrimination should be integral to their leadership approach.

vii. Continuous Feedback and Development: Instill a culture of regular feedback where employees consistently receive insights on their cross-cultural interactions and behaviors. Encourage self-reflection and actively promote individuals to seek feedback from colleagues, subordinates, and superiors. Provide resources for ongoing development, including workshops, webinars, and online materials specifically centered around cross-cultural management.

3. Case Study 1: Global Expansion in a Technology Company
    3.1 Background:
    A technology company based in the United States decides to expand its operations into Southeast Asia. The company aims to tap into the growing market and establish a local presence. However, the company

faces several cross-cultural management challenges during the expansion process.

3.2 Challenges:

3.2.1 Communication: Language barriers and cultural differences impact effective communication between the headquarters and the new branch in Southeast Asia. Misinterpretation of messages and misunderstandings hinder collaboration and decision-making.

3.2.2 Work Culture: The work culture in Southeast Asia differs significantly from that in the United States. The company struggles to align the local work practices with its established organizational culture, resulting in conflicts and resistance to change.

3.3. Strategies and Solutions:

3.3.1 Cultural Sensitivity Training: The company conducts cultural sensitivity training for both the headquarters and the new branch employees. The training sessions focus on understanding cultural nuances, communication styles, and work practices in Southeast Asia, fostering better cross-cultural understanding.

3.3.2 Local Hiring and Knowledge Transfer: The company hires local talent and provides opportunities for knowledge transfer between the headquarters and the new branch. This facilitates cultural integration, enhances local expertise, and promotes collaboration between the two locations.

3.4. Adaptive Leadership: Leaders from the headquarters visit the new branch regularly to gain firsthand experience of the local culture and engage with employees. They adapt their leadership style, taking into account the cultural context, and provide guidance and support to local managers.

4) Case Study 2: Merging Two Multinational Companies

4.1. Background: A global organization emerges from the merger of two multinational companies – one rooted in Europe and the other in Asia. This union introduces cross-cultural management challenges, given the distinct cultural backgrounds and work practices of the employees involved.

4.2. Challenges:

4.2.1. Cultural Clash: Employees from different cultural backgrounds struggle to adapt to each other's work styles, communication norms, and decision-making processes. This results in misunderstandings, conflicts, and decreased collaboration.

4.2.2. Leadership Integration: The merger requires aligning leadership practices and establishing a unified leadership approach. Differences in leadership styles and expectations pose challenges in creating a cohesive leadership team.
4.3. Strategies and Solutions:
4.3.1. Cultural Integration Workshops: The merged organization conducts cultural integration workshops to enhance understanding and appreciation of diverse cultures. These workshops provide a platform for employees to share their perspectives, address misconceptions, and evolve plans for successful fraternization explained by Laurent (1983).
4.3.2. Leadership Exchange Program: The organization implements a leadership exchange program where leaders from each region spend time in the other region. This program allows leaders to gain firsthand exposure to local work environments, grasp cultural subtleties, and establish connections with employees from diverse backgrounds.
4.3.3. Team-building Initiatives: Organized team-building activities aim to encourage collaboration and cultivate relationships among employees with varied cultural backgrounds. These initiatives create spaces for informal interactions, enhance mutual understanding, and establish common ground.
4.3.4. Platforms for Cross-cultural Communication: The organization introduces tools for cross-cultural communication, including multilingual communication tools and cultural competence training for employees. These platforms support efficient communication, overcome language barriers, and promote cultural understanding.

Conclusions

Admonishing a global workforce in presents a unique set of challenges that require thoughtful approaches for successful outcomes. To navigate these challenges successfully, organizations must prioritize key areas such as effective communication, cultural comprehension, trust-building, cross-cultural leadership, and managing differences in time zones and work-life balance.

Clear and open communication is essential to bridge language barriers and ensure effective collaboration. Providing language training, utilizing translation services, and leveraging common communication platforms can enhance fostering comprehension among team members with diverse cultural backgrounds.

Diversity in cultures can influence how teams work together and make decisions. Cascading cultural sensitivity, propone for training, and composing an inclusive workplace proffer to employees understanding and concerning different viewpoints, ultimately improving collaboration and productivity. Establishing trust and collaboration within a global workforce is essential. Whether through face-to-face or virtual interactions, along with transparent communication, trust among team members can be cultivated. Acknowledging and valuing diverse perspectives, coupled with opportunities for cross-functional and cross-cultural team projects, further strengthens collaboration.

Implementing leadership development programs that specifically address cross-cultural management enables leaders to acquire the essential competencies needed for adeptly navigating diverse work environments.

Addressing temporal disparities and equilibrium between professional and personal life is crucial for maintaining employee well-being and productivity. Implementing flexible work arrangements and utilizing technology tools for project management and collaboration can help accommodate diverse time zones and advocate for a harmonious work-life balance.

Through the application of these strategies, organizations can effectively address the complexities of managing a global workforce in the realm of cross-cultural management. By cultivating a culture rooted in understanding, collaboration, and productivity, organizations can leverage the advantages of diversity to propel success within the global business landscape.

References

Adler, N.J. and Gundersen, A. 2007. *International dimensions of organizational behavior*. 5th ed. Cengage Learning Publisher.

Bortolotti, T., Boscari, S. and Danese, P. 2015. Successful lean implementation: Organizational culture and soft lean practices. *International Journal of Production Economics*, 160, pp. 182–201. https://doi.org/10.1016/j.ijpe.2014.10.013

Chang, C.L.H. and Lin, T.C. 2015. The role of organizational culture in the knowledge management process. *Journal of Knowledge Management*, 19(3), pp. 433–455. https://doi.org/10.1108/JKM-08-2014-0353.

Cullen, J. B. and Parboteeah, K. P. 2013. *Multinational management*. Cengage Learning.

Diamantidis, A.D. and Chatzoglou, P. 2018. Factors affecting employee performance: An empirical approach. *International Journal of Productivity and Performance Management*, 68(1), pp. 171–193. https://doi.org/10.1108/IJPPM-01-2018-0012.

Earley, P.C. and Mosakowski, E. 2004. Cultural intelligence. *Harvard Business Review*, 82(10), pp. 139–146.

Elsbach, K.D. and Stigliani, I. 2018. Design thinking and organizational culture: A review and framework for future research. *Journal of Management*, 44(6), pp. 2274–2306.

Hofstede, G. 1980. *Culture's consequences: International differences in work-related values.* Beverly Hills/London: Sage.

House, R. J., Hanges, P. J., Javidan, M., Dorfman, P. W. and Gupta, V., eds. 2004. *Culture, leadership, and organizations: The GLOBE study of 62 societies.* Thousand Oaks: Sage.

Javidan, M., Stahl, G.K., Brodbeck, F. and Wilderom, C.P.M. 2005. Cross-border transfer of knowledge: Cultural lessons from project GLOBE. *Academy of Management Executive*, 19(2), pp. 59–76.

Laurent, A. 1983. The cultural diversity of Western conceptions of management. *International Studies of Management & Organization*, 13(1–2), pp. 75–96.

Lee, Y.T. and Gyamfi, N.Y.A. 2023. Cultural contingencies of resources: (Re)conceptualizing domestic employees in the context of globalization. *Academy of Management Review*, 48(1), pp. 165–168. https://doi.org/10.5465/amr.2021.0339.

Meyer, E. 2014. *The culture map: Breaking through the invisible boundaries of global business.* New York: Public Affairs.

Mills, A.J. 2017. Organization, Gender, and CultureIn: A.J. Milss, ed. *Insights and research on the study of gender and intersectionality in international airline cultures.* Leeds: Emerald Publishing Limited, pp. 15–33. https://doi.org/10.1108/978-1-78714-545-020171002.

Moran, R. T., Abramson, N. R. and Moran, S. V. 2014. *Managing cultural differences.* 9th ed. London: Routledge.

Oberg, K. 1960. Culture shock: Adjustment to new cultural environments. *Practical Anthropology*, 7(4), pp. 177–182. https://doi.org/10.1177/009182966000700405

Pudelko, M. and Harzing, A.W. 2007. Country-of-origin, localization, or dominance effect? An empirical investigation of HRM practices in foreign subsidiaries. *Human Resource Management*, 46(4), pp. 535–559. https://doi.org/10.1002/hrm.20181.

Søderberg, A.M. and Holden, N. 2002. Rethinking cross cultural management in a globalizing business world. *International Journal of Cross Cultural Management*, 2(1), pp. 103–121.

Stoermer, S., Hildisch, A.K. and Froese, F.J. 2016. Culture matters: The influence of national culture on inclusion climate. *Cross Cultural & Strategic Management*, 23(2).

Tarique, I. and Schuler, R. S. 2010. Global talent management: Literature review, integrative framework, and suggestions for further research. *Journal of World Business*, 45(2), pp. 122–133. https://doi.org/10.1016/j.jwb.2009.09.019.

Terjesen, S., Hessels, J. and Li, D. 2016. Comparative international entrepreneurship: A review and research agenda. *Journal of Management*, 42(1), pp. 299–344.

Thomas, D. C. and Peterson, M. F. 2017. *Cross-cultural management: Essential concepts*. 3rd ed. Sage Publications Inc.

Trompenaars, F. and Hampden-Turner, C. 2011. *Riding the waves of culture: Understanding cultural diversity in business*. Nicholas Brealey Publishing.

Xing, F. 1995. The Chinese cultural system: Implications for cross-cultural management. *SAM Advanced Management Journal*, 60(1), pp. 14–21.

Yeganeh, H. and Su, Z. 2006. Conceptual foundations of cultural management research. *International Journal of Cross Cultural Management*, 6(3), pp. 361–376.

Vishal Singh Varma, Reshu Gupta Singh, Ravi Kumar Goyal, Kritika Tekwani, Ramnika Kaur, Liana Mihaela Ferritsean

The economic potential of millet farming in Rajasthan (India): Opportunities and challenges

Abstract: *Millet farming has been an integral part of Rajasthan's agricultural landscape for centuries. With the increasing demand for healthy and sustainable food, millets have gained renewed attention as a nutritious and climate-resilient crop. This research paper aims to assess the economic potential of millet farming in Rajasthan, considering its contribution to rural livelihoods, food security, and sustainable agriculture. The study employs a mixed-methods approach, combining primary data from field surveys with secondary data from government reports and academic literature. The findings suggest that millet farming has enormous economic potential in Rajasthan, particularly in the context of smallholder agriculture. The crop's low-input requirements, high yield potential, and resilience to drought make it an attractive option for resource-poor farmers. However, several challenges hinder the expansion of millet cultivation, including low market demand, lack of processing infrastructure, and limited access to credit and markets. The study recommends policy interventions to address these challenges and promote millet farming as a viable source of income and food security in rural Rajasthan.*

Keywords: Millets, Economic Growth, Pearl Millets, Rajasthan, India, Nutritional Values, Sustainable Agriculture, Policy Implications.

Introduction

Rajasthan, India's largest state, is known for its arid and semi-arid climate, which poses significant challenges to agriculture. The state's agriculture is predominantly rain-fed, and farmers face frequent droughts and water scarcity. In recent years, climate change has exacerbated these challenges, leading to increased vulnerability for smallholder farmers. Millet farming, however, offers a promising solution to these challenges, this view has been supported by Choudhary and Kumar (2023). Millets are a group of cereal crops that are hardy, drought-tolerant, and require low inputs, making them well-suited to Rajasthan's climate and soil conditions. Moreover, millets are copious in nutriments, gluten exempt, and have a crouched glycemic indicant, making them a vigorous lieutenant to rice and wheat. In defiance of many perquisites, millets have received little attention from policymakers and markets, leading to their marginalization and underutilization. Millet is an important crop for the state of Rajasthan, particularly in the semi-arid regions where other crops struggle to grow. The state has a rich tradition of millet cultivation, and it remains an important staple crop for millions of people. Despite this, the potential of millet farming in Rajasthan has not been fully realized, and there is significant scope for increasing production and improving the economic viability of millet farming. This research paper aims to explore the economic potential of millet farming in Rajasthan, considering its role in promoting rural livelihoods, food security, and sustainable agriculture, explained by Padulosi et al. (2015).

Literature review

A review of existing research on millet farming in Rajasthan would be conducted to gain an understanding of the current state of knowledge, identify research gaps, and inform the research questions and hypotheses.

Choudhary and Kumar (2023) in an economic analysis of pearl millet production in Jaipur District, Rajasthan, reveal promising prospects. With favorable agro-climatic conditions and low input costs, pearl millet cultivation demonstrates strong potential for profitable returns. Additionally, its drought-resistant nature aligns with the region's water scarcity concerns, making it a resilient and sustainable crop choice for local farmers. Lenka et al. (2020) stated that millets are promising crops due to their resilience and adaptability to changing plight of climate. Their inferior water and injunction exigency make them eco-friendly choices, contributing to food security and mitigating climate change. Millets play a crucial role in promoting sustainable and resilient farming practices.

Devi et al. (2014) in "Health benefits of finger millet (Eleusine coracana L.) polyphenols and dietary fiber: a review" explain that Finger millet (Eleusine coracana L.) is a nutrient-dense grain renowned for its polyphenols and dietary fiber content, conferring various health advantages. Polyphenols in finger millet, such as flavonoids and tannins, exhibit potent antioxidant vista. These commixture succor brush oxidative crunch, curtail inflammation, and lower the peril of enduring diseases, withal cardiovascular indisposition and assertive cancers. Furthermore, the high dietary fiber content in finger millet supports digestive health by promoting regular bowel movements, preventing constipation, and improving nutrient absorption. It also aids in weight management by inducing a feeling of fullness and stabilizing blood sugar levels, making finger millet a valuable addition to a health-conscious diet.

Wilson et al. (2022) significantly impact millets stated about functional characteristics and nutritional accessibility. Techniques such as milling, dehulling, and fermentation can alter their physiochemical properties. Dehulling may enhance nutrient bioavailability by removing anti-nutrients, but also reduce fiber content. Fermentation can improve mineral bioavailability and reduce phytate levels, making minerals like iron and zinc more accessible. Milling can result in flour with varied particle sizes, influencing the texture and sensory attributes of millet-based products. Overall, processing methods must be chosen carefully to preserve the nutritional benefits of millets while meeting specific product requirements, highlighting the importance of balancing nutrient retention and product quality. Sharma et al. (2013) in "Economic analysis of pearl millet marketing in Rajasthan" reveals a complex picture. Pearl millet, a staple crop, plays a critical role in the state's agriculture. The market is influenced by factors like production levels, quality standards, and seasonal demand fluctuations. Small-scale farmers often face challenges in accessing markets and obtaining fair prices. Government initiatives and cooperatives can help improve market access. Efficient supply chain management and the promotion of value-added products like millet-based snacks can enhance profitability for farmers and contribute to food security. A comprehensive approach is needed to ensure sustainable pearl millet marketing in Rajasthan. Sharma et al. (2022) have identified a number of factors that influence the profitability of millet farming, including yields, prices, costs, and government policies.

Several studies have found that millet yields in Rajasthan are significantly lower than the national average, and that there is potential for proliferating turnout through the ratification of bettered cultivation conduct and technologies. E. g., the use of ameliorated seed species, fertilizers, and irrigation techniques can significantly increase yields and improve profitability.

Studies have also identified significant price fluctuations for millet in Rajasthan, which can affect the economic viability of millet farming. These price fluctuations are driven by a number of factors, including changes in demand, supply, and government policies. Some studies have suggested that there is potential for improving the marketing and distribution of millet in Rajasthan, which could help to stabilize prices and improve the economic viability of millet farming. In terms of costs, studies have found that millet cultivation in Rajasthan is proportionately squat cost as to other crops, distinctly those that compel immense exhortation of water and fertilizers. However, there are still opportunities for reducing costs through the adoption of more efficient cultivation practices and technologies.

The literature suggests that there is significant potential for increasing the economic viability of millet farming in Rajasthan. However, this will require a coordinated effort by farmers, researchers, policymakers, and other stakeholders to address the challenges facing the sector and to take advantage of the opportunities that exist. There are significant challenges facing the sector, including low yields, price fluctuations, and inadequate infrastructure. Addressing these challenges will require a coordinated effort by farmers, researchers, policymakers, and other stakeholders.

Methodology

The study employs a mixed-methods approach, depending with secondary data from government reports and academic literature. The secondary data analysis involved a review of government policies, market trends, and academic literature on millet farming in Rajasthan.

The methodology for assessing the economic potential of millet farming in Rajasthan would typically involve the following steps:

- Data analysis: The secondary data would be analyzed using statistical and econometric methods to estimate the costs, revenues, profits, and other economic indicators associated with millet farming in Rajasthan. Regression models would be used to identify the factors that contribute to variations in millet yields, costs, and profits across different regions and farming systems.
- Economic modeling: The results of the data analysis would be used to construct economic models that simulate the impacts of changes in market conditions, policy interventions, and technological innovations on millet farming in Rajasthan. These models would provide insights into the potential benefits and costs of different scenarios for millet production, marketing, and consumption.
- Policy implications: The findings from the analysis and modeling would be used to draw policy implications for promoting millet farming in Rajasthan. The policy recommendations would be based on an assessment of the trade-offs between economic, social, and environmental objectives, and would take into account the preferences and constraints of different stakeholders in the millet value chain.

The methodology would involve a combination of quantitative and qualitative methods, and would be guided by an interdisciplinary and participatory approach that involves farmers, researchers, policymakers, and other stakeholders.

Results and discussions

The study found that millet farming has enormous economic potential in Rajasthan, particularly in the context of smallholder agriculture. Millet crops such as pearl millet, finger millet, and sorghum have high yield potential and require low inputs, making them an attractive option for resource-poor farmers. Millet farming also has a low carbon footprint, as the crops do not require much fertilizer or irrigation. The study found that millet farming contributed to the food security of smallholder farmers, as millets provide a source of nutritious food and fodder for livestock. However, several challenges hinder the expansion of millet cultivation in Rajasthan. The study found that the demand for millets was low, and farmers often struggled to find buyers. The lack of processing infrastructure was another challenge, as most millets were sold in their raw form, which fetched lower prices. Moreover, limited access to credit and markets prevented many farmers from investing in millet farming.

I. Nutritional Values:

Table 1. Nutrients and energy values for commonly consumed millets

Millet Type	Energy Value per 100g (kcal)	Carbohydrates (g)	Protein (g)	Fat (g)	Fiber (g)	Composition (g)
Foxtail Millet	351	73.8	11.2	3.6	6.7	Iron: 2.8, Calcium:31,Magnesium:76, Phosphorus: 290, Potassium: 119, Zinc: 1.9
Finger Millet	336	72.9	7.3	1.3	3.6	Iron: 3.9, Calcium: 344, Magnesium: 114, Phosphorus: 290, Potassium:408,Zinc:2.9
Pearl Millet	378	67.5	11.0	5.0	1.3	Iron: 3.9, Calcium:42,Magnesium:114, Phosphorus: 290, Potassium: 342, Zinc: 1.4

(continued on next page)

Table 1. Continued

Millet Type	Energy Value per 100g (kcal)	Carbohydrates (g)	Protein (g)	Fat (g)	Fiber (g)	Composition (g)
Barnyard Millet	360	69.3	11.2	4.3	10.1	Iron: 15.2, Calcium:11,Magnesium:16, Phosphorus: 283, Potassium: 195, Zinc: 1.5
Little Millet	341	68.0	7.7	4.3	7.6	Iron: 9.3, Calcium:17,Magnesium:76, Phosphorus: 215, Potassium: 205, Zinc: 1.2
Kodo Millet	329	65.8	8.3	1.4	9.0	Iron: 1.9, Calcium:27,Magnesium:36, Phosphorus: 350, Potassium: 165, Zinc: 1.2
Proso Millet	378	72.9	10.6	3.1	3.9	Iron: 3.0, Calcium:20,Magnesium:87, Phosphorus: 284, Potassium: 195, Zinc: 1.2

Note: The nutrient and energy values can vary slightly depending on the variety, growing conditions, and processing method.
Source: USDA Food Data Central, 2021.

II. Global Scenario of Millets

Millet production varies from year to year and depends on a range of factors, including weather conditions, soil quality, and government policies. However, according to data from the Food and Agriculture Organization (FAO, 2023), millets were grown on around 35 million hectares of land, and the global formulation of millets was nearly 32 million metric tons. All over, India is the bulkiest farmer of millets, reckoning almost 40 % of the world's production. Other major producers include Nigeria, Niger, China, Mali, Burkina Faso, Sudan, and Ethiopia.

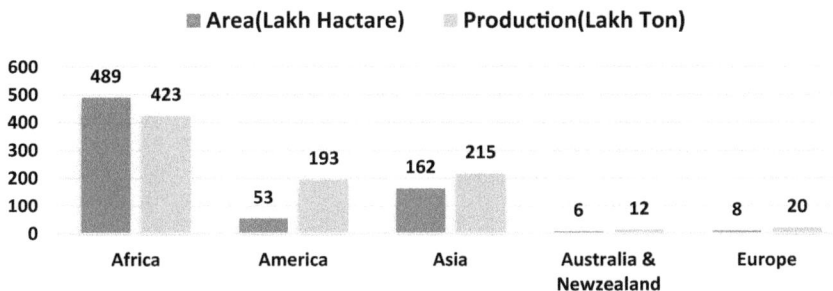

Figure 1. Global Millets Production and Sown Area Region-wise (2019)
Source: Ministry of Agriculture & Farmer's Welfare, 2022.

III. India & Millets' Trend

According to data from the Ministry of Agriculture and Farmers' Welfare (Government of India), in recent years, there has been growing interest in millets as a healthy and nutritious food, stated by Bhat, Nandini, and Tippeswamy (2018). In India, there are number of millets grown, inclusive of bajra, ragi, kangni, kutki, kodo millet, sanwa, and chena. Pearl millet (bajra) is an important crop in India, and it is well-suited to the country's climate and soil conditions. Pearl millet production is good in India due to its adaptability, nutritional value, versatility, high yield potential, and cultural importance.

- Adaptability to arid and semi-arid regions: Pearl millet is a highly resilient crop that can tolerate drought and heat stress. It is well-suited to the arid and semi-arid regions of India, where other crops may not grow well.
- Nutritional value: Pearl millet is a good source of protein, fiber, and micronutrients such as iron and zinc. It is an important staple crop in many parts of India, particularly in regions where other sources of protein and nutrients may be scarce (Kaushik et al., 2021).
- Versatility: Pearl millet can be used for a variety of purposes, including human consumption, animal feed, and as a raw material for the production of alcoholic beverages. This versatility makes it an attractive crop for farmers, as they can choose the best use for their particular crop
- High yield potential: Pearl millet has a high yield potential, and with proper management, it can produce good yields even in marginal lands. This makes it an important crop for small and marginal farmers, who may not have access to high-quality land, explained by Malathi et at. (2016).

◻ Cultural importance: Pearl millet has been an important crop in India for centuries, and it is deeply embedded in the country's cultural and culinary traditions. This has helped to sustain its production and consumption, even in the face of competition from other crops, as evaluated by Anbukkani Balaji and Nithyashree (2017).

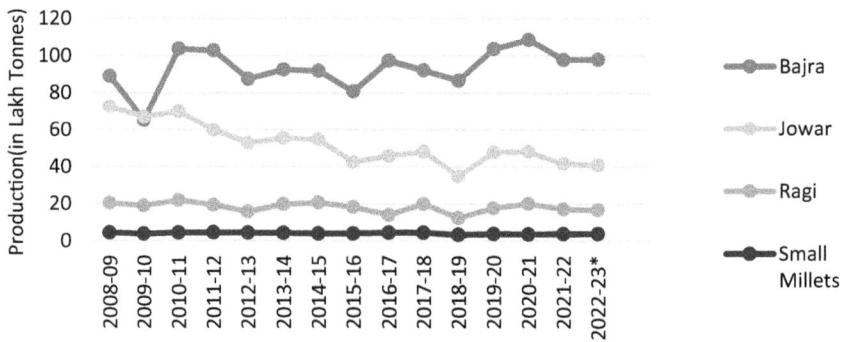

Figure 2. Trends in production of millets in India (2008–2023*)
Source: Ministry of Agriculture & Farmer's Welfare, 2023 (2022–23* 3rd Advanced Estimate).

Table 2. Millet production data for Rajasthan, India, for the last five years (2017–2018 to 2021–2022*)

Year	Pearl Millet Production (in lakh tonnes)	Jowar Production (in lakh tonnes)
2021–2022*	37.51	5.90
2020–2021	45.61	5.90
2019–2020	46.86	4.56
2018–2019	38.08	4.70
2017–2018	37.53	3.01

*4th Advanced Estimate
Source: APEDA, 2023.

Discussions

I. Millet's production is good in Rajasthan for several reasons:
- Climate: Rajasthan has an arid and semi-arid climate that is well-suited to the cultivation of millets. Millets are drought-resistant and can tolerate high temperatures, making them an ideal crop for the state.
- Soil: The soil in Rajasthan is generally sandy and low in organic matter. Millets are well-suited to these soil conditions and can thrive in poor soils where other crops struggle.
- Water conservation: Millets require less water than many other crops, making them an ideal choice for farmers in Rajasthan who may face water scarcity. The cultivation of millets can also contribute to water conservation efforts by reducing the demand for irrigation.
- Nutritional value: Millets are highly nutritious and can provide an important source of food for the population of Rajasthan. By promoting millet cultivation, the state can improve food security and provide a source of healthy food for its citizens, as evaluated by Nithiyanantham et al. (2019).
- Market demand: Millets are gaining popularity in India and other countries due to their nutritional value and health benefits. This has created a growing market for millets and provides an opportunity for farmers in Rajasthan to earn a higher income by growing this crop (Singh et al., 2023).

The climate, soil, water availability, nutritional value, and market demand make millets production a good option for farmers in Rajasthan. By promoting millets cultivation, the state can enhance its agricultural sector and contribute to its economic growth and development.

II. Millets have the potential to enhance Rajasthan's economy in several ways:
- Increase in agricultural income: Millets can provide an additional source of income for farmers in Rajasthan who may currently rely on traditional crops such as wheat and rice. By diversifying their crops and growing millets, farmers can increase their income and reduce their reliance on a single crop (NITI Aayog, 2018).
- Employment opportunities: The cultivation of millets can create employment opportunities for rural communities in Rajasthan. This includes both on-farm employment in activities such as cultivation and harvesting, as well as off-farm employment in processing and marketing activities (National Policy for Farmers, 2020. Ministry of Agriculture & Farmers' Welfare, GoI).

- Value addition: Millets can be processed into a variety of products, including flour, flakes, and ready-to-eat foods. This provides an opportunity for value addition and can increase the economic returns from millet cultivation (Mishra et al., 2021).
- Improved food security: Millets are highly nutritious and can provide an important source of food for households in Rajasthan. By promoting millet cultivation, the state can enhance food security for its population and reduce dependence on imported food, published by Dwivedi et al. (2016).
- Sustainable agriculture: Millets are well-suited to the arid climate of Rajasthan and require less water than many other crops. By stimulating millet produces, the state can spur sustainable agriculture practices that hoard water and lessen the concussion of climate change on agriculture.

The exaltation of millet cultivation in Rajasthan can contribute to the state's economic growth and development while also promoting sustainable agriculture and improving food security, evaluated by Chiffoleau et al. (2019).

III. Opportunities:
- Growing demand: Millets are gaining popularity as a healthy food option and have a thriving market bid both in India and internationally. This floats a space for farmers in Rajasthan to upsurge their income by millets, as stated by Birol, Munasib and Roy (2015).
- Drought-resistant: Millets are well-suited to the arid climate of Rajasthan and require less water than many other crops. This makes them a viable option for farmers in regions with limited access to water, evaluated by Grovermann et al. (2018).
- Crop diversification: Millets offer an opportunity for farmers in Rajasthan to diversify their crops and reduce their dependence on traditional crops such as wheat and rice, published by Shaktawat et al. (2012).
- Government initiatives: The Government of Rajasthan has launched several initiatives to promote the cultivation of millets, including subsidies and training programs for farmers.

IV. Challenges:
- Low productivity: Millets have lower yields compared to other crops such as wheat and rice, which makes them less profitable for farmers. There is a need for improved cultivation techniques and better seed varieties to increase productivity, published by Saxena et al. (2018).

- Lack of processing facilities: The lack of processing facilities for millets in Rajasthan is a major challenge. This limits the potential for value addition and marketing of millet products.
- Limited market linkages: The market for millets in Rajasthan is not well developed, which limits the opportunities for farmers to sell their crops at a fair price. There is a need for better market linkages and access to markets beyond the state.
- Limited awareness: The awareness of the nutritional benefits of millets is still limited among purchaser in Rajasthan. This presents a challenge for marketing millet products and increasing demand.

Conclusions

The agrology of millets in Rajasthan has enormous potential to enrich into the state's economy. However, there are also challenges that must be overcome to fully realize this potential.

Government policies also play an important role in the economic viability of millet farming in Rajasthan. Several studies have identified policies that can support the development of the millet sector, including investments in research and development, extension services, marketing, and infrastructure. In addition, policies that provide incentives for farmers to adopt improved cultivation practices and technologies can help to improve yields and profitability. While there are opportunities for millet farming to contribute to the economy of Rajasthan, there are also several challenges that need to be addressed. Improving productivity, developing processing facilities, creating market linkages, and increasing awareness of the benefits of millets are all key areas that need attention to unlock the full potential of millets in Rajasthan. The study concludes that millet farming has enormous economic potential in Rajasthan, but several challenges need to be addressed to realize this potential fully. The study recommends policy interventions to promote millets.

References

Agricultural and Processed Food Products Export Development Authority (APEDA), Ministry of Commerce and Industry, Govt of India. 2023. *State wise millet production* [online] Available at: <https://apeda.gov.in/milletportal/files/Statewise_Millet_Production.pdf> [Accessed 5 September 2023].

Anbukkani, P., Balaji, S.J. and Nithyashree, M.L. 2017. Production and consumption of minor millets in India-A structural break analysis. *Ann. Agricultural Research.* New Series, 38(4), pp. 1–8.

Bhat, S., Nandini, C. and Tippeswamy, V. 2018. Significance of small millets in nutrition and health-A review. *Asian Journal of Dairy and Food Research*, 37(1), pp. 35–40.

Birol, E., Munasib, A. and Roy, D. 2015. Networks and low adoption of modern technology: The case of pearl millet in Rajasthan, India. *Indian Growth and Development Review*, 8(2), pp. 142–162. https://doi.org/10.1108/IGDR-07-2014-0025.

Chiffoleau, Y., Millet-Amrani, S., Rossi, A., Rivera-Ferre, M.G. and Merino, P.L. 2019. The participatory construction of new economic models in short food supply chains. *Journal of Rural Studies*, 68, pp. 182–190. https://doi.org/10.1016/j.jrurstud.2019.01.019.

Choudhary, R.K. and Kumar, S. 2023. An economic analysis of production of pearl millet in Jaipur District of Rajasthan, India. *Journal of Experimental Agriculture International*, 45(9), pp. 20–25.

Devi, P.B., Vijayabharathi, R., Sathyabama, S., Malleshi, N.G. and Priyadarisini, V.B. 2014. Health benefits of finger millet (Eleusine coracana L.) polyphenols and dietary fiber: A review. *Journal of Food Science and Technology*, 51, pp. 1021–1040. https://doi.org/10.1007/s13197-011-0584-9

Dwivedi, B.S., Rawat, A.K., Dixit, B.K. and Thakur, R.K. 2016. Effect of inputs integration on yield, uptake and economics of Kodo Millet (Paspalum scrobiculatum L.), *Economic Affairs*, 61(3), pp. 519–524.

FAO. 2023. *Unleashing the potential if millets. International Year of Millets 2023. Background Paper*. Rome: FAO. [online] Available at: <https://fdc.nal.usda.gov/fdc-app.html#/?query=Millet> [Accessed 5 September 2023].

Grovermann, C., Umesh, K.B., Quiédeville, S., Kumar, B.G. and Moakes, S. 2018. The economic reality of underutilised crops for climate resilience, food security and nutrition: Assessing finger millet productivity in India, *Agriculture*, 8(9), p. 131. https://doi.org/10.3390/agriculture8090131.

Kaushik, N., Yadav, P., Khandal, R.K. and Aggarwal, M. 2021. Review of ways to enhance the nutritional properties of millets for their value-addition. *Journal of Food Processing and Preservation*, 45(6), p. e15550. https://doi.org/10.1111/jfpp.15550.

Lenka, B., Kulkarni, G.U., Moharana, A., Singh, A.P., Pradhan, G.S. and Muduli, L. 2020. Millets: Promising crops for climate-smart agriculture. *International Journal of Current Microbiology and Applied Sciences*, 9(11), pp. 656–668. https://doi.org/10.20546/ijcmas.2020.911.081.

Malathi, B., Appaji, C., Reddy, G.R., Dattatri, K. and Sudhakar, N. 2016. Growth pattern of millets in India. *Indian Journal of Agricultural Research*, 50(4), pp. 382–386.

Ministry of Agriculture & Farmers' Welfare. 2020. *National policy for farmers*, <https://agriwelfare.gov.in/en/DigiAgriDiv> [Accessed 5 September 2023].

Ministry of Agriculture & Farmer's Welfare. 2022. *International year of millets: India leading the way*. Press release, 26 December 2022. [online] Available at: <https://static.pib.gov.in/WriteReadData/specificdocs/documents/2022/dec/doc20221226147401.pdf> [Accessed 5 September 2023].

Ministry of Agriculture & Farmer's Welfare. 2023. Press release, 28 July 2023. [online] Available at: <https://sansad.in/getFile/annex/260/AS94.pdf?source=pqars> [Accessed 5 September 2023].

Mishra, P., Prakash, H.G., Devi, S., Sonkar, S., Yadav, S., Singh, C.H. and Singh, R.D. 2021. Nutritional quality of millets and their value-added products with the potential health benefits: A review, *International Journal of Current Microbiology and Applied Sciences*, 10(10), pp. 163–175. https://doi.org/10.20546/ijcmas.2021.1010.019.

Nithiyanantham, S., Kalaiselvi, P., Mahomoodally, M.F., Zengin, G., Abirami, A. and Srinivasan, G. 2019. Nutritional and functional roles of millets – A review. *Journal of food biochemistry*, 43(7), p. e12859. https://doi.org/10.1111/jfbc.12859

NITI Aayog. 2018. *Doubling farmers' income: Rationale, strategy, prospects and action plan*. [online] Available at: <https://www.niti.gov.in/sites/default/files/2018-09/Doubling-Farmers-Income-Rationale-Strategy-Prospects-and-Action-Plan.pdf>

Padulosi, S., Mal, B., King, O.I. and Gotor, E. 2015. Minor millets as a central element for sustainably enhanced incomes, empowerment, and nutrition in rural India. *Sustainability*, 7(7), pp. 8904–8933. https://doi.org/10.3390/su7078904.

Saxena, R., Vanga, S.K., Wang, J., Orsat, V. and Raghavan, V. 2018. Millets for food security in the context of climate change: A review. *Sustainability*, 10(7), p. 2228. https://doi.org/10.3390/su10072228.

Shaktawat, S., Singh, I.P., Nagaraj, N. and Sharma, S. 2012. Resource use efficiency and returns to investment in research of pearl millet in Rajasthan. *Indian Journal of Economics and Development*, 8(4), pp. 1–12.

Sharma, S., Singh, I.P., RAO, P.P., Basavaraj, G. and Nagaraj, N. 2013. Economic analysis of pearl millet marketing in Rajasthan. *International Journal of Commerce and Business Management*, 6(1), pp. 66–75.

Sharma, S., Sharma, G.L.M.L., Singh, H. and Chaudhary, R.S. 2022. Growth and instability in area, production and yield of major millets in Rajasthan. *The Pharma Innovation Journal*, SP-11(2), pp. 1536–1543.

Singh, S., Yadav, R.N., Tripathi, A.K., Kumar, M., Kumar, M., Yadav, S. and Yadav, R. 2023. Current status and promotional strategies of millets: A review. *International Journal of Environment and Climate Change*, 13(9), pp. 3088–3095.

USDA Food Data Central. 2021. [online] Available at: <https://fdc.nal.usda.gov/fdc-app.html#/?query=Millet> [Accessed 5 September 2023].

Wilson, A., Elumalai, A., Moses, J.A. and Anandharamakrishnan, C. 2022. Effect of processing on functional characteristics, physiochemical properties, and nutritional accessibility of millets, In: *Handbook of millets-processing, quality, and nutrition status*. Singapore: Springer, pp. 205–229.

www.ingramcontent.com/pod-product-compliance
Ingram Content Group UK Ltd.
Pitfield, Milton Keynes, MK11 3LW, UK
UKHW021830210426
5322IPUK00004B/111